The Definitive Guide to the ARM Cortex-M0

Joseph Yiu

ELSEVIER

AMSTERDAM • BOSTON • HEIDELBERG • LONDON • NEW YORK • OXFORD
PARIS • SAN DIEGO • SAN FRANCISCO • SINGAPORE • SYDNEY • TOKYO

Newnes is an imprint of Elsevier

Newnes

Newnes is an imprint of Elsevier

The Boulevard, Langford Lane, Kidlington, Oxford, OX5 1GB, UK

30 Corporate Drive, Suite 400, Burlington, MA 01803, USA

First published 2011

Notices

Knowledge and best practice in this field are constantly changing. As new research and experience broaden our understanding, changes in research methods, professional practices, or medical treatment may become necessary.

Practitioners and researchers must always rely on their own experience and knowledge in evaluating and using any information, methods, compounds, or experiments described herein. In using such information or methods they should be mindful of their own safety and the safety of others, including parties for whom they have a professional responsibility.

To the fullest extent of the law, neither the Publisher nor the authors, contributors, or editors, assume any liability for any injury and/or damage to persons or property as a matter of products liability, negligence or otherwise, or from any use or operation of any methods, products, instructions, or ideas contained in the material herein.

British Library Cataloguing in Publication Data
A catalogue record for this book is available from the British Library

Library of Congress Control Number: 2010940590

ISBN: 978-0-12-385477-3

For information on all Newnes publications
visit our website at www.elsevierdirect.com

Transferred to Digital Printing in 2012

Working together to grow
libraries in developing countries

www.elsevier.com | www.bookaid.org | www.sabre.org

ELSEVIER   BOOK AID International   Sabre Foundation

Contents

Foreword

It is an exciting time to be a microcontroller user. A growing range of ARM Cortex-M devices are available from many vendors, spanning a broad range of applications. Users who are familiar with 8-bit and 16-bit microcontrollers and are moving to ARM Cortex-M—based devices are surprised at just how easy they are to use and, with the introduction of ARM Cortex-M0 devices, how low-cost and efficient 32-bit microcontrollers have become.

So I was delighted that Joseph Yiu chose to write a guide for the users of these ARM Cortex-M0 devices. As a technical authority on the ARM Cortex-M family and a formative 8-bit user, Joseph is uniquely placed to guide users new to ARM Cortex microcontrollers on their first transition steps, and to impart detailed knowledge about the Cortex-M0 to the advanced user.

Dr Dominic Pajak
ARM Cortex-M0 Product Manager

Preface

I started learning about microcontrollers when I was studying at university. At that time, some of the single board computers I was using had an 8-bit microcontroller, and my programs were stored in external Eraseable Programmable Read Only Memory (EPROM) chips. The EPROM chips were in relatively large dual in line (DIP) packages and could be erased by shining ultraviolet light through the glass window. Since then, microcontroller technology has changed a lot: external EPROMs have been replaced by on-chip flash memories, DIP packages have been replaced by surface mount packages, and most microcontrollers have become in-system reprogrammable. More and more peripherals have been added to microcontrollers, and the complexity of the software has increased dramatically.

Since 2004, the microcontroller market has made some dramatic changes. Previously, the microcontrollers on the market were mostly 8-bit and 16-bit types, and 32-bit microcontroller applications were limited to high-end products, mainly because of the cost. Although most of the 8-bit and 16-bit microcontrollers could be programmed in C, trying to squeeze all the required functionalities into a small microcontroller was becoming more and more difficult. You might spend one day writing a program in C, and then find that you needed to spend another two days rewriting part of the program in assembly because the processing speed of the microcontroller was too slow for all the required processing tasks.

Even if you are developing simple applications and do not require high processing power in the microcontrollers, occasionally you might need to switch to a different microcontroller architecture because of project requirements, and this can take a tremendous effort. Not only will you need to spend the money to buy new tools, but it can take weeks to learn to use the tools and months to become familiar with a new architecture.

In October 2004, the price of an ARM7 microcontroller dropped to below 3 US dollars. This was very good news for many developers who needed to develop complex embedded software. Since then, with the availability of the Cortex-M3, the price of ARM microcontrollers has dropped much further and you can get an ARM microcontroller for less than 1 US dollar. As a result, the use of ARM microcontrollers is gaining acceptance. In addition to providing excellent performance, the modern ARM microcontrollers require very little power. The use of ARM microcontrollers is no longer limited to high-end products.

Like many good ideas, the concept for the Cortex-M0 started life as a conversation between engineers in a bar. A small and growing number of ARM partners were looking for a 32-bit processor that was small, really small. The concept quickly became a full-blown engineering project (codenamed "Swift"). In 2009, the Cortex-M0 design was completed, and it quickly became one of the most successful ARM processor products.

Through the examples in this book, you will find that the Cortex-M0 microcontrollers are easy to use. In some aspects, they are even easier to use than some 8-bit microcontrollers because of the simplicity of the linear memory architecture, an uncomplicated and yet flexible exception model, comprehensive debug features, and the software infrastructures provided by ARM, microcontroller vendors, and software solution providers.

Because the Cortex-M processors are extremely C friendly, there is no need to optimize the applications with assembly. There is no need to learn lots of special C directives just to get the interrupt handlers working. For some embedded developers, the switch to ARM micro-controllers also means it is much easier to switch between different microcontroller products without the need to buy new tools and learn a new architecture. On the Internet you can find that many people are already using Cortex-M0 microcontrollers on a number of interesting projects.

After working on a number of ARM processor projects, I gained a lot of experience in the use of the Cortex-M processors (and possibly some gray hairs as well). With encouragement from some friends and help from lots of people, I decided to put those experiences into a book to share with numerous embedded developers who are interested in using the ARM Cortex-M processors. I learned a lot while writing my first book, which is about the Cortex-M3 processor. Many thanks to those people who gave me useful feedback when the first book was published, both inside and outside ARM. I know it wasn't perfect, but at least it is encouraging to find that many readers found my Cortex-M3 book useful. I hope this book, *The Definitive Guide to the ARM Cortex-M0*, will be even better.

This book targets a wide range of audiences: from students, hobbyists, and electronic enthu-siasts, to professional embedded software developers, researchers, and even semiconductor product designers. As a result, it covers a wide range of information, including many advanced technical details that most embedded developers might never need. At the same time, it contains many examples, making it easy for novice embedded software developers to use.

I hope that you find this book helpful and that you enjoy using the Cortex-M0 in your next embedded project.

Acknowledgments

A number of people have assisted me in researching for and writing this book.

First of all, a big thank you to my colleagues in ARM for their help in reviewing and suggesting improvements, especially Edmund Player, Nick Sampays, and Dominic Pajak, and also Drew Barbier, Colin Jones, Simon Craske, Jon Marsh, and Robert Boys.

I would also like to thank a number of external reviewers whose input in terms of feedback has enabled me to hone and improve the quality of the book's technical information. They are Joe Yu and Kenneth Dwyer from NXP, John Davies from Glasgow University, and Jeffrey S. Mueller from Triad Semiconductor. In addition to those who reviewed the prepublication book, I would also thank the following who assisted me while I was writing it by answering my technical inquiries and by providing product information. They are Kenneth Dwyer, David Donley, and Amit Bhojraj from NXP; Drew Barbier from ARM; Milorad Cvjetkovic from Keil; Jeffrey S. Mueller, William Farlow, and Jim Kemerling from Triad Semiconductor; Jamie Brettle from National Instruments; Derek Morris from Code Red Technologies; Brian Barrera from CodeSourcery; and the sales teams from Rowley Associates, Steinert Technologies, IAR Systems, and TASKING.

I would also thank the staff from Elsevier for their professional work in getting this book published.

And finally, a big thank you to all my friends within and outside of my work environment whose unstinting encouragement enthused and enabled me (though the odd cup of coffee helped too) to start and then complete the book.

Conventions

Various typographical conventions have been used in this book, as follows:

- Normal assembly program codes:

    MOV R0, R1; Move data from Register R1 to Register R0

- Assembly code in generalized syntax; items inside "< >" must be replaced by real register names:

    MRS < reg >, < special_reg >;

- C program codes:

    for (i = 0; i < 3; i + +) { func1(); }

- Pseudo code:

    if (a > b) {...

- Values:
    1. 4'hC, 0x123 are both hexadecimal values.
    2. *#3* indicates item number 3 (e.g., IRQ #3 means IRQ number 3).
    3. *#immed_12* refers to 12-bit immediate data.
    4. Register bits are typically used to illustrate a part of a value based on bit position. For example, bit[15:12] means bit number 15 down to 12.
- Register access types:
    1. R is Read only.
    2. W is Write only.
    3. R/W is Read or Write accessible.
    4. R/Wc is Readable and cleared by a Write access.

## CHAPTER 1

Introduction

Why Cortex-M0?

The ARM Cortex-M0 processor is designed to meet the needs of modern ultra-low-power microcontroller units (MCUs) and mixed-signal devices. It is intended to satisfy the demand for ever-lower-cost applications with increasing connectivity (e.g., Ethernet, USB, low-power wireless) and uses of analog sensors (e.g., touch sensors and accelerometers). These applications require tight integration of analog and digital functionality to preprocess and communicate data. Existing 8-bit and 16-bit devices often can't support these applications without significant increases in code size and clock frequency, therefore increasing power. The Cortex-M0 addresses the need for increased performance efficiency while remaining low cost and extending battery life. It is unsurprising, therefore, that the Cortex-M0 processor is now available in a rapidly growing range of silicon products, with the processor being the fastest licensing ARM design to date.

The idea behind ARM Cortex-M0 was to create the smallest, lowest power processor possible, while remaining upward compatible with the higher-performance ARM Cortex-M3. ARM announced the Cortex-M0 processor in February 2009, and it achieved its goals. The resulting design is just 12,000 logic gates in minimum configuration, as small as an 8-bit or 16-bit processor, but it is a full 32-bit processor that incorporates advanced technologies with many compelling benefits over 8-bit or 16-bit devices.

Energy Efficiency

The performance efficiency of the Cortex-M0 (0.9 DMIPS/MHz) means it can get a task done in fewer cycles (even a 32-bit multiply can be completed in one cycle). This means Cortex-M0 devices can spend more time in a low-power sleep state, offering better energy efficiency. Alternately, they can get the same job done in fewer MHz, meaning lower active power and electromagnetic interference (EMI). The low gate count of the Cortex-M0 means that the leakage, and sleep current, is minimized. The highly efficient interrupt controller (NVIC) means that interrupt overhead is low, even when handling nested interrupts of different priorities.

Code Density

The code size offered by the Thumb-2-based instruction set is smaller than 8-bit or 16-bit architectures for many applications. This means users can choose devices with smaller flash

The Definitive Guide to the ARM Cortex-M0. DOI: 10.1016/B978-0-12-385477-3.10001-1

memory sizes for the same application. This saves both device cost and power, as flash accesses contribute significantly to the total device power.

Ease of Use

The Cortex-M0 is designed as an ideal C target, with many modern compilers supporting it, and its interrupt service routines are able to be coded directly as C functions without the need for an assembler. However, the instruction set is just 56 instructions, so assembler coding is also easy to learn. Although it is a high-performance pipelined processor, the instruction and interrupt timings are fully deterministic (zero jitter), allowing the designs to be used in applications that require deterministic timing behavior and allowing developers to predict or analyze system timing accurately.

As microcontrollers are a key application area for the Cortex-M0 processor, it is designed with a number of vital microcontroller features:

- A built-in interrupt controller with easy-to-use interrupt priority control
- Low interrupt latency, allowing higher interrupt processing throughput and predictable system responsiveness
- A highly efficient instruction set called Thumb that provides high code density
- A very low gate count (as small as 12k in minimum configuration)
- A number of power-saving support features and very high energy efficiency
- Various debug features

In addition to the processor itself, the Cortex-M0 also has the following features:

- A wide range of tools supports including the Keil Microcontroller Development Kit as well as a number of third-party tools
- A wide range of software written for ARM architecture, including many open source projects like embedded operating systems (OSs) and compressors/decompressors (codecs)

The Cortex-M0 processor delivers the best energy-efficient 32-bit processing capability on the market, alongside a complete ecosystem to provide all the needs for embedded developers. A number of semiconductor vendors are already developing different products based on the Cortex-M0 processor with a variety of peripherals, memory sizes, and speeds. Products based on the Cortex-M0 started appearing on the market by the end of 2009, with more products arriving afterward.

Application of the Cortex-M0 Processors

The Cortex-M0 processor is used with a wide range of products. The most common usage is microcontrollers. Many Cortex-M0 microcontrollers are low cost and are designed for low-power applications. They can be used in applications that include computer peripherals and accessories; toys; white goods; industrial, heating, ventilating, and air conditioning (HVAC) controls; and home automation.

By using Cortex-M0 microcontrollers, these products can be built with more features, a more sophisticated user interface, better performance, and often better energy efficiency. At the same time, software development with Cortex-M processors is just as easy as using 8-bit and 16-bit microcontrollers. The costs of Cortex-M0 microcontrollers are competitive too.

Another important group of the Cortex-M0 applications are application-specific standard products (ASSPs) and system-on-chip (SoC). For ASSPs like mixed-signal controllers, the low gate count advantage of the Cortex-M0 allows a 32-bit processing capability to be included in chip designs that traditionally only allow 8-bit or simple 16-bit processors to be used. One example is the touch screen controller, and the Cortex-M0 processor is already used in this type of product (see Chapter 22).

For complex system-on-chips, the designs are often divided into a main application processor system and a number of subsystems for I/O controls, communication protocol processing, and system management. In some cases, Cortex-M0 processors are used in part of the subsystems to offload some activities from the main application processor and to allow a small amount of processing be carried out while the main processor is in standby mode (e.g., in battery-powered products).

Background of ARM and ARM processors

ARM has a long and successful history of 32-bit microprocessor design. Nowadays most mobile phones use several ARM processors in their design, and the application of ARM processors has been extended to many home entertainment systems, electronic toys, household electrical products, mobile computing, and industrial applications. However, unlike most semiconductor companies, ARM does not manufacture or sell these microprocessors. Instead, the processors designed by ARM are used by other semiconductor companies through a licensing business model. ARM provides a number of different processor designs, and the Cortex-M0 is one of the products in the Cortex-M processor family that is designed for microcontroller applications.

The Cortex-M0 is not the first processor ARM developed for the microcontroller market. ARM processors have been around for more than 20 years. ARM was originally formed in 1990 as Advanced RISC Machine Ltd, a joint venture of Acorn Computer Group, Apple Computer, and VLSI Technology. In 1991, ARM released the ARM6 processor family and VLSI Technology became the initial licensee. Since then, a number of well-known semiconductor companies including Texas Instruments, NXP (formerly Philips Semiconductors), ST Microelectronics, NEC, and Toshiba have also become ARM licensees. The use of the ARM processor also extended to various consumer products as well as industrial applications.

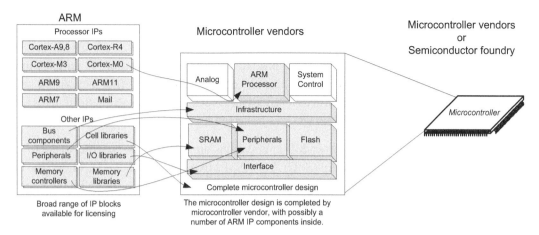

Figure 1.1:
Use of intellectual property (IP) in microcontroller design.

Unlike most microcontroller vendors, ARM does not manufacture microcontroller products. The ARM business is based on intellectual property (IP) licensing (Figure 1.1). With this business model, ARM provides the design of processors to microcontroller designers, and these companies integrate the processor design with the rest of the chip design. Apart from the processors, ARM also designs system infrastructure blocks, peripherals, and silicon process libraries, which microcontroller vendors may choose to integrate into their chips.

In addition, from IP licensing ARM also provides software development tools including C compilers, debug interface hardware, and hardware platforms, as well as services such as consultancy and technical training courses. All these activities support microcontroller vendors as well as software developers in using ARM technology.

One of the most successful processor products from ARM is the ARM7TDMI processor, which is used in many 32-bit microcontrollers around the world. Unlike traditional 32-bit processors, the ARM7TDMI supports two instruction sets, one called the ARM instruction set with 32-bit instructions and another 16-bit instruction set called Thumb. By allowing both instruction sets to be used on the processor, the code density is greatly increased, hence reducing memory footprint. At the same time, critical tasks can still execute with good speed. This enables ARM processors to be used in many portable devices that require low power and small memory. As a result, ARM processors are the first choice for mobile devices like mobile phones.

After the success of ARM7TDMI, ARM continues to develop faster and more powerful processors (Figure 1.2). For example, the ARM9 processor family is used in a large number of 32-bit microcontrollers and ARM11 devices, which are popular for use in smart phones and personal digital assistants (PDAs). Many new technologies have been introduced in these processors, like the Jazelle Java acceleration support and TrustZone, a feature that provides

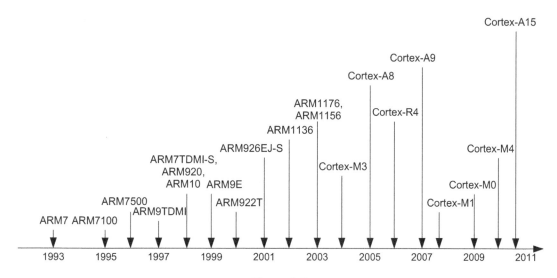

Figure 1.2:
Timeline of popular ARM processors.

enhanced on-chip level system security. Nowadays, most mobile phones contain at least one ARM processor, like the ARM11, and the recent Cortex-A8 and Cortex-A9 processor products, which are used in mobile Internet devices. In September 2010, ARM announced the introduction of the Cortex-A15 MPCore processor, which can be used in an even wider range of computing platforms from mobile applications, to high-end home entertainment systems, to server applications.

Besides the high-end processor products, ARM also increased the product portfolio for low-end products like microcontrollers. In 2004, ARM introduced the Cortex-M3 processor design; products started shipping in 2006, unleashing a new trend in the microcontroller market. Following the success of the Cortex-M3, the entry-level Cortex-M0 and the Cortex-M4, featuring floating point and DSP capabilities, joined the Cortex-M family. Today there are more than 60 licensees using Cortex-M family processors, and many microcontroller vendors are already shipping Cortex-M-based microcontrollers. In the third quarter of 2008, ARM's partners shipped more than 1 billion ARM processors. According to analysis[1] in 2009, the worldwide CPU core shipments (with CPU licensing) reached 5.3 billion in 2008, and they are expected to increase to 10 billion in 2012.

Cortex-M0 Processor Specification and ARM Architecture

The specification of the Cortex-M0 is outlined in a number of ARM documents. The Cortex-M0 Devices Generic User Guide (reference 1) covers the programmer's model, instruction set,

---

[1] Market analysis from the Linley Group (Mountain View, California) (www.eetimes.eu/semi/214600305).

and general information about the architecture. The full details of the instruction set, programmer's model, and other topics are specified in a document called the ARMv6-M Architecture Reference Manual (reference 3). The timing information of the processor core, and implementation-related information are described in a document called the Cortex-M0 Technical Reference Manual (TRM) (reference 2). These documents are available from ARM web site (www.arm.com). Please note that the download of the ARMv6-M Architecture Reference Manual requires a registration process.

You might wonder what "ARMv6-M" means. Because there are several generations of ARM processors, the architectures of these processors are also divided into different versions. For example, the popular ARM7TDMI processor is based on ARM architecture version 4T, or ARMv4T (T for Thumb instruction support), whereas the Cortex-M3 processor design is based on architecture version 7-M. The design of the Cortex-M0 processor code is based on a version of the architecture called ARMv6-M. The instruction set defined in ARMv6-M is a superset of Thumb instruction set in ARM7TDMI, which provides a highly efficient instruction set and excellent code density. It has access to 4GB of linear memory address space but does not need to use memory paging as in some 8-bit and 16-bit microcontrollers.

For each version, the architecture document covers the following:

- Programmer's model
- Instruction set details
- Exception mechanism
- Memory model
- Debug architecture

For each generation of ARM processors, new instructions and architectural features are added to the processor architecture specification, which results in various versions of architecture. The version number of the architecture is separated from the processor naming, and it is possible for a processor family to contain more than one architecture version. For example, early versions of ARM9 processors (ARM920T and ARM922T) are both architecture version 4T, whereas newer versions of ARM9 processors (ARM926EJ-S, ARM946E, ARM966E, etc.) are based on architecture version 5TE. Table 1.1 shows some of the commonly used ARM processors and their architecture versions.

You might notice that in the past, most ARM processors used a number of suffixes to specify the features available on the processor. Some of the feature suffixes are no longer in use as they have become standard features in newer ARM processors or have been replaced by newer technologies.

After the release of the ARM11 processor family, it was decided that some of the new features and technologies used in the most advanced ARM processors are just as useful to the lower-cost or deeply embedded processor devices. For example, the Thumb-2 Instruction Set

Table 1.1: Examples of ARM Processors and Their Architecture

Processor	Suffixes in Processor Names	Architecture
ARM7TDMI	T = Thumb instruction support D = JTAG debugging M = fast multiplier I = Embedded ICE module	ARMv4T
ARM920T	T = Thumb instruction support	ARMv4T
ARM946E, ARM966E	E = Enhanced digital signal processing instructions	ARMv5TE
ARM926EJ-S	E = Enhanced digital signal processing instructions, J = Jazelle (Java accelerator) S = Synthesizable design	ARMv5TE
ARM1136J(F)-S	(F) = Optional floating point	ARMv6
ARM1176JZ(F)-S	Z = TrustZone security support	ARMv6
ARM1156T2(F)-S	T2 = Thumb-2 Instruction Set support	ARMv6
Cortex-A8, Cortex-A9, Cortex-A15	A = Application	ARMv7-A
Cortex-R4(F)	R = RealTime, with optional floating point support	ARMv7-R
Cortex-M3	M = Microcontroller	ARMv7-M
Cortex-M1	M = Microcontroller (for FPGA)	ARMv6-M
Cortex-M0	M = Microcontroller	ARMv6-M
Cortex-M4	M = Microcontroller with optional floating point support	ARMv7E-M

provides a performance boost to ARM processors running in Thumb state, and CoreSight Debug architecture provides scalable debug technologies that give better debug capability than previous solutions. As a result, starting from version 7, the architecture is divided into three profiles, targeted at different product ranges. This new generation of processors is called "Cortex," with a suffix to identify individual designs and indicate which architectural profile they belong to (Figure 1.3 and Table 1.2).

Table 1.2: Three Profiles in the ARMv7 Architecture

Architecture	Targets
ARMv7-A(e.g., Cortex-A9)	Application processors that are required to support complex applications like smart phones, PDAs, and GPSs
ARMv7-R(e.g., Cortex-R4)	Real-time, high-performance processors to support highly demanding applications like hard disk controller and automotive control systems
ARMv7-M(e.g., Cortex-M3)	Microcontroller processors for industrial control like generic microcontrollers or cost-sensitive embedded systems like low-cost consumer products

So how does this relate to the Cortex-M0 processor? Following the success of the Cortex-M3 processor release, ARM decided to further expand its product range in the microcontroller applications. The first step is to allow users to implement their ARM processor on

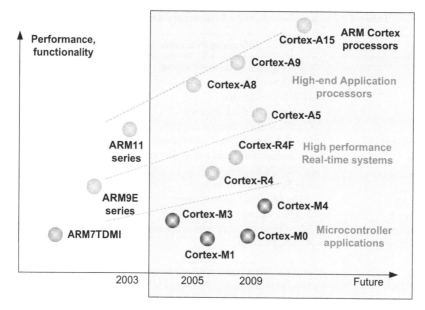

Figure 1.3:
Diversity of processor architecture to three areas in the Cortex processor family.

a field programmable gate array (FPGA) easily, and the second step is to address the ultra-low-power embedded processor. To do this, ARM took the Thumb instruction set from the existing ARMv6 architecture and developed a new architecture based on the exception and debug features in the ARMv7-M architecture. As a result, ARMv6-M was formed, and the processors based on this architecture are the Cortex-M0 processor (for microcontroller and ASICs) and the Cortex-M1 processor (for FPGA) (Figure 1.4).

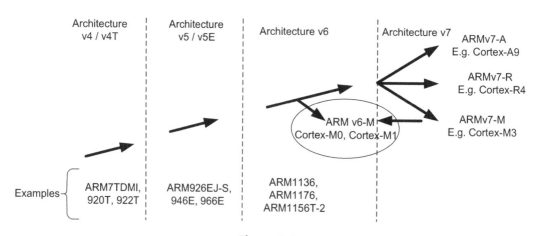

Figure 1.4:
The evolution of ARM processor architecture.

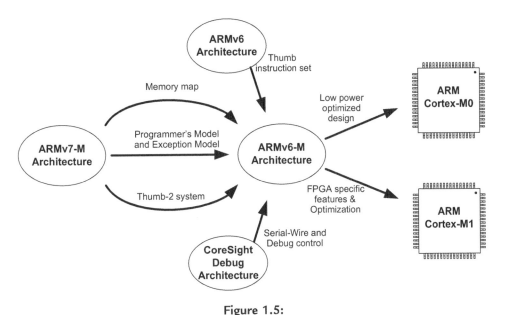

Figure 1.5:
ARMv6-M architecture provides attractive features from various ARM architectures.

This development results in a processor architecture that is very small and efficient and yet is easy to use and can achieve a high performance. Similar to the Cortex-M3 processor, both the Cortex-M0 and the Cortex-M1 processors include a nested vectored interrupt controller (NVIC) and use the same exception/interrupt mechanism. They also use a programmer's mode similar to ARMv7-M, which defines Thread mode and Exception mode (Figure 1.5). They also support the CoreSight Debug architecture, which makes it easy for users to develop and test applications.

The rest of the book will only focus on the Cortex-M0 processor and not the Cortex-M1 processor.

ARM Processors and the ARM Ecosystem

What makes the ARM architecture special compared to proprietary architectures? Aside from the processor technology, the ecosystem surrounding ARM development plays a very important role.

As well as working directly with the microcontroller vendors that offer ARM processor-based devices, ARM works closely with vendors that provide the ecosystem supporting those devices. These include vendors providing compilers, middleware, operating systems, and development tools, as well as training and design services companies, distributors, academic researchers, and so on (Figure 1.6).

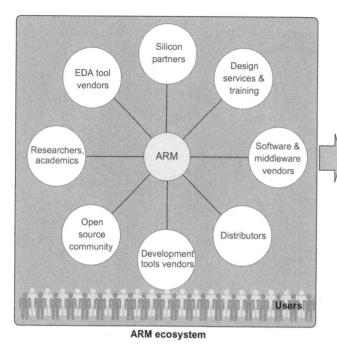

Choices
- More choices of microcontrollers
- More choice on development tools
- More development boards
- More open source project support
- More OS support
- More middleware and software solutions

Knowledge sharing
- Resources on the Internet
- Large user community
- Technical forums
- Seminars and webinars (many free)
- Strong supports

Figure 1.6:
The ARM ecosystem.

The ARM ecosystem allows a lot more choices. Apart from choice of microcontroller devices from different vendors, the user also has a greater choice of software tools. For example, you can get development tools from Keil, IAR Systems, TASKING, CodeSourcery, Rowley Associates, GNU C compiler, and the like. As a result, you have much more freedom in project development. Information on some of the compiler products is presented in Chapter 22.

ARM also invests in various open source projects to help the open source community to develop software on ARM platforms. The combined effort of all these parties not only makes the ARM products better, it also results in more choices of hardware and software solutions.

The ARM ecosystem also enables better knowledge sharing, which helps developers build products on ARM microcontrollers quicker and more effectively. Aside from the many Internet resources available, you can also find expert advices on web-based technical forums from ARM (some links are shown at the end of this chapter), ARM microcontroller vendors, and others. Microcontroller vendors, distributors, and other training service providers also organize regular ARM microcontroller training courses. The open nature of the ARM ecosystem also enables healthy competition. As a result, users are getting high-quality products at competitive prices.

Any company that develops ARM products or uses ARM technologies can become an ARM partner by becoming a member of the ARM Connected Community. The ARM Connected Community is a global network of companies aligned to provide a complete solution, from design to manufacture and end use, for products based on the ARM architecture. ARM offers a variety of resources to Connected Community members, including promotional programs and peer-networking opportunities that enable a variety of ARM partners to come together to provide end-to-end customer solutions. Today, the ARM Connected Community has more than 700 corporate members. Joining the ARM Connected Community is easy; details are presented on the ARM web site (http://cc.arm.com).

Getting Started with the Cortex-M0 Processor

The Cortex-M0 is easy to use and is supported by various microcontroller vendors and development tools vendors. For example, software for the Cortex-M0 processor can be developed with the ARM Keil Microcontroller Development Kit (MDK, and sometimes referred as MDK-ARM in ARM/Keil documentation), the ARM RealView Development Suite (RVDS), various GNU tool chains (e.g., CodeSourcery G++), and a number of other embedded development packages.

Because the Cortex-M0 processor is extremely C friendly, you can reuse a majority of your existing C programs. Also, a number of embedded operating systems supports the Cortex-M0 processor, including the RTX kernel from Keil, µC/OS-II/III from Micriµm, embOS from SEGGER, ThreadX from Express Logic, and µClinux from the open source community. In addition, the Cortex-M0 can also reuse most of the Thumb assembly code written for the ARM7TDMI processor.

Organization of This Book and Resources

The contents of this book can be divided into the areas outlined in Table 1.3.

Table 1.3: Organization of This Book

Chapters	Descriptions
1	Introduction
2-7	Cortex-M0 processor features, architecture, programmer's model, instruction set, and memory map
8-13	Exceptions, interrupt, and various features
14-21	Software development
22	Product information
Appendix	Instruction set summary, quick references

Apart from this book and documentation from the ARM web site, you can get additional information from the following sources:

- Documentation from microcontroller vendors
- ARM forum (www.arm.com/forums)
- Keil forum (www.keil.com/forum)
- OnARM website (www.onarm.com)
- ARM Connected Community web page (www.arm.com/community)
- Forums of various microcontroller vendors.

## CHAPTER 2

Cortex-M0 Technical Overview

General Information on the Cortex-M0 Processor

The Cortex-M0 processor is a 32-bit Reduced Instruction Set Computing (RISC) processor with a von Neumann architecture (single bus interface). It uses an instruction set called Thumb, which was first supported in the ARM7TDMI processor; however, several newer instructions from the ARMv6 architecture and a few instructions from the Thumb-2 technology are also included. Thumb-2 technology extended the previous Thumb instruction set to allow all operations to be carried out in one CPU state. The instruction set in Thumb-2 included both 16-bit and 32-bit instructions; most instructions generated by the C compiler use the 16-bit instructions, and the 32-bit instructions are used when the 16-bit version cannot carry out the required operations. This results in high code density and avoids the overhead of switching between two instruction sets.

In total, the Cortex-M0 processor supports only 56 base instructions, although some instructions can have more than one form. Although the instruction set is small, the Cortex-M0 processor is highly capable because the Thumb instruction set is highly optimized. Academically, the Cortex-M0 processor is classified as load-store architecture, as it has separate instructions for reading and writing to memory, and instructions for arithmetic or logical operations that use registers.

A simplified block diagram of the Cortex-M0 is shown in Figure 2.1.

The processor core contains the register banks, ALU, data path, and control logic. It is a three-stage pipeline design with fetch stage, decode stage, and execution stage. The register bank has sixteen 32-bit registers. A few registers have special usages.

The Nested Vectored Interrupt Controller (NVIC) accepts up to 32 interrupt request signals and a nonmaskable interrupt (NMI) input. It contains the functionality required for comparing priority between interrupt requests and the current priority level so that nested interrupts can be handled automatically. If an interrupt is accepted, it communicates with the processor so that the processor can execute the correct interrupt handler.

The Wakeup Interrupt Controller (WIC) is an optional unit. In low-power applications, the microcontroller can enter standby state with most of the processor powered down. In this situation, the WIC can perform the function of interrupt masking while the NVIC and the

The Definitive Guide to the ARM Cortex-M0. DOI: 10.1016/B978-0-12-385477-3.10002-3
Copyright © 2011 Man Cheung Joseph Yiu. Published by Elsevier Inc. All rights reserved.

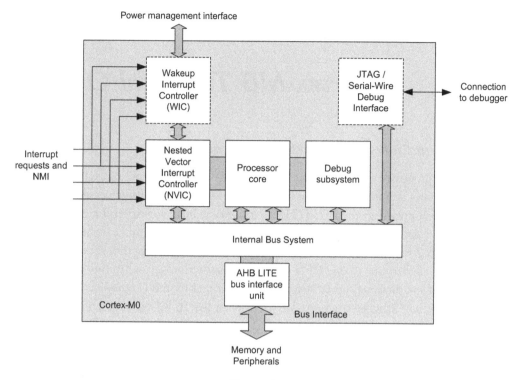

Figure 2.1:
Simplified block diagram of the Cortex-M0 processor.

processor core are inactive. When an interrupt request is detected, the WIC informs the power management to power up the system so that the NVIC and the processor core can then handle the rest of the interrupt processing.

The debug subsystem contains various functional blocks to handle debug control, program breakpoints, and data watchpoints. When a debug event occurs, it can put the processor core in a halted state so that embedded developers can examine the status of the processor at that point.

The JTAG or serial wire interface units provide access to the bus system and debugging functionalities. The JTAG protocol is a popular five-pin communication protocol commonly used for testing. The serial wire protocol is a newer communication protocol that only requires two wires, but it can handle the same debug functionalities as JTAG.

The internal bus system, the data path in the processor core, and the AHB LITE bus interface are all 32 bits wide. AHB-Lite is an on-chip bus protocol used in many ARM processors. This bus protocol is part of the Advanced Microcontroller Bus Architecture (AMBA) specification, a bus architecture developed by ARM that is widely used in the IC design industry.

The ARM Cortex-M0 Processor Features

The ARM Cortex-M0 processor contains many features. Some are visible system features, and others are not visible to embedded developers.

System Features

- Thumb instruction set. Highly efficient, high code density and able to execute all Thumb instructions from the ARM7TDMI processor.
- High performance. Up to 0.9 DMIPS/MHz (Dhrystone 2.1) with fast multiplier or 0.85 DMIPS/MHz with smaller multiplier.
- Built-in Nested Vectored Interrupt Controller (NVIC). This makes interrupt configuration and coding of exception handlers easy. When an interrupt request is taken, the corresponding interrupt handler is executed automatically without the need to determine the exception vector in software.
- Interrupts can have four different programmable priority levels. The NVIC automatically handles nested interrupts.
- Deterministic exception response timing. The design can be set up to respond to exceptions (e.g., interrupts) with a fixed number of cycles (constant interrupt latency arrangement) or to respond to the exception as soon as possible (minimum 16 clock cycles).
- Nonmaskable interrupt (NMI) input for safety critical systems.
- Architectural predefined memory map. The memory space of the Cortex-M0 processor is architecturally predefined to make software porting easier and to allow easier optimization of chip design. However, the arrangement is very flexible. The memory space is linear and there is no memory paging required like in a number of other processor architectures.
- Easy to use and C friendly. There are only two modes (Thread mode and Handler mode). The whole application, including exception handlers, can be written in C without any assembler.
- Built-in optional System Tick timer for OS support. A 24-bit timer with a dedicated exception type is included in the architecture, which the OS can use as a tick timer or as a general timer in other applications without an OS.
- SuperVisor Call (SVC) instruction with a dedicated SVC exception and PendSV (Pendable Supervisor service) to support various operations in an embedded OS.
- Architecturally defined sleep modes and instructions to enter sleep. The sleep features allow power consumption to be reduced dramatically. Defining sleep modes as an architectural feature makes porting of software easier because sleep is entered by a specific instruction rather than implementation defined control registers.
- Fault handling exception to catch various sources of errors in the system.

Implementation Features

- Configurable number of interrupts (1 to 32)
- Fast multiplier (single cycle) or small multiplier (for a smaller chip area and lower power, 32 cycles)
- Little endian or big endian memory support
- Optional Wakeup Interrupt Controller (WIC) to allow the processor to be powered down during sleep, while still allowing interrupt sources to wake up the system
- Very low gate count, which allows the design to be implemented in mixed signal semiconductor processes

Debug Features

- Halt mode debug. Allows the processor activity to stop completely so that register values can be accessed and modified. No overhead in code size and stack memory size.
- CoreSight technology. Allows memories and peripherals to be accessed from the debugger without halting the processor. It also allows a system-on-chip design with multiple processors to share a single debug connection.
- Supports JTAG connection and serial wire debug connections. The serial wire debug protocol can handle the same debug features as the JTAG, but it only requires two wires and is already supported by a number of debug solutions from various tools vendors.
- Configurable number of hardware breakpoints (from 0 to maximum of 4) and watchpoints (from 0 to maximum of 2). The chip manufacturer defines this during implementation.
- Breakpoint instruction support for an unlimited number of software breakpoints.
- All debug features can be omitted by chip vendors to allow minimum size implementations.

Others

- Programmer's model similar to the ARM7TDMI processor. Most existing Thumb code for the ARM7TDMI processor can be reused. This also makes it easy for ARM7TDMI users, as there is no need to learn a new instruction set.
- Compatible with the Cortex-M1 processor. This allows users of the Cortex-M1 processor to migrate their FPGA designs to an ASICs easily.
- Forward compatibility with the ARM Cortex-M3 and Cortex-M4 processors. All instructions supported in the Cortex-M0 processor are supported on the Cortex-M3 processor, which allows an easy upgrade path.
- Easy porting from the ARM Cortex-M3/M4. Because of the similarities between the architectures, many C applications for the Cortex-M3/M4 can be ported to the Cortex-M0 processor easily. This is great news for middleware vendors and embedded OS vendors, as it is straightforward to port their existing software products for Cortex-M3 microcontrollers to Cortex-M0 microcontrollers.

- Supported by various development suites including the ARM Keil Microcontroller Development Kit (MDK), the ARM RealView Development Suite (RVDS), the IAR C compiler, and the open source GNU C compiler, including tool chains based on gcc (e.g., CodeSourcery G++ development suite).
- Support of various embedded operating systems (OSs). A number of OS for the Cortex-M0 processor are available, including some free OSs. For example, the Keil MDK toolkit includes a free embedded OS called the RTX kernel. Examples of using the RTX are covered in Chapter 18.

Advantages of the Cortex-M0 Processor

With all these features on the Cortex-M0 processor, what does it really mean for an embedded developer? And why should embedded developers moved from 8-bit and 16-bit architectures?

Energy Efficiency

The most significant benefit of the Cortex-M0 processor over other 8-bit and 16-bit processors is its energy efficiency. The Cortex-M0 processor is about the same size as a typical 16-bit processor and possibly several times bigger than some of the 8-bit processors. However, it has much better performance than 16-bit and 8-bit architectures. As a result, you can put the processor into sleep mode for the majority of the time to reduce power to a minimum, yet you will still be able to get the processing task done.

For comparison, the DMIPS figures of some popular architectures are shown in Figure 2.2 and Table 2.1.

Note: You might wonder why the Dhrystone 2.1 is used for comparison while there are other well-established benchmarks like the EEMBC. However, the EEMBC has restrictions on the use of its benchmark results and therefore cannot be openly published.

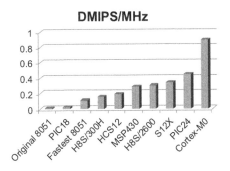

Figure 2.2:
Dhrystone comparison.

Table 2.1: Dhrystone Performance Data Based on Information Available on the Internet

Architecture	Estimated DMIPS/MHz with Dhrystone 2.1
Original 80C51	0.0094
PIC18	0.01966
Fastest 8051	0.113
H8S/300H	0.16
HCS12	0.19
MSP430	0.288
H8S/2600	0.303
S12X	0.34
PIC24	0.445
Cortex-M0	0.896 (if a small multiplier is used, the performance is 0.85)

As you can see, the Cortex-M0 processor is significantly faster than all popular 16-bit microcontrollers and eight times faster than the fastest 8051 implementation. This advantage can be used in conjunction with the sleep mode feature in the Cortex-M0 processor so that an embedded system can stay in low-power mode more often to reduce the average power consumption without losing performance. For example, Figure 2.3 illustrates that in an interrupt-driven application, the Cortex-M0 processor can have much lower average power consumption compared to 8-bit and 16-bit microcontrollers.

Although some 8-bit microcontrollers having a very low gate count, which can reduce the sleep mode current consumption, the average current consumed by the processor can be much larger than that for the Cortex-M0. The comparison is even more significant at the chip level, when including the power consumption of the memory system and the peripherals.

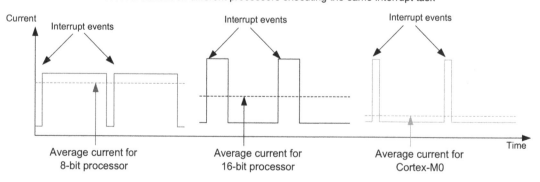

Figure 2.3:
The Cortex-M0 provides better energy saving at the same processing performance.

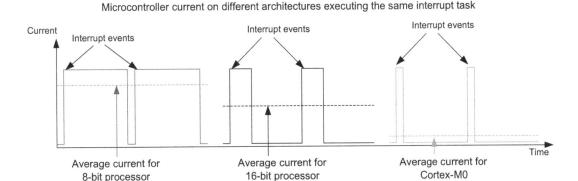

Figure 2.4:
At the chip level, the duty cycle of processor activity becomes more significant.

In a microcontroller design, the processor core only takes a small amount of the chip area, whereas a large portion of the power is consumed by other parts of the chip. As a result, the duty cycle (portion of time where the processor is active) dominates the power calculation at chip component level, as shown in Figure 2.4.

When running other applications that are not interrupt driven, the clock frequency for the Cortex-M0 processor can be reduced significantly, compared to 8-bit/16-bit processors, to lower the power consumption. Even if an 8-bit or 16-bit microcontroller has a lower operating current than the Cortex-M0 at the same clock frequency, you can still achieve lower power consumption on the Cortex-M0 by reducing the clock speed without losing the performance level compared to 8-bit/16-bit solutions (Figure 2.5).

Although other 32-bit microcontrollers are available that some of them have a higher performance than the Cortex-M0, their processor sizes are a number of times larger than the Cortex-M0 processor. As a result, the average power consumptions of these microcontrollers are higher than the Cortex-M0 microcontrollers.

Limitations in 8-Bit and 16-Bit Architectures

Another important reason to use the 32-bit Cortex-M0 processor rather than the traditional 16-bit or 8-bit architectures is that it does not have many architectural limitations found in these architectures.

The first obvious limitation of 8-bit and 16-bit architectures is memory size. Whereas program size and data RAM size can directly limit the capability of an embedded product, other less obvious limitations like stack memory size (e.g., 8051 stack is located in the internal RAM, which is limited to 256 bytes, including the register bank space) can also affect what you can

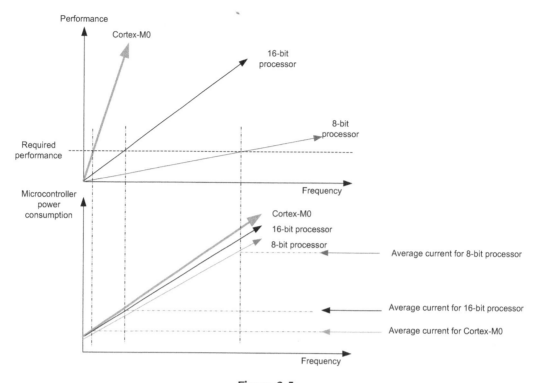

Figure 2.5:
The Cortex-M0 can provide lower power consumption by running at lower clock frequencies.

develop. With the ARM architecture, the memory space is much larger and the stack is located in system memory, making it much more flexible.

Many 8-bit and 16-bit microcontrollers allow access to a larger memory range by dividing memory space into memory pages. By doing so, development of software can become difficult because accessing addresses in a different memory page is not straightforward. It also increases code size and reduces performance because of the overhead in switching memory pages. For example, a processing task with a program size larger than one memory page might need page-switching code to be inserted within it, or it might need to be partitioned into multiple parts. ARM microcontrollers use 32-bit linear addresses and do not require memory paging; therefore, they are easier to use and provide better efficiency.

Another limitation of 8-bit microcontroller architectures can be the limitations of their instruction sets. For example, 8051 heavily relies on the accumulator register to handle data processing and memory transfers. This increases the code size because you need to keep transferring data into the accumulator and taking it out before and after operations. For instance, when processing integer multiplications on an 8051, a lot of data transfer is required to move data in and out of the ACC (Accumulator) register and B register.

Addressing modes account for another factor that limits performance in many 8-bit and 16-bit microcontrollers. A number of addressing modes are available in the Cortex-M0, allowing better code density and making it easier to use.

The instruction set limitations on 8-bit and 16-bit architectures not only reduce the performance of the embedded system, but they also increase code size and hence increase power consumption, as a larger program memory is required.

Easy to Use, Software Portability

When compared to other processors, including many 32-bit processors, the ARM Cortex microcontrollers are much easier to use. All the software code for the ARM Cortex microcontrollers can be written in C, allowing shorter software development time as well as improving software portability. Even if a software developer decided to use assembly code, the instruction set is fairly easy to understand. Furthermore, because the programmer's model is similar to ARM7TDMI, those who are already familiar with ARM processors will quickly become familiar with the Cortex microcontrollers.

The architecture of the Cortex-M0 also allows an embedded OS to be implemented efficiently. In complex applications, use of an embedded OS can make it easier to handle parallel tasks.

Wide Range of Choices

Because ARM operates as an intellectual property (IP) supplier, and ARM processors are adopted by most of the microcontroller vendors, you can easily find the right ARM microcontroller for your application. Also, you do not need to change your development tools if you change the target microcontroller between different vendors.

Apart from the hardware, you can also find a wide range of choices of embedded OS, code libraries, development tools, and other resources. This ecosystem allows you to focus on product development and get your product ready faster.

Low-Power Applications

One of the key targets of the Cortex-M0 processor is low power. The result is that the processor consumes only 12μW/MHz with a 65nm semiconductor process or 85μW/MHz with a 180nm semiconductor process. This is very low power consumption for a 32-bit processor. How was this target achieved?

ARM put a lot of effort into various areas to ensure the Cortex-M0 processor could reach its low power-consumption target. These areas included the following:

- Small gate count
- High efficiency

- Low-power features (sleep modes)
- Logic cell enhancement

Let us take a look at these areas one by one.

Small Gate Count

The Cortex-M0 processor's small gate count characteristic directly reduces the active current and leakage current of the processor. During the development of the Cortex-M0 processor, various design techniques and optimizations were used to make the circuit size as small as possible. Each part of the design was carefully developed and reviewed to ensure that the circuit size is small (it is a bit like writing an application program in assembly to achieve the best optimization). This allows the gate count to be 12k gates at minimum configuration. Typically, the gate count could be 17k to 25k gates when including more features. This is about the same size or smaller than typical 16-bit microprocessors, with more than double the system performance.

High Efficiency

By having a highly efficient architecture, embedded system designers can develop their product so that it has a lower clock frequency while still being able to provide the required performance, reducing the active current of the product. With a performance of 0.9DMIPS/MHz, despite not being very high compared with some modern 32-bit processors, the Dhrystone benchmark result of Cortex-M0 is still higher than the older generation of 32-bit desktop processors like the 80486DX2 (0.81DMIPS/MHz), and it is a lot smaller. The high efficiency of the Cortex-M0 processor is mostly due to the efficiency of the Thumb instruction set, as well as highly optimized hardware implementation.

Low-Power Features

The Cortex-M0 processors have a number of low-power features that allow embedded product developers to reduce the product's power consumption. First, the processor provides two sleep modes and they can be entered easily with "Wait-for-Interrupt" (WFI) or "Wait-for-Event" (WFE) instructions. The power management unit on the chip can use the sleep status of the processor to reduce the power consumption of the system. The Cortex-M0 processor also provides a "Sleep-on-Exit" feature, which causes the processor to run only when an interrupt service is required. In addition, the Cortex-M0 processor has been carefully developed so that some parts of the processor, like the debug system, can be switched off when not required.

Apart from these normal sleep features, the Cortex-M0 processor also supports a unique feature called the Wakeup Interrupt Controller (WIC). This allows the processor to be powered down while still allowing interrupt events to power up the system and resume operation almost instantaneously when required. This greatly reduces the leakage current (static power consumption) of the system during sleep.

Logic Cell Enhancement

In recent years, there have been enhancements in logic cell designs. Apart from pushing logic gate designs to smaller transistor sizes, the Physical IP (intellectual property) division in ARM has also been working hard to find innovative ways to reduce power consumption in embedded systems. One of the major developments is the introduction of the Ultra Low Leakage (ULL) logic cell library. The first ULL cell library has been developed with a 0.18um process. Apart from reducing the leakage current, the new cell library also supports special state retention cells that can hold state information while the rest of the system is powered down. ARM also works with leading EDA tools vendors to allow chip vendors to make use of these new technologies in their chip designs.

Cortex-M0 Software Portability

The Cortex-M0 is the third processor released from the Cortex-M family. The Cortex-M processors are developed to target microcontroller products and other products that require a processor architecture that is easy to use and has flexible interrupt support. The first Cortex-M processor released was the Cortex-M3 processor, a high-performance processor with many advanced features. The second processor released was the Cortex-M1, a processor developed for FPGA applications. Despite being developed for different types of applications, they all have a consistent architecture, similar programmer's models, and use a compatible instruction set.

Both Cortex-M0 and Cortex-M1 processors are based on the ARMv6-M architecture. Therefore, they have exactly the same instruction set and programmer's model. However, they have different physical characteristics like instruction timing and have different system features.

The Cortex-M3 processor is based on the ARMv7-M architecture, and its Thumb-2 instruction set is a superset of the instruction set used in ARMv6-M. The programmer's model is also similar to ARMv6-M. As a result, software developed for the Cortex-M0 can run on the Cortex-M3 processor without changes (Figure 2.6).

The similarity between the Cortex-M processors provides various benefits. First, it provides better software portability. In most cases, C programs can be transferred between these processors without changes. And binary images from Cortex-M0 or Cortex-M1 processors can run on a Cortex-M3 processor because of its upward compatibility.

The second benefit is that the similarities between Cortex-M processors allow development tool chains to support multiple processors easily. Apart from similarities on the instruction set and programmer's model, the debug architecture is also similar.

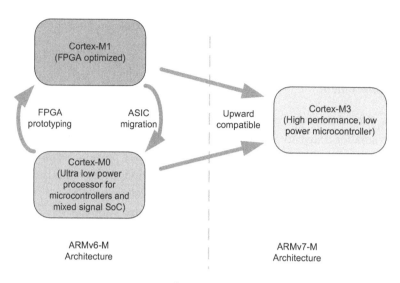

Figure 2.6:
Cortex-M0 compatibility.

The consistency of instruction set and programmer's model also make it easier for embedded programmers to migrate between different products and projects without facing a sharp learning curve.

## CHAPTER 3

Architecture

Overview

The ARMv6-M architecture that the Cortex-M0 processor implemented covers a number of different areas. The complete details of the ARMv6-M architecture are documented in the ARMv6-M Architecture Reference Manual [reference 3]. This document is available from the ARM web site via a registration process. However, you do not have to know the complete details of the architecture to start using a Cortex-M0 microcontroller. To use a Cortex-M0 device with C language, you only need to know the memory map, the peripheral programming information, the exception handling mechanism, and part of the programmer's model.

In this chapter, we will cover the programmer's model and a basic overview of the memory map and exceptions. Most users of the Cortex-M0 processor will work in C language; as a result, the underlying programmer's model will not be visible in the program code. However, it is still useful to know about the details, as this information is often needed during debugging and it will also help readers to understand the rest of this book.

Programmer's Model

Operation Modes and States

The Cortex-M0 processor has two operation modes and two states (Figure 3.1).

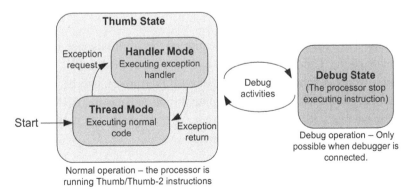

Figure 3.1:
Processor modes and states in the Cortex-M0 processor.

The Definitive Guide to the ARM Cortex-M0. DOI: 10.1016/B978-0-12-385477-3.10003-5

When the processor is running a program, it is in the Thumb state. In this state, it can be either in the Thread mode or the Handler mode. In the ARMv6-M architecture, the programmer's model of Thread mode and Handler mode are almost completely the same. The only difference is that Thread mode can use a shadowed stack pointer (Figure 3.7, presented later in the chapter) by configuring a special register called CONTROL. Details of stack pointer selection are covered later in this chapter.

The Debug state is used for debugging operation only. Halting the processor stops the instruction execution and enter debug state. This state allows the debugger to access or change the processor register values. The debugger can access system memory locations in either the Thumb state or the Debug state.

When the processor is powered up, it will be running in the Thumb state and Thread mode by default.

Registers and Special Registers

To perform data processing and controls, a number of registers are required inside the processor core. If data from memory are to be processed, they have to be loaded from the memory to a register in the register bank, processed inside the processor, and then written back to the memory if needed. This is commonly called a "load-store architecture." By having a sufficient number of registers in the register bank, this mechanism is easy to use and is C friendly. It is easy for C compilers to compile a C program into machine code with good performance. By using internal registers for short-term data storage, the amount of memory accesses can be reduced.

The Cortex-M0 processor provides a register bank of 13 general-purpose 32-bit registers and a number of special registers (Figure 3.2).

The register bank contains sixteen 32-bit registers. Most of them are general-purpose registers, but some have special uses. The detailed descriptions for these registers are as follows.

R0–R12

Registers R0 to R12 are for general uses. Because of the limited space in the 16-bit Thumb instructions, many of the Thumb instructions can only access R0 to R7, which are also called the low registers, whereas some instructions, like MOV (move), can be used on all registers. When using these registers with ARM development tools such as the ARM assembler, you can use either uppercase (e.g., R0) or lowercase (e.g., r0) to specify the register to be used. The initial values of R0 to R12 at reset are undefined.

R13, Stack Pointer (SP)

R13 is the stack pointer. It is used for accessing the stack memory via PUSH and POP operations. There are physically two different stack pointers in Cortex-M0. The main stack

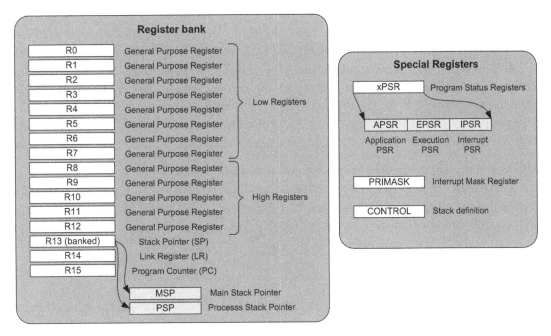

Figure 3.2:
Registers in the Cortex-M0 processor.

pointer (MSP, or SP_main in ARM documentation) is the default stack pointer after reset, and it is used when running exception handlers. The process stack pointer (PSP, or SP_process in ARM documentation) can only be used in Thread mode (when not handling exceptions). The stack pointer selection is determined by the CONTROL register, one of the special registers that will be introduced later.

When using ARM development tools, you can access the stack pointer using either "R13" or "SP." Both uppercase and lowercase (e.g., "r13" or "sp") can be used. Only one of the stack pointers is visible at a given time. However, you can access to the MSP or PSP directly when using the special register access instructions MRS and MSR. In such cases, the register names "MSP" or "PSP" should be used.

The lowest two bits of the stack pointers are always zero, and writes to these two bits are ignored. In ARM processors, PUSH and POP are always 32-bit accesses because the registers are 32-bit, and the transfers in stack operations must be aligned to a 32-bit word boundary. The initial value of MSP is loaded from the first 32-bit word of the vector table from the program memory during the startup sequence. The initial value of PSP is undefined.

It is not necessary to use the PSP. In many applications, the system can completely rely on the MSP. The PSP is normally used in designs with an OS, where the stack memory for OS Kernel and the thread level application code must be separated.

R14, Link Register (LR)

R14 is the Link Register. The Link Register is used for storing the return address of a subroutine or function call. At the end of the subroutine or function, the return address stored in LR is loaded into the program counter so that the execution of the calling program can be resumed. In the case where an exception occurs, the LR also provides a special code value, which is used by the exception return mechanism. When using ARM development tools, you can access to the Link Register using either "R14" or "LR." Both upper and lowercase (e.g., "r14" or "lr") can be used.

Although the return address in the Cortex-M0 processor is always an even address (bit[0] is zero because the smallest instructions are 16-bit and must be half-word aligned), bit zero of LR is readable and writeable. In the ARMv6-M architecture, some instructions require bit zero of a function address set to 1 to indicate Thumb state.

R15, Program Counter (PC)

R15 is the Program Counter. It is readable and writeable. A read returns the current instruction address plus four (this is caused by the pipeline nature of the design). Writing to R15 will cause a branch to take place (but unlike a function call, the Link Register does not get updated).

In the ARM assembler, you can access the Program Counter, using either "R15" or "PC," in either upper or lower case (e.g., "r15" or "pc"). Instruction addresses in the Cortex-M0 processor must be aligned to half-word address, which means the actual bit zero of the PC should be zero all the time. However, when attempting to carry out a branch using the branch instructions (BX or BLX), the LSB of the PC should be set to 1.[2] This is to indicate that the branch target is a Thumb program region. Otherwise, it can imply trying to switch the processor to ARM state (depending on the instruction used), which is not supported and will cause a fault exception.

xPSR, combined Program Status Register

The combined Program Status Register provides information about program execution and the ALU flags. It is consists of the following three Program Status Registers (PSRs) (Figure 3.3):

- Application PSR (APSR)
- Interrupt PSR (IPSR)
- Execution PSR (EPSR)

The APSR contains the ALU flags: N (negative flag), Z (zero flag), C (carry or borrow flag), and V (overflow flag). These bits are at the top 4 bits of the APSR. The common use of these flags is to control conditional branches.

---

[2] Not required when a move (MOV) or add (ADD) instruction is used to modify the program counter.

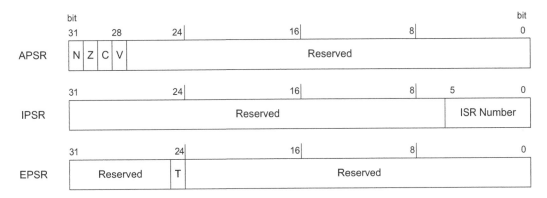

Figure 3.3:
APSR, IPSR, and EPSR.

The IPSR contains the current executing interrupt service routine (ISR) number. Each exception on the Cortex-M0 processor has a unique associated ISR number (exception type). This is useful for identifying the current interrupt type during debugging and allows an exception handler that is shared by several exceptions to know what exception it is serving.

The EPSR on the Cortex-M0 processor contains the T-bit, which indicates that the processor is in the Thumb state. On the Cortex-M0 processor, this bit is normally set to 1 because the Cortex-M0 only supports the Thumb state. If this bit is cleared, a hard fault exception will be generated in the next instruction execution.

These three registers can be accessed as one register called xPSR (Figure 3.4). For example, when an interrupt takes place, the xPSR is one of the registers that is stored onto the stack memory automatically and is restored automatically after returning from an exception. During the stack store and restore, the xPSR is treated as one register.

Figure 3.4:
xPSR.

Direct access to the Program Status Registers is only possible through special register access instructions. However, the value of the APSR can affect conditional branches and the carry flag in the APSR can also be used in some data processing instructions.

Behaviors of the Application Program Status Register (APSR)

Data processing instructions can affect destination registers as well as the APSR, which is commonly known as ALU status flags in other processor architectures. The APSR is essential for controlling conditional branches. In addition, one of the APSR flags, the C (Carry) bit, can also be used in add and subtract operations.

There are four APSR flags in the Cortex-M0 processor, and they are identified in Table 3.1.

Table 3.1: ALU Flags on the Cortex-M0 Processor

Flag	Descriptions
N (bit 31)	Set to bit [31] of the result of the executed instruction. When it is "1", the result has a negative value (when interpreted as a signed integer). When it is "0", the result has a positive value or equal zero.
Z (bit 30)	Set to "1" if the result of the executed instruction is zero. It can also be set to "1" after a compare instruction is executed if the two values are the same.
C (bit 29)	Carry flag of the result. For unsigned addition, this bit is set to "1" if an unsigned overflow occurred. For unsigned subtract operations, this bit is the inverse of the borrow output status.
V (bit 28)	Overflow of the result. For signed addition or subtraction, this bit is set to "1" if a signed overflow occurred.

A few examples of the ALU flag results are shown in Table 3.2.

Table 3.2: ALU Flags Example

Operation	Results, Flags
0x70000000 + 0x70000000	Result = 0xE0000000, N = 1, Z = 0, C = 0, V = 1
0x90000000 + 0x90000000	Result = 0x30000000, N = 0, Z = 0, C = 1, V = 1
0x80000000 + 0x80000000	Result = 0x00000000, N = 0, Z = 1, C = 1, V = 1
0x00001234 − 0x00001000	Result = 0x00000234, N = 0, Z = 0, C = 1, V = 0
0x00000004 − 0x00000005	Result = 0xFFFFFFFF, N = 1, Z = 0, C = 0, V = 0
0xFFFFFFFF − 0xFFFFFFFC	Result = 0x00000003, N = 0, Z = 0, C = 1, V = 0
0x80000005 − 0x80000004	Result = 0x00000001, N = 0, Z = 0, C = 1, V = 0
0x70000000 − 0xF0000000	Result = 0x80000000, N = 1, Z = 0, C = 0, V = 1
0xA0000000 − 0xA0000000	Result = 0x00000000, N = 0, Z = 1, C = 1, V = 0

In the Cortex-M0, almost all of the data processing instructions modify the APSR; however, some of these instructions do not update the V flag or the C flag. For example, the MULS (multiply) instruction only changes the N flag and the Z flag.

The ALU flags can be used for handling data that is larger than 32 bits. For example, we can perform a 64-bit addition by splitting the operation into two 32-bit additions. The pseudo form of the operation can be written as follows:

```
// Calculating Z = X + Y, where X, Y and Z are all 64-bit
Z[ 31:0]  = X[ 31:0] + Z[ 31:0] ; // Calculate lower word addition, carry flag get updated
Z[ 63:32] = X[ 63:32] + Z[ 63:32] + Carry; // Calculate upper word addition
```

An example of carry out such 64-bit add operation in assembly code can be found in Chapter 6.

Figure 3.5:
PRIMASK.

The other common usage of APSR flag is to control branching. This topic is addressed in Chapter 4, where the details of the condition branch instruction will be covered.

PRIMASK: Interrupt Mask Special Register

The PRIMASK register is a 1-bit-wide interrupt mask register (Figure 3.5). When set, it blocks all interrupts apart from the nonmaskable interrupt (NMI) and the hard fault exception. Effectively it raises the current interrupt priority level to 0, which is the highest value for a programmable exception.

The PRIMASK register can be accessed using special register access instructions (MSR, MRS) as well as using an instruction called the Change Processor State (CPS). This is commonly used for handling time-critical routines.

CONTROL: Special Register

As mentioned earlier, there are two stack pointers in the Cortex-M0 processor. The stack pointer selection is determined by the processor mode as well as the configuration of the CONTROL register (Figure 3.6).

After reset, the main stack pointer (MSP) is used, but can be switched to the process stack pointer (PSP) in Thread mode (when not running an exception handler) by setting bit [1] in the CONTROL register (Figure 3.7). During running of an exception handler (when the processor

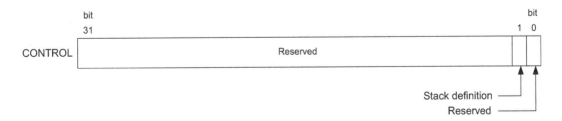

Figure 3.6:
CONTROL.

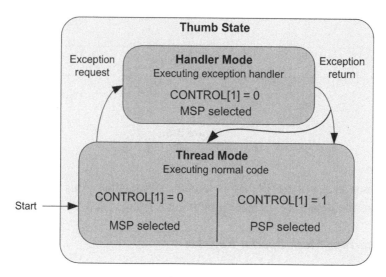

Figure 3.7:
Stack pointer selection.

is in Handler mode), only the MSP is used, and the CONTROL register reads as zero. The CONTROL register can only be changed in Thread mode or via the exception entrance and return mechanism.

Bit 0 of the CONTROL register is reserved to maintain compatibility with the Cortex-M3 processor. In the Cortex-M3 processor, bit 0 can be used to switch the processor to User mode (non-privileged mode). This feature is not available in the Cortex-M0 processor.

Memory System Overview

The Cortex-M0 processor has 4 GB of memory address space (Figure 3.8). The memory space is architecturally defined as a number of regions, with each region having a recommended usage to help software porting between different devices.

The Cortex-M0 processor contains a number of built-in components like the NVIC and a number of debug components. These are in fixed memory locations within the system region of the memory map. As a result, all the devices based on the Cortex-M0 have the same programming model for interrupt control and debug. This makes it convenient for software porting and helps debug tool vendors to develop debug solutions for the Cortex-M0 based microcontroller or system-on-chip (SoC) products.

In most cases, the memories connected to the Cortex-M0 are 32-bits, but it is also possible to connect memory of different data widths to the Cortex-M0 processor with suitable memory interface hardware. The Cortex-M0 memory system supports memory transfers of different sizes such as byte (8-bit), half word (16-bit), and word (32-bit). The Cortex-M0 design can be

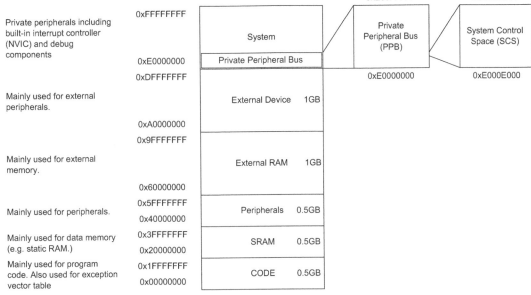

Figure 3.8:
Memory map.

configured to support either little endian or big endian memory systems, but it cannot switch from one to another in an implemented design.

Because the memory system and peripherals connected to the Cortex-M0 are developed by microcontroller vendors or system-on-chip (SoC) designers, different memory sizes and memory types can be found in different Cortex-M0 based products.

Stack Memory Operations

Stack memory is a memory usage mechanism that allows the system memory to be used as temporary data storage that behaves as a first-in, last-out buffer. One of the essential elements of stack memory operation is a register called the stack pointer. The stack pointer is adjusted automatically each time a stack operation is carried out. In the Cortex-M0 processor, the stack pointer is register R13 in the register bank. Physically there are two stack pointers in the Cortex-M0 processor, but only one of them is used at one time, depending on the current value of the CONTROL register and the state of the processor (see Figure 3.7).

In common terms, storing data to the stack is called pushing (using the PUSH instruction) and restoring data from the stack is called popping (using the POP instruction). Depending on processor architecture, some processors perform storing of new data to stack memory using incremental address indexing and some use decrement address indexing. In the Cortex-M0 processor, the stack operation is based on a "full-descending" stack model. This means the

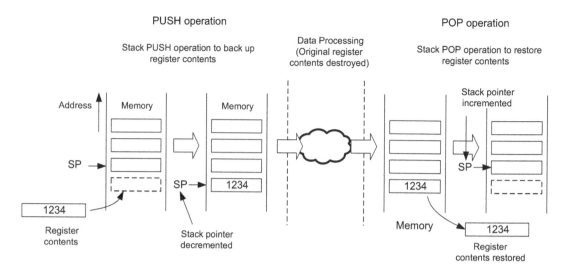

Figure 3.9:
Stack PUSH and POP in the Cortex-M0 processor.

stack pointer always points to the last filled data in the stack memory, and the stack pointer predecrements for each new data store (PUSH) (Figure 3.9).

PUSH and POP are commonly used at the beginning and end of a function or subroutine. At the beginning of a function, the current contents of the registers used by the calling program are stored onto the stack memory using a PUSH operation, and at the end of the function, the data on the stack memory is restored to the registers using a POP operation. Typically, each register PUSH operation should have a corresponding register POP operation, otherwise the stack pointer will not be able to restore registers to their original values. This can result in unpredictable behavior, for example, stack overflow.

The minimum data size to be transferred for each push and pop operations is one word (32-bit), and multiple registers can be pushed or popped in one instruction. The stack memory accesses in the Cortex-M0 processor are designed to be always word aligned (address values must be a multiple of 4, for example, 0x0, 0x4, 0x8, etc.), as this gives the best efficiency for minimum design complexity. For this reason, bits [1:0] of both stack pointers in the Cortex-M0 processor are hardwired to zeros and read as zeros.

The stack pointer can be accessed as either R13 or SP. Depending on the processor state and the CONTROL register value, the stack pointer accessed can either be the main stack pointer (MSP) or the process stack pointer (PSP). In many simple applications, only one stack pointer is needed and by default the main stack pointer (MSP) is used. The process stack pointer (PSP) is usually only required when an operating system (OS) is used in the embedded application (Table 3.3).

Table 3.3: Stack Pointer Usage Definition

Processor State	CONTROL[1] = 0 (Default Setting)	CONTROL[1] = 1 (OS Has Started)
Thread mode	Use MSP (R13 is MSP)	Use PSP (R13 is PSP)
Handler mode	Use MSP (R13 is MSP)	Use MSP (R13 is MSP)

In a typical embedded application with an OS, the OS kernel uses the MSP and the application processes use the PSP. This allows the stack for the kernel to be separate from stack memory for the application processes. This allows the OS to carry out context switching quickly (switching from execution of one application process to another). Even though the OS kernel only uses the MSP as its stack pointer, it can still access the value in PSP by using special register access instructions (MRS and MSR).

Because the stack grows downward (full-descending), it is common for the initial value of the stack pointer to be set to the upper boundary of SRAM. For example, if the SRAM memory range is from 0x20000000 to 0x20007FFF, we can start the stack pointer at 0x20008000. In this case, the first stack PUSH will take place at address 0x20007FFC, the top word of the SRAM.

The initial value of MSP is stored at the beginning of the program memory. Here we will find the exception vector table, which is introduced in the next section. The initial value of PSP is undefined, and therefore the PSP must be initialized by software before using it.

Exceptions and Interrupts

Exceptions are events that cause change to program control: instead of continuing program execution, the processor suspends the current executing task and executes a part of the program code called the exception handler. After the exception handler is completed, it will then resume the normal program execution. There are various types of exceptions, and interrupts are a subset of exceptions. The Cortex-M0 processor supports up to 32 external interrupts (commonly referred as IRQs) and an additional special interrupt called the nonmaskable interrupt (NMI). The exception handlers for interrupt events are commonly known as interrupt service routines (ISRs). Interrupts are usually generated by on-chip peripherals, or by external input through I/O ports. The number of available interrupts on the Cortex-M0 processor depends on the microcontroller product you use. In systems with more peripherals, it is possible for multiple interrupt sources to share one interrupt connection.

Besides the NMI and IRQ, there are a number of system exceptions in the Cortex-M0 processor, primarily for OS use and fault handling (Table 3.4).

Each exception has an exception number. This number is reflected in various registers including the IPSR and is used to define the exception vector addresses. Note that exception numbers are separated from interrupt numbers used in device driver libraries. In most device

Table 3.4: Exception Types

Exception Type	Exception Number	Description
Reset	1	Power on reset or system reset.
NMI	2	Nonmaskable interrupt—highest priority exception that cannot be disabled. For safety critical events.
Hard fault	3	For fault handling—activated when a system error is detected.
SVCall	11	Supervisor call—activated when SVC instruction is executed. Primarily for OS applications.
PendSV	14	Pendable service (system) call—activated by writing to an interrupt control and status register. Primarily for OS applications.
SysTick	15	System Tick Timer exception—typically used by an OS for a regular system tick exception. The system tick timer (SysTick) is an optional timer unit inside the Cortex-M0 processor.
IRQ0 to IRQ31	16 - 47	Interrupts—can be from external sources or from on-chip peripherals.

driver libraries, system exceptions are defined using negative numbers, and interrupts are defined as positive numbers from 0 to 31.

Reset is a special type of exception. When the Cortex-M0 processor exits from a reset, it executes the reset handler in Thread mode (no need to return from handler to thread). Also, the exception number of 1 is not visible in the IPSR.

Apart from NMI, hard fault, and reset, all other exceptions have a programmable priority level. The priority level for NMI and hard fault are fixed and both have a higher priority than the rest of the exceptions. More details will be covered in Chapter 8.

Nested Vectored Interrupt Controller (NVIC)

To prioritize the interrupt requests and handle other exceptions, the Cortex-M0 processor has a built-in interrupt controller called the Nested Vectored Interrupt Controller (NVIC). The interrupt management function is controlled by a number of programmable registers in the NVIC. These registers are memory mapped, with the addresses located within the System Control Space (SCS) as illustrated in Figure 3.8.

The NVIC supports a number of features:

- Flexible interrupt management
- Nested interrupt support
- Vectored exception entry
- Interrupt masking

Flexible Interrupt Management

In the Cortex-M0 processor, each external interrupt can be enabled or disabled and can have its pending status set or clear by software. It can also accept exception requests at signal level (interrupt request from a peripheral remain asserted until the interrupt service routine clears the interrupt request), as well as an exception request pulse (minimum 1 clock cycle). This allows the interrupt controller to be used with any interrupt source.

Nested Interrupt Support

In the Cortex-M0 processor, each exception has a priority level. The priority level can be fixed or programmable. When an exception occurs, such as an external interrupt, the NVIC will compare the priority of this exception to the current level. If the new exception has a higher priority, the current running task will be suspended. Some of the registers will be stored on to the stack memory, and the processor will start executing the exception handler of the new exception. This process is called "preemption." When the higher priority exception handler is complete, it is terminated with an exception return operation and the processor automatically restores the registers from the stack and resumes the task that was running previously. This mechanism allows nesting of exception services without any software overhead.

Vectored Exception Entry

When an exception occurs, the processor will need to locate the starting point of the corresponding exception handler. Traditionally, in ARM processors such as the ARM7TDMI, software usually handles this step. The Cortex-M0 automatically locates the starting point of the exception handler from a vector table in the memory. As a result, the delay from the start of the exception to the execution of the exception handlers is reduced.

Interrupt Masking

The NVIC in the Cortex-M0 processor provides an interrupt masking feature via the PRIMASK special register. This can disable all exceptions except hard fault and NMI. This masking is useful for operations that should not be interrupted such as time critical control tasks or real-time multimedia codecs.

The above NVIC features help makes the Cortex-M0 processor easier to use, provides better response times, and reduces program code size by managing the exceptions in the NVIC hardware.

System Control Block (SCB)

Apart from the NVIC, the System Control Space (SCS) also contains a number of other registers for system management. This is called the System Control Block (SCB). It contains registers for sleep mode features and system exception configurations, as well as a register containing the processor identification code (which can be used by in circuit debuggers for detection of the processor type).

Debug System

Although it is currently the smallest processor in the ARM processor family, the Cortex-M0 processor supports various debug features. The processor core provides halt mode debug, stepping, and register accesses, and additional debug blocks provide debug features like the Breakpoint Unit (BPU) and Data Watchpoint (DWT) units. The BPU supports up to four hardware breakpoints, and the DWT supports up to two watchpoints.

To allow a debugger to control the aforementioned debug components and carry out debug operations, the Cortex-M0 processor provides a debug interface unit. This debug interface unit can either use the JTAG protocol or the serial wire debug (SWD) protocol. In some Cortex-M0 products, the microcontroller vendors can also choose to use a debug interface unit, which supports both JTAG and serial wire debug protocol. However, typical Cortex-M0 implementations are likely to support only one protocol with SWD probably being preferred because fewer pins are required.

The serial wire debug protocol is a new standard developed by ARM, and it can reduce the number connection to just two signals. It can handle all the same debug features as JTAG without any loss of performance. The serial wire debug interface shares the same connector as JTAG: The Serial clock signal is shared with JTAG TCK signal, and the serial wire data are shared with the JTAG TMS signal (Figure 3.10). There are many debug emulators for ARM microcontrollers, including ULINK2 (from Keil) and JLink (from SEGGER), that already support the serial wire debug protocol.

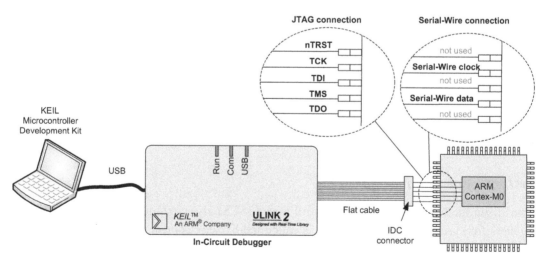

Figure 3.10:
Debug interface connections can be JTAG or the serial wire debug protocol.

Program Image and Startup Sequence

To understand the startup sequence of the Cortex-M0 processor, we need to have a quick overview on the program image first. Normally, the program image for the Cortex-M0 processor is located from address 0x00000000.

The beginning of the program image contains the vector table (Figure 3.11). It contains the starting addresses (vectors) of exceptions. Each vector is located in address of "Exception_ Number × 4." For example, external IRQ #0 is exception type #16, therefore the address of the vector for IRQ#0 is in $16 \times 4 = 0 \times 40$. These vectors have LSB set to 1 to indicate that the exceptions handlers are to be executed with Thumb instructions. The size of the vector table depends on how many interrupts are implemented.

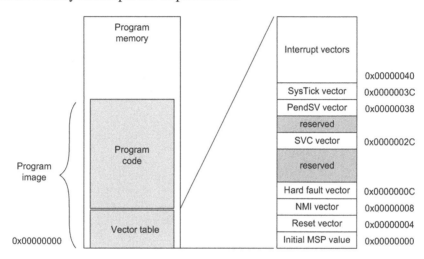

Figure 3.11:
Vector table in a program image.

The vector table also defines the initial value of the main stack pointer (MSP). This is stored in the first word of the vector table.

When the processor exits from reset, it will first read the first two-word addresses in the vector table. The first word is the initial MSP value, and the second word is the reset vector (Figure 3.12), which determines the starting of the program execution address (reset handler).

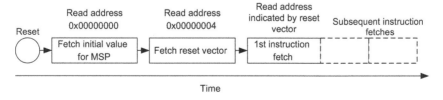

Figure 3.12:
Reset sequence.

For example, if we have boot code starting from address 0x000000C0, we need to put this address value in the reset vector location with the LSB set to 1 to indicate that it is Thumb code. Therefore, the value in address 0x00000004 is set to 0x000000C1 (Figure 3.13). After the processor fetches the reset vector, it will start executing program code from the address found there. This behavior is different from traditional ARM processors (e.g., ARM7TDMI), where the processor executes the program starting from address 0x00000000, and the vectors in the vector table are instructions as oppose to address values in the Cortex-M processors.

The reset sequence also initializes the main stack pointer (MSP). Assume we have SRAM located from 0x20000000 to 0x20007FFF, and we want to put the main stack at the top of the SRAM; we can set this up by putting 0x20008000 in address 0x00000000 (Figure 3.13).

Because the Cortex-M0 processor will first decrement the stack pointer before pushing the data on to the stack, the first stacked item will be located in 0x200007FFC, which is just at the top of the SRAM, whereas the second stacked item will be in 0x20007FF8, below the first stacked item.

This behavior differs from that of traditional ARM processors and many other microcontroller architectures where the stack pointer has to be initialized by software code rather than a value in a fixed address.

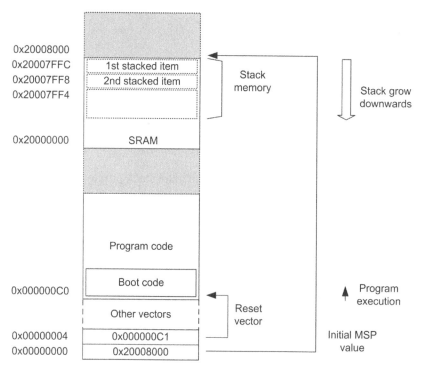

Figure 3.13:
Example of MSP and PC initialization.

If the process stack pointer (PSP) is to be used, it must be initialized by software code before writing to the CONTROL register to switch the stack pointer. The reset sequence only initializes the MSP and not the PSP.

Different software development tools have different ways of specifying the initial stack pointer value and the values for the reset and exception vectors. Most of the development tools come with code examples demonstrating how this can be done with their development flow. In most compilation tools, the vector table can be defined completely using C.

## CHAPTER 4

Introduction to Cortex-M0 Programming

Introduction to Embedded System Programming

All microcontrollers need program code to enable them to perform their intended tasks. If your only experience comes from developing programs for personal computers, you might find the software development for microcontrollers very different. Many embedded systems do not have any operating systems (sometimes these systems are referred as bare metal targets) and do not have the same user interface as a personal computer. If you are completely new to microcontroller programming, do not worry. Programming the Cortex-M0 is easy. As long as you have a basic understanding of the C language, you will soon be able develop simple applications on the Cortex-M0.

What Happens When a Microcontroller Starts?

Most modern microcontrollers have on-chip flash memory to hold the compiled program. The flash memory holds the program in binary machine code format, and therefore programs written in C must be compiled before programmed to the flash memory. Some of these microcontrollers might also have a separate boot ROM, which contains a small boot loader program that is executed when the microcontroller starts, before executing the user program in the flash memory. In most cases, only the program code in the flash memory can be changed and the boot loader is fixed.

After the flash memory (or other types of program memory) is programmed, the program is then accessible by the processor. After the processor is reset, it carries out the reset sequence, as outlined at the end of the previous chapter (Figure 4.1).

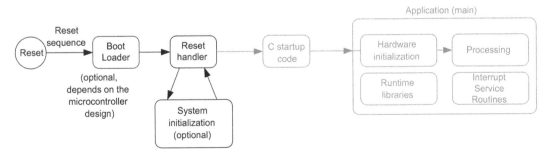

Figure 4.1:
What happens when a microcontroller starts—the Reset handler.

The Definitive Guide to the ARM Cortex-M0. DOI: 10.1016/B978-0-12-385477-3.10004-7

In the reset sequence, the processor obtains the initial MSP value and reset vector, and then it executes the reset handler. All of this required information is usually stored in a program file called startup code. The reset handler in the startup code might also perform system initialization (e.g., clock control circuitry and Phase Locked Loop [PLL]), although in some cases system initialization is carried out later when the C program "main()" starts. Example startup code can usually be found in the installation of the development suite or from software packages available from the microcontroller vendors. For example, if the Keil Microcontroller Development Kit (MDK) is used for development, the project creation wizard can optionally copy a default startup code file into your project that matches the microcontroller you selected.

For applications developed in C, the C startup code is executed before entering the main application code. The C startup code initializes variables and memory used by the application, and they are inserted to the program image by the C development suite (Figure 4.2).

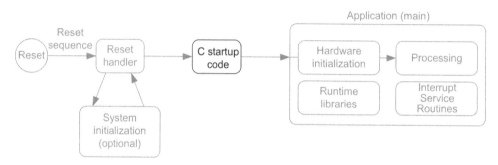

Figure 4.2:
What happens when a microcontroller starts—C startup code.

After the C startup code is executed, the application starts. The application program often contains the following elements:

- Initialization of hardware (e.g., clock, PLL, peripherals)
- The processing part of the application
- Interrupt service routines

In addition, the application might also use C library functions (Figure 4.3). In such cases, the C compiler/linker will include the required library functions into the compiled program image.

The hardware initialization might involve a number of peripherals, some system control registers, and interrupt control registers inside the Cortex-M0 processors. The initialization of the system clock control and the PLL might also take place if this were not carried out in the reset handler. After the peripherals are initialized, the program execution can then proceed to the application processing part.

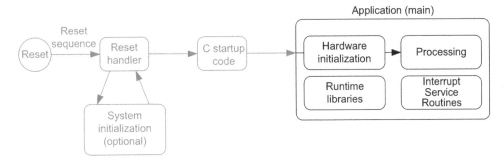

Figure 4.3:
What happens when a microcontroller starts—application.

Designing Embedded Programs

There are many ways to structure the flow of the application processing. Here we will cover a few fundamental concepts.

Polling

For simple applications, polling (sometimes also called super loop) is easy to set up and works fairly well for simple tasks (Figure 4.4).

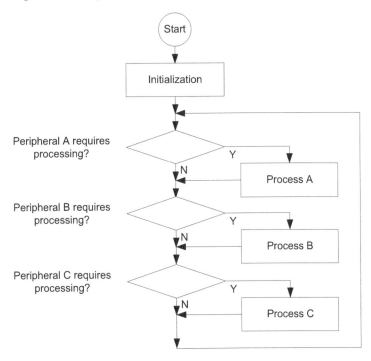

Figure 4.4:
Polling method for simple application processing.

However, when the application gets complicated and demands higher processing performance, polling is not suitable. For example, if one of the processes takes a long time, other peripherals will not receive any service for some time. Another disadvantage of using the polling method is that the processor has to run the polling program all the time, even if it requires no processing.

Interrupt Driven

In applications that require lower power, processing can be carried out in interrupt service routines so that the processor can enter sleep mode when no processing is required. Interrupts are usually generated by external sources or on chip peripherals to wake up the processor.

In interrupt-driven applications (Figure 4.5), the interrupts from different devices can be set at different priorities. In this way a high-priority interrupt request can obtain service even when a lower-priority interrupt service is running, which will be temporarily stopped. As a result, the latency for the higher-priority interrupt is reduced.

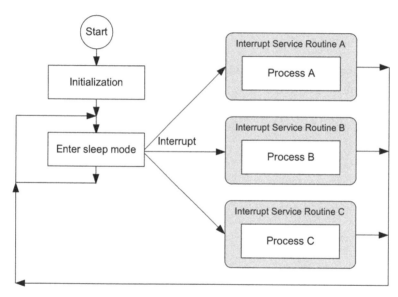

Figure 4.5:
An interrupt-driven application.

Combination of Polling and Interrupt Driven

In many cases, applications can use a combination of polling and interrupt methods (Figure 4.6). By using software variables, information can be transferred between interrupt service routines and the application processes.

By dividing a peripheral processing task into an interrupt service routine and a process running in the main program, we can reduce the duration of interrupt services so that even lower-priority interrupt services gain a better chance of getting serviced. At the same time, the system

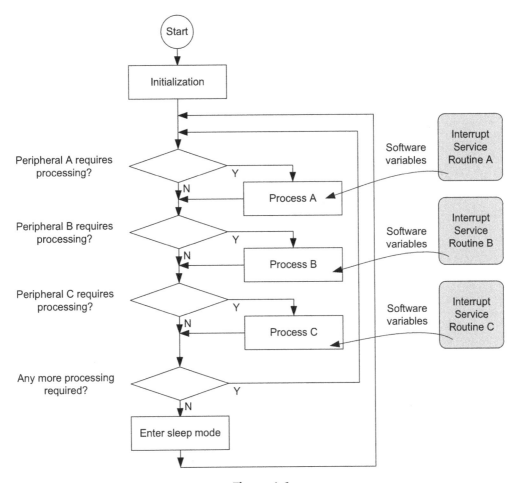

Figure 4.6:
Combination of polling and interrupt-driven application.

can still enter sleep mode when no processing task is required. In Figure 4.6, the application is partitioned into processes A, B, and C, but in some cases, an application cannot be partitioned into individual parts easily and needs to be written as a large combined process.

Handling Concurrent Processes

In some cases, an application process could take a significant amount of time to complete and therefore it is undesirable to handle it in a big loop as shown in Figure 4.6. If process A takes too long to complete, processes B and C will not able to respond to peripheral requests fast enough, resulting in system failure. Common solutions are as follows:

1. Breaking down a long processing task to a sequence of states. Each time the process is accessed, only one state is executed.
2. Using a real-time operating system (RTOS) to manage multiple tasks.

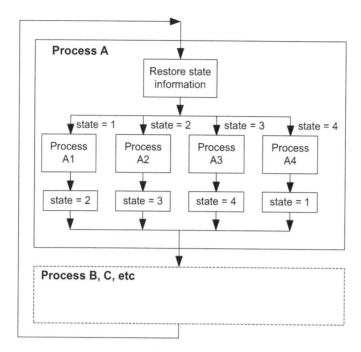

Figure 4.7:
Partitioning a process into multiple parts in the application loop.

For method 1, a process is divided into a number of parts, and software variables are used to track the state of the process (Figure 4.7). Each time the process is executed, the state information is updated so that next time the process is executed, the processing can resume correctly.

Because the execution path of the process is shortened, other processes in the main loop can be reached quicker inside the big loop. Although the total processing time required for the processing remains unchanged (or increases slightly because of the overhead of state saving and restoring), the system is more responsive. However, when the application tasks become more complex, partitioning the application task manually can become impractical.

For more complex applications, a real-time operating system (RTOS) can be used (Figure 4.8). An RTOS allows multiple application processes to be executed by dividing processor execution time into time slots and allocating one to each task. To use an RTOS, a timer is needed to generate regular interrupt requests. When each time slot ends, the timer generates an interrupt that triggers the RTOS task scheduler, which determines if context switching should be carried out. If context switching should be carried out, the task schedule suspends the current executing task and then switches to the next task that is ready to be executed.

Using an RTOS improves the responsiveness of a system by ensuring that all tasks are reached within a certain amount of time. Examples of using an RTOS are covered in Chapter 18.

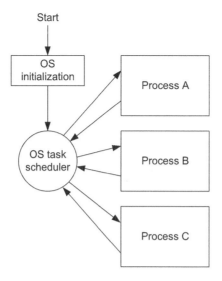

Figure 4.8:
Using an RTOS to handle multiple concurrent application processes.

Inputs and Outputs

On many embedded systems, the available inputs and outputs can be limited to simple electronic interfaces like digital and analog inputs and outputs (I/Os), UARTs, I2C, SPI, and so on. Many microcontrollers also offer USB, Ethernet, CAN, graphics LCD, and SD card interfaces. These interfaces are handled by peripherals in the microcontrollers.

On Cortex-M0 microcontrollers, peripherals are controlled by memory-mapped registers (examples of accessing peripherals are presented later in this chapter). Some of these peripherals are more sophisticated than peripherals available on 8-bit and 16-bit microcontrollers, and there might be more registers to program during the peripheral setup.

Typically, the initialization process for peripherals may consist of the following steps:

1. *Programming the clock control circuitry to enable the clock signal to the peripheral and the corresponding I/O pins if necessary.* In many low-power microcontrollers, the clock signals reaching different parts of the chip can be turned on or off individually to save power. By default, most of the clock signals are usually turned off and need to be enabled before the peripherals are programmed. In some cases you also need to enable the clock signals for the peripherals bus system.
2. *Programming of I/O configurations.* Most microcontrollers multiplex their I/O pins for multiple uses. For a peripheral interface to work correctly, the I/O pin assignment might need to be programmed. In addition, some microcontrollers also offer configurable electrical characteristics for the I/O pins. This can result in additional steps in I/O configurations.

3. *Peripheral configuration.* Most interface peripherals contain a number of programmable registers to control their operations, and therefore a programming sequence is usually needed to allow the peripheral to work correctly.
4. *Interrupt configuration.* If a peripheral operation requires interrupt processing, additional steps are required for the interrupt controller (e.g., the NVIC in the Cortex-M0).

Most microcontroller vendors provide device driver libraries for peripheral programming to simplify software development. Unlike programming on personal computers, you might need to develop your own user interface functions to design a user-friendly standalone embedded system. However, the device driver libraries provided by the microcontroller vendors will make the development of your user interface easier.

For the development of most deeply embedded systems, it is not necessary to have a rich user interface. However, basic interfaces like LEDs, DIP switches, and push buttons can deliver only a limited amount of information. For debugging software, a simple text input/output console is often sufficient. This can be handled by a simple RS-232 connection through a UART interface on the microcontroller to a UART interface on a personal computer (or via a USB adaptor) so that we can display the text messages and enter user inputs using a terminal application (Figure 4.9).

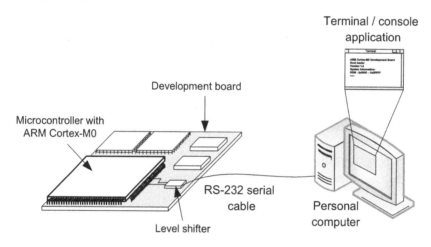

Figure 4.9:
Using UART interface for user input and output.

The technique to redirect text messages from a "printf" (in C language) to a UART (or another interface) is commonly referred to as "retargeting." Retargeting can also handle user inputs and system functions. Examples of simple retargeting will be presented in later chapters of this book.

Typically, microcontrollers also provide a number of general-purpose input and output ports (GPIOs) that are suitable for simple control, user buttons or switches, LEDs, and the like. You can also develop an embedded system with a full feature graphics display using a microcontroller with built-in LCD controllers or using an external LCD module with a parallel or SPI interface. Although microcontroller vendors usually provide device driver libraries for the peripheral blocks, you might still need to develop your own user input and output functions.

Development Flow

Many development tool chains are available for ARM microcontrollers. The majority of them support C and assembly language. Embedded projects can be developed in either C or assembly language, or a mixture of both. In most cases, the program-generation flow can be summarized in a diagram, as shown in Figure 4.10.

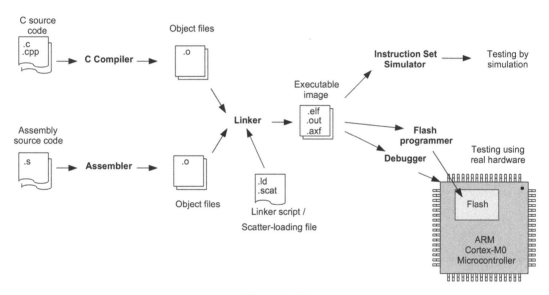

Figure 4.10:
Typical program-generation flow.

In most simple applications, the programs can be completely written in the C language. The C compiler compiles the C program code into object files and then generates the executable program image file using the linker. In the case of GNU C compilers, the compile and linking stages are often merged into one step.

Projects that require assembly programming use the assembler to generate object code from assembly source code. The object files can then be linked with other object files in the project to produce an executable image. Besides the program code, the object files and the executable image may also contain various debug information.

Depending on the development tools, it is possible to specify the memory layout for the linker using command line options. However, in projects using GNU C compilers, a linker script is normally required to specify the memory layout. A linker script is also required for other development tools when the memory layout gets complicated. In ARM development tools, the linker scripts are often called scatter-loading files. If you are using the Keil Microcontroller Development Kit (MDK), the scatter-loading file is generated automatically from the memory layout window. You can use your own scatter file if you prefer.

After the executable image is generated, we can test it by downloading it to the flash memory or internal RAM of the microcontroller. The whole process can be quite easy; most development suites come with a user-friendly integrated development environment (IDE). When working together with an in-circuit debugger (sometimes referred to as an in-circuit emulator [ICE], debug probe, or USB-JTAG adaptor), you can create a project, build your application, and download your embedded application to the microcontroller in a few steps (Figure 4.11).

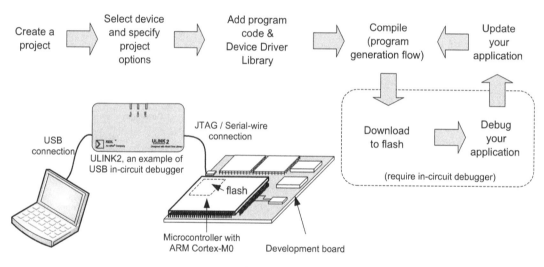

Figure 4.11:
An example of development flow.

In many cases, an in-circuit debugger is needed to connect the debug host (personal computer) to the target board. The Keil U-LINK2 is one of the products available and can be used with Keil MDK and CodeSourcery g++ (Figure 4.12).

The flash programming function can be carried out by the debugger software in the development suite (Figure 4.13) or in some cases by a flash programming utility downloadable from microcontroller vendor web site. The program can then be tested by running it on the microcontroller, and by connecting the debugger to the microcontroller, the program execution

Figure 4.12:
ULINK 2 USB-JTAG adaptor.

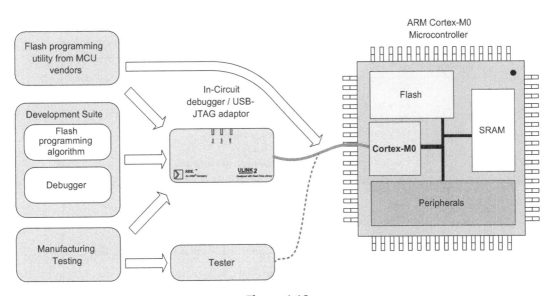

Figure 4.13:
Various usages of the debug interface on the Cortex-M0 processor.

can be controlled and the operations can be observed. All these actions can be carried out via the debug interface of the Cortex-M0 processor.

For simple program codes, we can also test the program using a simulator. This allows us to have full visibility to the program execution sequence and allows testing without actual hardware. Some development suites provide simulators that can also imitate peripheral behavior. For example, Keil MDK provides device simulation for many ARM Cortex microcontrollers.

Apart from the fact that different C compilers perform differently, different development suites also provide different C language extension features, as well as different syntax and directives in assembly programming. Chapters 5, 6, and 16 provide assembly syntax information for ARM development tools (including ARM RealView Development Suite [RVDS] and Keil MDK) and GNU C compilers. In addition, different development suites also provide different features in debug, utilities, and support different debug hardware product range.

C Programming and Assembly Programming

The Cortex-M0 processor can be programmed using C language, assembly language, or a mix of both. For beginners, C language is usually the best choice as it is easier to learn and most modern C compilers are very good at generating efficient code for the Cortex microcontrollers. Table 4.1 compares the use of C language and assembly language.

Table 4.1: Comparison between C Programming and Assembly Language Programming

Language	Pros and Cons
C	**Pros** Easy to learn Portable Easy handling of complex data structures **Cons** Limited/no direct access to core register and stack No direct control over instruction sequence generation No direct control over stack usage
Assembly	**Pros** Allows direct control to each instruction step and all memory operations Allows direct access to instructions that cannot be generated with C **Cons** Take longer time to learn Difficult to manage data structure Less portable (syntax of assembly language in different tool chains can be different)

Most C compilers provide workarounds to allow assembly code to be used within C program code. For example, ARM C compilers provide an *Embedded Assembler* so that assembly functions can be included in C program code easily. Similarly, most other C compilers provide an *Inline Assembler* for inlining assembly code within a C program file. However, the assembly syntax for using an Embedded Assembler and Inline Assembler are tool specific (not portable). Note that the ARM C compiler has an Inline Assembler feature as well, but this is only available for 32-bit ARM instructions (e.g., for ARM7TDMI). Because the Cortex-M0 processor supports the Thumb instruction set only, the Embedded Assembler is used.

Some C compilers (including ARM C compilers in RealView Development Suite and Keil MDK) also provide intrinsic functions to allow special instructions to be used that cannot be generated using normal C code. Intrinsic functions are normally tool dependent. However, a tool-independent version of similar functions for Cortex-M0 is also available via the Cortex Microcontroller Software Interface Standard (CMSIS). This will be covered later in the chapter.

As Figure 4.10 shows, you can mix C and assembly code together in a project. This allows most parts of the program to be written in C, and some parts that cannot be handled in C can be written in assembly code. To do this, the interface between functions must be handled in a consistent manner to allow input parameters and returned results to be transferred correctly. In ARM software development, the interface between functions is specified by a specification document called the ARM Architecture Procedure Call Standard (AAPCS, reference 4). The AAPCS is part of the Embedded Application Binary Interface (EABI). When using the Embedded Assembler, you should follow the guidelines set by the AAPCS. The AAPCS document and the EABI document can be downloaded from the ARM web site.

More details in this area are covered in Chapter 16.

What Is in a Program Image?

At the end of Chapter 3 we covered the reset sequence of the Cortex-M0 and briefly introduced the vector table. Now we will look at the program image in more detail.

A program image for the Cortex-M0 microcontroller often contains the following components:

- Vector table
- C startup routine
- Program code (application code and data)
- C library code (program codes for C library functions, inserted at link time)

Vector Table

The vector table can be programmed in either C language or assembly language. The exact details of the vector table code are tool chain dependent because vector table entries require symbols created by the compiler and linker. For example, the initial stack pointer value is linked to stack region address symbols generated by the linker, and the reset vector is linked to C startup code address symbols, which are compiler dependent. For example, in the RealView Development Suite (RVDS), you can define the vector table with the following C code:

Example of vector table in C language

```
/* Stack and heap settings  */
#define STACK_BASE  0x20020000      /* Stack start address */
#define STACK_SIZE  0x8000          /* length stack grows downwards */
#define HEAP_BASE   0x20010000      /* Heap starts address */
#define HEAP_SIZE   0x10000-0x8000  /* Heap Length */

/* Linker-generated Stack Base addresses  */
extern unsigned int Image$$ARM_LIB_STACK$$ZI$$Limit;
extern unsigned int Image$$ARM_LIB_STACKHEAP$$ZI$$Limit;
typedef void(* const ExecFuncPtr)(void) __irq;
extern int __main(void);
/*
 * Exception Table, in separate section so it can be correctly placed at 0x0
 */
#pragma arm section rodata="exceptions_area"

ExecFuncPtr exception_table[] = {
    /* Configure Initial Stack Pointer, using linker-generated symbols*/
    #pragma import(__use_two_region_memory)
    (ExecFuncPtr)&Image$$ARM_LIB_STACK$$ZI$$Limit,
                               /* Initial Main Stack Pointer */
    (ExecFuncPtr) Reset_Handler, /* Initial PC, set to entry point.
                                 Branch to __main */
    NMI_Handler,                 /* Non-maskable Interrupt handler */
    HardFault_Handler,           /* Hard fault handler */
    0, 0, 0, 0, 0, 0, 0,         /* Reserved */
    SVC_Handler,                 /* SVC handler */
    0, 0,                        /* Reserved */
    PendSV_Handler,              /* PendSV handler */
    SysTick_Handler,             /* SysTick Handler */

    /* Device specific configurable interrupts start here...*/
    Interrupt0_Handler,
    Interrupt1_Handler,        /* dummy default interrupt handlers */
    Interrupt2_Handler
    /*
    :
    */
};
#pragma arm section
```

Some development tools, including Keil MDK, create the vector table as part of the assembly startup code. In this case, the Define Constant Data (DCD) directive is used to create the vector table.

Example of vector table in assembly

```
        AREA    STACK, NOINIT, READWRITE, ALIGN=3
StackMem
        SPACE   0x8000   ; Allocate space for the stack.
__initial_sp
        AREA    HEAP, NOINIT, READWRITE, ALIGN=3
__heap_base
HeapMem
        SPACE   0x8000   ; Allocate space for the heap.
__heap_limit
        PRESERVE8          ; Indicate that the code in this file
                           ; preserves 8-byte alignment of the stack.
;
; The vector table.
        AREA    RESET, CODE, READONLY
        THUMB
        EXPORT  __Vectors
__Vectors
        DCD     __initial_sp              ; Top of Stack
        DCD     Reset_Handler             ; Reset Handler (branch to __main)
        DCD     NMI_Handler               ; NMI Handler
        DCD     HardFault_Handler         ; Hard Fault Handler
        DCD     0                         ; Reserved
        DCD     0                         ; Reserved
        DCD     0                         ; Reserved
        DCD     0                         ; Reserved
        DCD     0                         ; Reserved
        DCD     0                         ; Reserved
        DCD     0                         ; Reserved
        DCD     SVC_Handler               ; SVCall Handler
        DCD     0                         ; Reserved
        DCD     0                         ; Reserved
        DCD     PendSV_Handler            ; PendSV Handler
        DCD     SysTick_Handler           ; SysTick Handler

        ; Device specific configurable interrupts start here...
        DCD     Interrupt0_Handler        ;
        DCD     Interrupt1_Handler        ;  dummy default interrupt handlers
        DCD     Interrupt2_Handler        ;
```

You might notice that in both examples, the vector tables are given section names (*exceptions_area* in the C example and *RESET* in the assembly example). The vector table needs to be placed at the beginning of the system memory map (address 0x00000000). This can be done by a linker script or command line option, which requires a section name so that the contents of the vector table can be identified and mapped correctly by the linker.

In normal applications, the reset vector can point to the beginning of the C startup code. However, you can also define a reset handler to carry out additional initialization before branching to the C startup code.

C Startup Code

The C startup code is used to set up data memory such as global data variables. It also zero initializes part of the data memory for variables that are uninitialized at load time. For applications that use C functions like malloc(), the C startup code also needs to initialize the data variables controlling the heap memory. After this initialization, the C startup code branches to the beginning of the main() program.

The C startup code is inserted by the compiler/linker automatically and is tool chain specific; it might not be present if you are writing a program purely in assembly. For ARM compilers, the C startup code is labeled as "__main," whereas the startup code generated by GNU C compilers is normally labeled as "_start."

Program Code

The instructions generated from your application program code carry out the tasks you specify. Apart from the instruction sequence, there are also various types of data:

- Initial values of variables. Local variables in functions or subroutines need to be initialized, and these initial values are set up during program execution.
- Constants in program code. Constant data are used in application codes in many ways: data values, addresses of peripheral registers, constant strings, and so on. These data are sometimes grouped together within the program images as a number of data blocks called literal pools.
- Some applications can also contain additional constant data like lookup tables and graphics image data (e.g., bit map) that are merged into the program images.

C Library Code

C library code is injected in to the program image by the linker when certain C/C++ functions are used. In addition, C library code can also be included because of data processing tasks such as floating point operations and divide. The Cortex-M0 does not have a divide instruction, and this function typically needs to be carried out by a C library divide function.

Some development tools offer various versions of C libraries for different purposes. For example, in Keil MDK or ARM RVDS there is an option to use a special version of C library called Microlib. The Microlib is targeted for microcontrollers and is very small, but it does not offer all features of the standard C library. In embedded applications that do not require high data processing capability and have tight program memory requirement, the Microlib offers a good way to reduce code size.

Depending on the application, C library code might not be present in simple C applications (no C library function calls) or pure assembly language projects.

Apart from the vector table, which must be placed at the beginning of the memory map, there are no other constraints on the placement of the rest of the elements inside a program image. In some cases, if the layout of the items in the program memory is important, the layout of the program image can be controlled by a linker script.

Data in RAM

Like program ROM, the RAM of microcontrollers is used in different ways. Typically, the RAM usage is divided into data, stack, and heap regions (Figure 4.14).

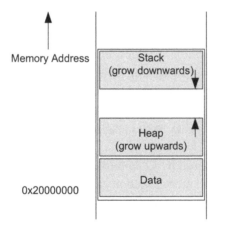

Example RAM usage in systems without OS

Figure 4.14:
Example of RAM usage in single task systems (without OS).

For microcontroller systems with an embedded OS (e.g., μClinux) or RTOS (e.g., Keil RTX), the stacks for each task are separate. Some OSs allow a user-defined stack for tasks that require larger stack memory. Some OSs divide the RAM into a number of segments, and each segment is assigned to a task, each containing individual data, stack, and heap regions (Figure 4.15).

So what is stored inside these data, stack, and heap regions?

- *Data.* Data stored in the bottom of RAM usually contain global variables and static variables. (Note: Local variables can be spilled onto the stack to reduce RAM usage. Local variables that belong to a function that is not in use do not take up memory space.)

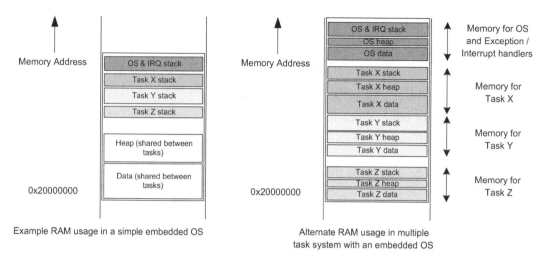

Memory Address

OS & IRQ stack

Task X stack

Task Y stack

Task Z stack

Heap (shared between tasks)

Data (shared between tasks)

0x20000000

Example RAM usage in a simple embedded OS

Memory Address

OS & IRQ stack
OS heap
OS data

Task X stack
Task X heap
Task X data

Task Y stack
Task Y heap
Task Y data

Task Z stack
Task Z heap
Task Z data

0x20000000

Memory for OS and Exception / Interrupt handlers

Memory for Task X

Memory for Task Y

Memory for Task Z

Alternate RAM usage in multiple task system with an embedded OS

Figure 4.15:
Example of RAM usage in multiple task systems (with an OS).

- *Stack.* The role of stack memory includes temporary data storage (normal stack PUSH and POP operations), memory space for local variables, parameter passing in function calls, register saving during an exception sequence, and so on. The Thumb instruction set is very efficient in handling data accesses that use a stack pointer (SP) related addressing mode and allows data in the stack memory to be accessed with very low instruction overhead.
- *Heap.* The heap memory is used by C functions that dynamically reserve memory space, like "alloc()," "malloc()," and other function calls that use these functions. To allow these functions to allocate memory correctly, the C startup code needs to initialize the heap memory and its control variables.

Usually, the stack is placed at the top of the memory space and the heap memory is placed underneath. This gives the best flexibility for the RAM usage. In an OS environment, there can be multiple regions of data, stack, and heap in the RAM.

C Programming: Data Types

The C language supports a number of "standard" data types. However, the implementation of data type can be processor architecture dependent and C compiler dependent. In ARM processors including the Cortex-M0, the data type implementations shown in Table 4.2 are supported by all C compilers.

When porting applications from other processor architectures to ARM processors, if the data types have different sizes, it might be necessary to modify the C program code in order to

Table 4.2: Size of Data Types in Cortex-M Processors

C and C99 (stdint.h) Data Type	Number of Bits	Range (Signed)	Range (Unsigned)
char, int8_t, uint8_t	8	−128 to 127	0 to 255
short int16_t, uint16_t	16	−32768 to 32767	0 to 65535
int, int32_t, uint32_t	32	−2147483648 to 2147483647	0 to 4294967295
long	32	−2147483648 to 2147483647	0 to 4294967295
long long, int64_t, uint64_t	64	$-(2^{63})$ to $(2^{63} - 1)$	0 to $(2^{64} - 1)$
float	32	$-3.4028234 \times 10^{38}$ to 3.4028234×10^{38}	
double	64	$-1.7976931348623157 \times 10^{308}$ to $1.7976931348623157 \times 10^{308}$	
long double	64	$-1.7976931348623157 \times 10^{308}$ to $1.7976931348623157 \times 10^{308}$	
pointers	32	0x0 to 0xFFFFFFFF	
enum	8/16/32	Smallest possible data type, except when overridden by compiler option	
bool (C++ only), _Bool (C only)	8	True or false	
wchar_t	16	0 to 65535	

ensure the program operates correctly. More details on porting software from 8-bit and 16-bit architecture are covered in Chapter 21.

In Cortex-M0 programming, the data variables stored in memory need to be stored at an address location that is a multiple of its size. More details on this area are covered in Chapter 7 (the data alignment section).

In ARM programming, we also refer to data size as word, half word, and byte (Table 4.3).

Table 4.3: Data Size Definition in ARM Processor

Terms	Size
Byte	8-bit
Half word	16-bit
Word	32-bit
Double word	64-bit

These terms are commonly found in ARM documentation, such as in the instruction set details.

Accessing Peripherals in C

Apart from data variables, a C program for microcontroller applications normally needs to access peripherals. In ARM Cortex-M0 microcontrollers, peripheral registers are memory mapped and can be accessed by memory pointers. In most cases, you can use the device drivers provided by the microcontroller vendors to simplify the software development task and make it easier to port software between different microcontrollers. If it is necessary to access the peripheral registers directly, the following methods can be used.

In simple cases of accessing a few registers, you can define a peripheral register as a pointer as follows:

Example registers definition for a UART using pointers and accessing the registers

```c
#define UART_BASE  0x40003000 // Base of ARM Primecell PL011
#define UART_DATA  (*((volatile unsigned long *)(UART_BASE + 0x00)))
#define UART_RSR   (*((volatile unsigned long *)(UART_BASE + 0x04)))
#define UART_FLAG  (*((volatile unsigned long *)(UART_BASE + 0x18)))
#define UART_LPR   (*((volatile unsigned long *)(UART_BASE + 0x20)))
#define UART_IBRD  (*((volatile unsigned long *)(UART_BASE + 0x24)))
#define UART_FBRD  (*((volatile unsigned long *)(UART_BASE + 0x28)))
#define UART_LCR_H (*((volatile unsigned long *)(UART_BASE + 0x2C)))
#define UART_CR    (*((volatile unsigned long *)(UART_BASE + 0x30)))
#define UART_IFLS  (*((volatile unsigned long *)(UART_BASE + 0x34)))
#define UART_MSC   (*((volatile unsigned long *)(UART_BASE + 0x38)))
#define UART_RIS   (*((volatile unsigned long *)(UART_BASE + 0x3C)))
#define UART_MIS   (*((volatile unsigned long *)(UART_BASE + 0x40)))
#define UART_ICR   (*((volatile unsigned long *)(UART_BASE + 0x44)))
#define UART_DMACR (*((volatile unsigned long *)(UART_BASE + 0x48)))
/* ----- UART Initialization  ---- */
void uartinit(void) // Simple initialization for ARM Primecell PL011
{
  UART_IBRD  =40;   // ibrd : 25MHz/38400/16 = 40
  UART_FBRD  =11;   // fbrd : 25MHz/38400 - 16*ibrd = 11.04
  UART_LCR_H =0x60;  // Line control : 8N1
  UART_CR    =0x301; // cr : Enable TX and RX, UART enable
  UART_RSR   =0xA; // Clear buffer overrun if any
}
/* ----- Transmit a character ---- */
int sendchar(int ch)
{
  while (UART_FLAG & 0x20); // Busy, wait
  UART_DATA = ch; // write character
  return ch;
}
/* ----- Receive a character ---- */
int getkey(void)
{
  while ((UART_FLAG & 0x40)==0); // No data, wait
  return UART_DATA; // read character
}
```

This solution is fine for simple applications. However, when multiple units of the same peripherals are available in the system, defining registers will be required for each of these peripherals, which can make code maintenance difficult. In addition, defining each register as a separated pointer might result in larger program size, as each register access requires a 32-bit address constant to be stored in the program flash memory.

To simplify the code, we can define the peripheral register set as a data structure and define the peripheral as a memory pointer to this data structure.

Example registers definition for a UART using data structure and accessing the registers using pointer of structure

```
typedef struct { // Base on ARM Primecell PL011
  volatile unsigned long DATA;          // 0x00
  volatile unsigned long RSR;           // 0x04
          unsigned long RESERVED0[4];// 0x08 - 0x14
  volatile unsigned long FLAG;          // 0x18
          unsigned long RESERVED1;   // 0x1C
  volatile unsigned long LPR;           // 0x20
  volatile unsigned long IBRD;          // 0x24
  volatile unsigned long FBRD;          // 0x28
  volatile unsigned long LCR_H;         // 0x2C
  volatile unsigned long CR;            // 0x30
  volatile unsigned long IFLS;          // 0x34
  volatile unsigned long MSC;           // 0x38
  volatile unsigned long RIS;           // 0x3C
  volatile unsigned long MIS;           // 0x40
  volatile unsigned long ICR;           // 0x44
  volatile unsigned long DMACR;         // 0x48
} UART_TypeDef;
#define Uart0   ((   UART_TypeDef *)    0x40003000)
#define Uart1   ((   UART_TypeDef *)    0x40004000)
#define Uart2   ((   UART_TypeDef *)    0x40005000)

/* ----- UART Initialization  ---- */
void uartinit(void) // Simple initialization for Primecell PL011
{
 Uart0->IBRD  =40;  // ibrd : 25MHz/38400/16 = 40
 Uart0->FBRD  =11;  // fbrd : 25MHz/38400 - 16*ibrd = 11.04
 Uart0->LCR_H =0x60;   // Line control : 8N1
 Uart0->CR    =0x301;  // cr : Enable TX and RX, UART enable
 Uart0->RSR   =0xA; // Clear buffer overrun if any
}
/* ----- Transmit a character ---- */
int sendchar(int ch)
{
 while (Uart0->FLAG & 0x20); // Busy, wait
 Uart0->DATA = ch; // write character
 return ch;
}
/* ----- Receive a character ---- */
int getkey(void)
{
 while ((Uart0->FLAG & 0x40)==0); // No data, wait
 return Uart0->DATA; // read character
}
```

In this example, the Integer Baud Rate Divider (IBRD) register for UART #0 is accessed by the symbol Uart0->IBRD, and the same register for UART #1 is accessed by Uart1->IBRD.

With this arrangement, the same register data structure for the peripheral can be shared between multiple instantiations, making code maintenance easier. In addition, the compiled code could be smaller because of the reduced requirement of immediate data storage.

With further modification, a function developed for the peripherals can be shared between multiple units by passing the base pointer to the function:

Example registers definition for a UART and driver code that support multiple UART using pointer passing

```c
typedef struct { // Base on ARM Primecell PL011
  volatile unsigned long DATA;        // 0x00
  volatile unsigned long RSR;         // 0x04
           unsigned long RESERVED0[4];// 0x08 - 0x14
  volatile unsigned long FLAG;        // 0x18
           unsigned long RESERVED1;   // 0x1C
  volatile unsigned long LPR;         // 0x20
  volatile unsigned long IBRD;        // 0x24
  volatile unsigned long FBRD;        // 0x28
  volatile unsigned long LCR_H;       // 0x2C
  volatile unsigned long CR;          // 0x30
  volatile unsigned long IFLS;        // 0x34
  volatile unsigned long MSC;         // 0x38
  volatile unsigned long RIS;         // 0x3C
  volatile unsigned long MIS;         // 0x40
  volatile unsigned long ICR;         // 0x44
  volatile unsigned long DMACR;       // 0x48
} UART_TypeDef;
#define Uart0   ((  UART_TypeDef *)    0x40003000)
#define Uart1   ((  UART_TypeDef *)    0x40004000)
#define Uart2   ((  UART_TypeDef *)    0x40005000)

/* ----- UART Initialization  ---- */
void uartinit(UART_Typedef *uartptr) //
{
  uartptr->IBRD  =40;  // ibrd : 25MHz/38400/16 = 40
  uartptr->FBRD  =11;  // fbrd : 25MHz/38400 - 16*ibrd = 11.04
  uartptr->LCR_H =0x60;   // Line control : 8N1
  uartptr->CR    =0x301;  // cr : Enable TX and RX, UART enable
  uartptr->RSR   =0xA; // Clear buffer overrun if any
}
/* ----- Transmit a character ---- */
int sendchar(UART_Typedef *uartptr, int ch)
{
  while (uartptr->FLAG & 0x20); // Busy, wait
  uartptr->DATA = ch; // write character
  return ch;
}
/* ----- Receive a character ---- */
int getkey(UART_Typedef *uartptr)
{
  while ((uartptr ->FLAG & 0x40)==0); // No data, wait
  return uartptr ->DATA; // read character
}
```

In most cases, peripheral registers are defined as 32-bit words. This is because most peripherals are connected to a peripheral bus (using APB protocol; see Chapter 7) that handles all transfers as 32 bit. Some peripherals might be connected to the processor bus (with AHB protocol that supports various transfer sizes; see Chapter 7). In such cases, the registers might be accessed in other transfer sizes. Please refer to the user manual of the microcontroller to determine the supported transfer size for each peripheral.

Note that when defining memory pointers for peripheral accesses, the "volatile" keyword should be used.

Cortex Microcontroller Software Interface Standard (CMSIS)

Introduction of CMSIS

As the complexity of embedded systems increase, the compatibility and reusability of software code becomes more important. Having reusable software often reduces development time for subsequent projects and hence speeds up time to market, and software compatibility helps the use of third-party software components. For example, an embedded system project might involve the following software components:

- Software from in-house software developers
- Software reused from other projects
- Device driver libraries from microcontroller vendors
- Embedded OS
- Other third-party software products like a communication protocol stack and codec (compressor/decompressor)

The use of the third-party software components is becoming more and more common. With all these software components being used in one project, compatibility is becoming critical for many large-scale software projects. To allow a high level of compatibility between these software products and improve software portability, ARM worked with various microcontroller vendors and software solution providers to develop the CMSIS, a common software framework covering most Cortex-M processors and Cortex-M microcontroller products (Figure 4.16).

The CMSIS is implemented as part of device driver library from microcontroller vendors. It provides a standardized software interface to the processor features like NVIC control and system control functions. Many of these processors feature access functions are available in CMSIS for the Cortex-M0, Cortex-M3 and Cortex-M4, allowing easy software porting between these processors.

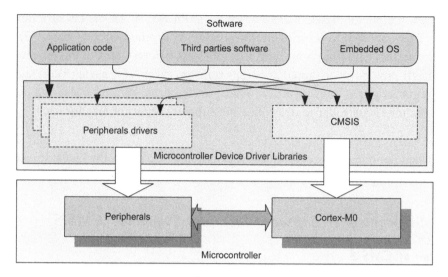

Figure 4.16:
CMSIS provides standardized access functions for processor features.

The CMSIS is standardized across multiple microcontroller vendors and is supported by multiple C compiler vendors. For example, it can be used with the Keil MDK, the ARM RealView Development Suite (RVDS), the IAR Embedded Workbench, the TASKING compiler, and various GNU-based C compiler suites including the CodeSourcery G++ tool chain.

What Is Standardized in CMSIS

The CMSIS standardized the following areas for embedded software:

- *Standardized access functions for accessing NVIC, System Control Block (SCB), and System Tick timer (SysTick) such as interrupt control and SysTick initialization.* These functions will be covered in various chapters of this book and in the CMSIS functions quick reference in Appendix C.
- *Standardized register definitions for NVIC, SCB, and SysTick registers.* For best software portability, we should use the standardized access functions. However, in some cases we need to directly access the registers in NVIC, SCB, or the SysTick. In such cases, the standardized register definitions help the software to be more portable.
- *Standardized functions for accessing special instructions in Cortex-M microcontrollers.* Some instructions on the Cortex-M microcontroller cannot be generated by normal C code. If they are needed, they can be generated by these functions provided in CMSIS. Otherwise, users will have to use intrinsic functions provided by the C compiler or embedded/inline assembly language, which are tool chain specific and less portable.

- *Standardized names for system exceptions handlers.* An embedded OS often requires system exceptions. By having standardized system exception handler names, supporting different device driver libraries in an embedded OS is much easier.
- *Standardized name for the system initialization function.* The common system initialization function "void SystemInit(void)" makes it easier for software developers to set up their system with minimum effort.
- *Standardize variable for clock speed information.* A standardized software variable called "SystemFreq" (CMSIS v1.00 to v1.20) or "SystemCoreClock" (CMSIS v1.30 or newer). This is used to determine the processor clock frequency.

The CMSIS also provides the following:

- A common platform for device driver libraries—each device driver library has the same look and feel, making it easier for beginners to learn and making it easier for software porting.
- In future release of CMSIS, it could also provide a set of common communication access functions so that middleware that has been developed can be reused on different devices without porting.

The CMSIS is developed to ensure compatibility for the basic operations. Microcontroller vendors can add functions to enhance their software solution so that CMSIS does not restrict the functionality and the capability of the embedded products.

Organization of the CMSIS

The CMSIS is divided into multiple layers:

Core Peripheral Access Layer
- Name definitions, address definitions, and helper functions to access core registers and core peripherals like the NVIC, SCB, and SysTick

Middleware Access Layer (work in progress)
- Common method to access peripherals for typical embedded systems
- Targeted at communication interfaces including UART, Ethernet, and SPI
- Allows embedded software to be used on any Cortex microcontrollers that support the required communication interface

Device Peripheral Access Layer (MCU specific)
- Register name definitions, address definitions, and device driver code to access peripherals

Access Functions for Peripherals (MCU specific)
- Optional helper functions for peripherals

The role of these layers is summarized in Figure 4.17.

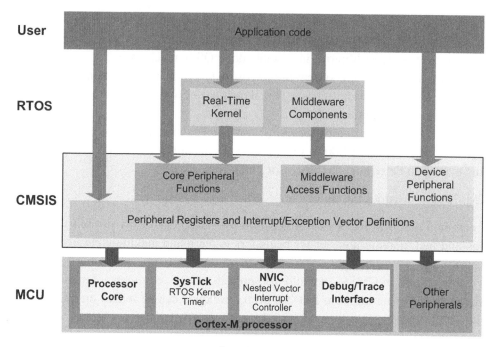

Figure 4.17:
CMSIS structure.

Using CMSIS

The CMSIS is an integrated part of the device driver package provided by the microcontroller vendors. If you are using the device driver libraries for software development, you are already using the CMSIS. If you are not using device driver libraries from microcontroller vendors, you can still use CMSIS by downloading the CMSIS package from OnARM web site (www.onarm.com), unpacking the files, and adding the required files for your project.

For C program code, normally you only need to include one header file provided in the device driver library from your microcontroller vendor. This header file then pulls in the all the required header files for CMSIS features as well as peripheral drivers.

You also need to include the CMSIS-compliant startup code, which can be either in C or assembly code. CMSIS provides various versions of startup code customized for different tool chains.

Figure 4.18 shows a simple project setup using the CMSIS package. The name of some the files depends on the actual microcontroller device name (indicated as <device> in Figure 4.18). When you use the header file provided in the device driver library, it automatically includes the other required header files for you (Table 4.4).

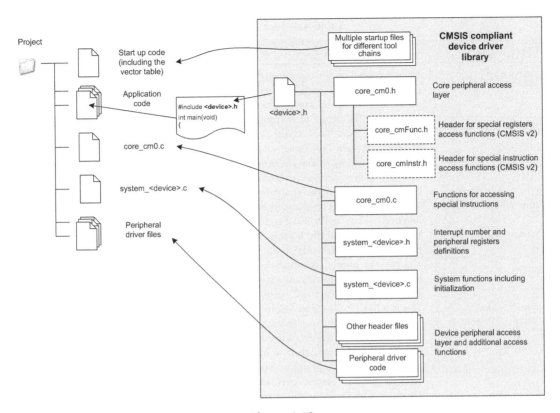

Figure 4.18:
Using CMSIS in a project.

Figure 4.19 shows a simple example of using CMSIS.

Typically, information and examples of using CMSIS can be found in the device driver libraries package from your microcontroller vendor. There are also some simple examples of using the CMSIS in the CMSIS package on the OnARM web site (www.onarm.com).

Benefits of CMSIS

For most users, CMSIS offer a number of key advantages.

Porting of applications from one Cortex-M microcontroller to another Cortex-M microcontroller is much easier. For example, most of the interrupt control functions are available for Cortex-M0, Cortex-M3, and Cortex-M4 (only a few functions for Cortex-M3/M4 are not available for Cortex-M0 because of the extra functionality of the Cortex-M3/M4 processors). This makes it straightforward to reuse the same application code for a different project. You can migrate a Cortex-M3 project to Cortex-M0 for lower cost, or you can move a Cortex-M0 project to Cortex-M3 if higher performance is required.

Table 4.4: Files in CMSIS

Files	Descriptions
<device>.h	A file provided by the microcontroller vendor that includes other header files and provides definitions for a number of constants required by CMSIS, definitions of device specific exception types, peripheral register definitions, and peripheral address definitions. The actual filtername depends on the device.
core_cm0.h	The file core_cm0.h contains the definitions of the registers for processor peripherals like NVIC, System Tick Timer, and System Control Block (SCB). It also provides the core access functions like interrupt control and system control. This file and the file core_cm0.c provide the core peripheral access layer of the CMSIS. In CMSIS version 2, this file is spitted into multiple files (see Figure 4.18).
core_cm0.c	The file core_cm0.c provides intrinsic functions of the CMSIS. The CMSIS intrinsic functions are compiler independent.
Startup code	Multiple versions of the startup code can be found in CMSIS because it is tools specific. The startup code contains a vector table and dummy definitions for a number of system exceptions handler, and from version 1.30 of the CMSIS, the reset handler also executes the system initialization function "void SystemInit(void)" before it branches to the C startup code.
system_<device>.h	This is a header file for functions implemented in system_<device>.c
system_<device>.c	This file contains the implementation of the system initialization function "void SystemInit(void)," the definition of the variable "SystemCoreClock" (processor clock speed) and a function called "void SystemCoreClockUpdate(void)" that is used after clock frequency changes to update "SystemCoreClock." The "SystemCoreClock" variable and the "SystemCoreClockUpdate" are available from CMSIS version 1.3.
Other files	There are additional files for peripheral control code and other helper functions. These files provide the device peripheral access layer of the CMSIS.

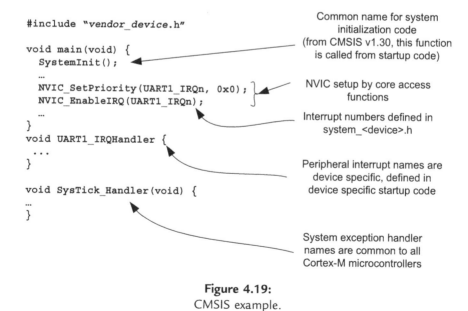

Figure 4.19:
CMSIS example.

Learning to use a new Cortex-M microcontroller is made easier. Once you have used one Cortex-M microcontroller, you can start using another quickly because all CMSIS device driver libraries have the same core functions and a similar look and feel.

The CMSIS also lowers the risk of incompatibility when integrating third-party software components. Because middleware and an embedded RTOS will be based on the same core peripheral register definitions and core access functions in CMSIS files, this reduces the chance of conflicting code. This can happen when multiple software components carry their own core access functions and register definitions. Without CMSIS, you might possibly find that different third-party software programs contain unique driver functions. This could lead to register name clashes, confusion because of multiple functions with similar names, and a waste of code space as a result of duplicated functions (Figure 4.20).

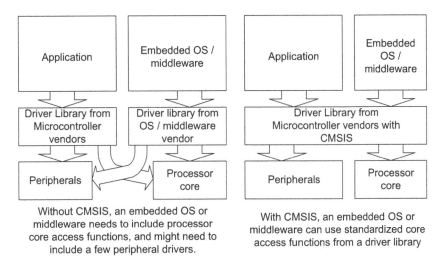

Figure 4.20:
CMSIS avoids overlapping of driver code.

CMSIS makes your software code future proof. Future Cortex-M microcontrollers will also have CMSIS support, so you can reuse your application code in future products.

The CMSIS core access functions have a small memory footprint. Multiple parties have tested CMSIS, and this helps reduce your software testing time. The CMSIS is Motor Industry Software Reliability Association (MISRA) compliant.

For companies developing an embedded OS or middleware products, the advantage of CMSIS is significant. Because CMSIS supports multiple compiler suites and is supported by multiple microcontroller vendors, the embedded OS or middleware developed with CMSIS can work on multiple complier products and can be used on multiple microcontroller families. Using CMSIS also means that these companies do not have to develop their own portable device drivers, which saves development time and verification efforts.

## CHAPTER 5

Instruction Set

Background of ARM and Thumb Instruction Set

The early ARM processors use a 32-bit instruction set called the ARM instructions. The 32-bit ARM instruction set is powerful and provides good performance, but at the same time it often requires larger program memory when compared to 8-bit and 16-bit processors. This was and still is an issue, as memory is expensive and could consume a considerable amount of power.

In 1995, ARM introduced the ARM7TDMI processor, adding a new 16-bit instruction set called the Thumb instruction set. The ARM7TDMI supports both ARM instructions and Thumb instructions, and a state-switching mechanism is used to allow the processor to decide which instruction decode scheme should be used (Figure 5.1). The Thumb instruction set provides a subset of the ARM instructions. By itself it can perform most of the normal functions, but interrupt entry sequence and boot code must still be in ARM state. Nevertheless, most processing can be carried out using Thumb instructions and interrupt handlers could switch themselves to use the Thumb state, so the ARM7TDMI processor provides excellent code density when compared to other 32-bit RISC architectures.

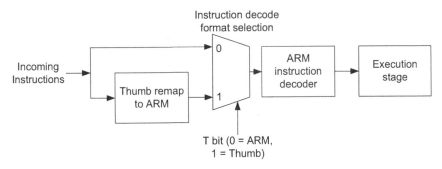

Figure 5.1:
ARM7TDMI design supports both ARM and the Thumb instruction set.

Thumb code provides a code size reduction of approximately 30% compared to the equivalent ARM code. However, it has some impact on the performance and can reduce the performance by 20%. On the other hand, in many applications, the reduction of program memory size and

The Definitive Guide to the ARM Cortex-M0. DOI: 10.1016/B978-0-12-385477-3.10005-9

the low-power nature of the ARM7TDMI processor made it extremely popular with portable electronic devices like mobile phones and microcontrollers.

In 2003, ARM introduced Thumb-2 technology. This technology provides a number of 32-bit Thumb instructions as well as the original 16-bit Thumb instructions. The new 32-bit Thumb instructions can carry out most operations that previously could only be done with the ARM instruction set. As a result, program code compiled for Thumb-2 is typically 74% of the size of the same code compiled for ARM, but it maintains similar performance.

The Cortex-M3 processor is the first ARM processor that supports only Thumb-2 instructions. It can deliver up to 1.25 DMIPS per MHz (measured with Dhrystone 2.1), and various microcontroller vendors are already shipping microcontroller products based on the Cortex-M3 processor. By implementing only one instruction set, the software development is made simpler and at the same time improves the energy efficiency because only one instruction decoder is required (Figure 5.2).

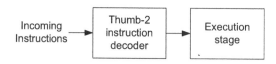

Figure 5.2:
Cortex-M processors do not have to remap instructions from Thumb to ARM.

In the ARMv6-M architecture used in the Cortex-M0 processor, in order to reduce the circuit size to a minimum, only the 16-bit Thumb instructions and a minimum subset of 32-bit Thumb instructions are supported. These 32-bit Thumb instructions are essential because the ARMv6-M architecture uses a number of features in the ARMv7-M architecture, which requires these instructions. For example, the accesses to the special registers require the MSR and MRS instructions. In addition, the Thumb-2 version of Branch and Link instruction (BL) is also included to provide a larger branch range.

Although the Cortex-M0 processor does not support many 32-bit Thumb instructions, the Thumb instruction set used in the Cortex-M0 processor is a superset of the original 16-bit Thumb instructions supported on the ARM7TDMI, which is based on ARMv4T architecture. Over the years, both ARM and Thumb instructions have gone through a number of enhancements as the architecture has evolved. For example, a number of instructions for data type conversions have been added to the Thumb instruction set for the ARMv6 and ARMv6-M architectures. These instruction set enhancements, along with various implementation optimizations, allow the Cortex-M0 processor to deliver the same level of performance as an ARM7TDMI running ARM instructions.

Table 5.1 shows the base 16-bit Thumb instructions supported in the Cortex-M0.

Table 5.1: 16-Bit Thumb Instructions Supported on the Cortex-M0 Processor

16-Bit Thumb Instructions Supported on Cortex-M0									
ADC	ADD	ADR	AND	ASR	B	BIC	BLX	BKPT	BX
CMN	CMP	CPS	EOR	LDM	LDR	LDRH	LDRSH	LDRB	LDRSB
LSL	LSR	MOV	MVN	MUL	NOP	ORR	POP	PUSH	REV
REV16	REVSH	ROR	RSB	SBC	SEV	STM	STR	STRH	STRB
SUB	SVC	SXTB	SXTH	TST	UXTB	UXTH	WFE	WFI	YIELD

The Cortex-M0 processor also supports a number of 32-bit Thumb instructions from Thumb-2 technology (Table 5.2):

- MRS and MSR special register access instructions
- ISB, DSB, and DMB memory synchronization instructions
- BL instruction (BL was supported in traditional Thumb instruction set, but the bit field definition was extended in Thumb-2)

Table 5.2: 32-Bit Thumb Instructions Supported on the Cortex-M0 Processor

32-Bit Thumb Instructions Supported on Cortex-M0					
BL	DSB	DMB	ISB	MRS	MSR

Assembly Basics

This chapter introduces the instruction set of the Cortex-M0 processor. In most situations, application code can be written entirely in C language and therefore it is not necessary to know the details of the instruction set. However, it is still useful to know what instructions are available and their usages; for example, this information might be needed during debugging.

The complete details of each instruction are documented in the ARMv6-M Architecture Reference Manual (reference 3). Here, the basic syntax and usage are introduced. First of all, to help explain the assembly instructions covered in this chapter, some of the basic information about assembly syntax is introduced here.

Quick Glance at Assembly Syntax

Most of the assembly examples in this book are written for the ARM assembler (armasm). Assembly tools from different vendors (e.g., GNU tool chain) have different assembly syntax. In most cases, the mnemonics of the assembly instructions are the same, but compile directives, definitions, labeling, and comment syntax can be different.

For ARM assembly (applies to ARM RealView Development Suite and Keil Microcontroller Development Kit), the following instruction formatting is used:

```
label
        mnemonic     operand1, operand2,...     ; Comments
```

The "label" is used as a reference to an address location. It is optional; some instructions might have a label in front of them so that the address of the instruction can be obtained using the label. Labels can also be used to reference data addresses. For example, you can put a label for a lookup table inside the program. After the "label," you can find the "mnemonic," which is the name of the instruction, followed by a number of operands. For data processing instructions written for the ARM assembler, the first operand is the destination of the operation. For a memory read or write, the first operand is the register that data are loaded into or the register that holds the write data (except for instructions that handle multiple loads and stores, which have a different syntax). The number of operands for each instruction depends on the instruction type. Some instructions do not need any operands, and some might need just one operand.

Note that some mnemonics can use different types of operands and can result in different instruction encodings. For example, the MOV (move) instruction can be used to transfer data between two registers, or it can be used to put an immediate constant value into a register.

The number of operands in an instruction depends on what type of instruction it is, and the syntax format can also be different. For example, immediate data are usually prefixed with "#":

```
        MOVS     R0, #0x12   ; Set R0 = 0x12 (hexadecimal)
        MOVS     R1, #'A'    ; Set R1 = ASCII character A
```

The text after each semicolon ";" is a comments. Comments do not affect the program operation but should make programs easier for humans to understand.

With GNU tool chain, the common assembly syntax is

```
label:
        mnemonic     operand1, operand2,...     /* Comments */
```

The opcode and operands are the same as the ARM assembler syntax, but the syntax for label and comments is different. For the same instructions as in the previous example, the GNU version is

```
        MOVS     R0, #0x12   /* Set R0 = 0x12 (hexadecimal) */
        MOVS     R1, #'A'    /* Set R1 = ASCII character A */
```

One of the commonly required features in assembly code is constant definitions. By using constant definitions, the program code can be more readable and can make code maintenance easier. In ARM assembly, an example of defining a constant is

```
        NVIC_IRQ_SETEN     EQU    0xE000E100
        NVIC_IRQ0_ENABLE   EQU    0x1
```

```
    ...
    LDR     R0,=NVIC_IRQ_SETEN    ; Put 0xE000E100 into R0
                    ; LDR here is a pseudo instruction that will be converted
                    ; to a PC relative literal data load by the assembler
    MOVS    R1, #NVIC_IRQ0_ENABLE ; Put immediate data (0x1) into
                                          ; register R1
    STR     R1, [ R0] ; Store 0x1 to 0xE000E100, this enable external
                    ; interrupt IRQ#0
```

Similarly, the same code can be written with GNU tool chain assembler syntax:

```
    .equ    NVIC_IRQ_SETEN,      0xE000E100
    .equ    NVIC_IRQ0_ENABLE,    0x1
    ...
    LDR     R0,=NVIC_IRQ_SETEN /* Put 0xE000E100 into R0
            LDR here is a pseudo instruction that will be
            converted to a PC relative load by the assembler */
    MOVS    R1, #NVIC_IRQ0_ENABLE /* Put immediate data (0x1) into
                                          register R1 */
    STR     R1, [ R0] /* Store 0x1 to 0xE000E100, this enable
                    external interrupt IRQ#0 */
```

Another typical feature in most assembly tools is allowing data to be inserted inside programs. For example, we can define data in a certain location in the program memory and access it with memory read instructions. In the ARM assembler, an example is

```
    LDR     R3,=MY_NUMBER  ; Get the memory location of MY_NUMBER
    LDR     R4, [ R3]      ; Read the value 0x12345678 into R4
    ...
    LDR     R0,=HELLO_TEXT ; Get the starting address of HELLO_TEXT
    BL      PrintText      ; Call a function called PrintText to
                                  ; display string

    ...
    ALIGN   4
MY_NUMBER   DCD  0x12345678
HELLO_TEXT  DCB  "Hello\n", 0 ; Null terminated string
```

In the preceding example, "DCD" is used to insert a word-size data, and "DCB" is used to insert byte-size data into the program. When inserting word-size data in program, we should use the "ALIGN" directive before the data. The number after the ALIGN directive determines the alignment size; in this case, the value is 4, which forces the following data to be aligned to a word boundary. Unaligned accesses are not supported in the Cortex-M0 processor. By ensuring the data following (MY_NUMBER) is word aligned, the program will be able to access the data correctly, avoiding any potential alignment faults.

Again, this example can be rewritten into GNU tool chain assembler syntax:

```
    LDR     R3,=MY_NUMBER   /* Get the memory location of MY_NUMBER */
    LDR     R4, [ R3]       /* Read the value 0x12345678 into R4 */
```

```
    ...
    LDR    R0,=HELLO_TEXT  /* Get the starting address of
                              HELLO_TEXT */
    BL     PrintText       /* Call a function called PrintText to
                              display string */
    ...
    .align 4
MY_NUMBER:
    .word  0x12345678
HELLO_TEXT:
    .asciz "Hello\n"        /* Null terminated string */
```

A number of different directives are available in both the ARM assembler and the GNU assembler for inserting data into a program. Table 5.3 presents a few commonly used examples.

Table 5.3: Commonly Used Directives for Inserting Data into a Program

Type of Data to Insert	ARM Assembler	GNU Assembler
Word	DCD (e.g., DCD 0x12345678)	.word / .4byte (e.g., .word 0x012345678)
Half word	DCW (e.g., DCW 0x1234)	.hword / .2byte (e.g., .hword 0x01234)
Byte	DCB (e.g., DCB 0x12)	.byte (e.g., .byte 0x012)
String	DCB (e.g., TXT DCB "Hello\n", 0)	.ascii /.asciz (with NULL termination) (e.g., .ascii "Hello\n" .byte 0 /* add NULL character */) (e.g., .asciz "Hello\n")
Instruction	DCI (e.g., DCI 0xBE00 ; Breakpoint-BKPT 0)	.word /.hword (e.g., .hword 0xBE00 /* Breakpoint (BKPT 0) */)

Use of a Suffix

In the assembler for ARM processors, some instructions can be followed by suffixes. For Cortex-M0, the available suffixes are shown in Table 5.4.

Table 5.4: Suffixes for Cortex-M0 Assembly Language

Suffix	Descriptions
S	Update APSR (flags); for example, `ADDS R0, R1; this ADD operation will update APSR`
EQ, NE, CS, CC, MI, PL, VS, VC, HI, LS, GE, LT, GT, LE	Conditional execution. EQ = Equal, NE = Not Equal, LT = Less Than, GT = Greater Than, etc. On the Cortex-M0 processor, these conditions can only be applied to conditional branches. For example, `BEQ label; branch to label if equal`

For the Cortex-M0 processor, most of the data processing instructions always update the APSR (flags); only a few of the data operations do not update the APSR. For example, when moving a piece of data from one register to another, it is possible to use

```
MOVS    R0, R1  ; Move R1 into R0 and update APSR
```

or

```
MOV     R0, R1  ; Move R1 into R0
```

The second group of suffixes is for conditional execution of instructions. In the Cortex-M0 processor, the only instruction that can be conditionally executed is a conditional branch. By updating the APSR using data operations, or instructions like test (TST) or compare (CMP), the program flow can be controlled. More details of this instruction will be covered later in this chapter when the conditional branch is introduced.

Thumb Code and Unified Assembler Language (UAL)

Traditionally, programming of the ARM processors in Thumb state is done with the Thumb Assembly syntax. To allow better portability between architectures and to use a single assembly language syntax between different ARM processors with various architectures, recent ARM development tools have been updated to support the Unified Assembler Language (UAL). For users who have used ARM7TDMI in the past, the most noticeable differences are the following:

- Some data operation instructions use three operands even when the destination register is the same as one of the source registers. In the past (pre-UAL), syntax might only use two operands for the same instructions.
- The "S" suffix becomes more explicit. In the past, when an assembly program file was assembled into Thumb code, most data operations were implied as instructions that updated the APSR; as a result, the "S" suffix was not essential. With the UAL syntax, instructions that update the APSR should have the "S" suffix to clearly indicate the expected operation. This prevents program code from failing when being ported from one architecture to another.

For example, a pre-UAL ADD instruction for 16-bit Thumb code is

```
ADD     R0, R1   ; R0 = R0 + R1, update APSR
```

With UAL syntax, this should be written as

```
ADDS    R0, R0, R1   ; R0 = R0 + R1, update APSR
```

But in most cases (dependent on tool chain being used), you can still write the instruction with a pre-UAL style (only two operands), but the use of "S" suffix will be more explicit:

```
ADDS    R0, R1    ; R0 = R0 + R1, update APSR
```

Most development tools still accept the pre-UAL syntax, including the ARM RealView Development Suite (RVDS) and the Keil Microcontroller Development Kit for ARM (MDK). However, the use of UAL is recommended for new projects. For assembly development with RVDS or Keil MDK, you can specify using UAL syntax with "THUMB" directives and pre-UAL syntax with "CODE16" directives. The choice of assembler syntax depends on which tool you use. Please refer to the documentation for your development suite to determine the suitable syntax.

Instruction List

The instructions in the Cortex-M0 processor can be divided into various groups based on functionality:

- Moving data within the processor
- Memory accesses
- Stack memory accesses
- Arithmetic operations
- Logic operations
- Shift and rotate operations
- Extend and reverse ordering operations
- Program flow control (branch, conditional branch, and function calls)
- Memory barrier instructions
- Exception-related instructions
- Other functions

In this section, the instructions are discussed in more detail. The syntax illustrated here uses symbols "Rd," "Rm," and the like. In real program code, these need to be substituted with register names R0, R1, R2, and so on.

Moving Data within the Processor

Transferring data is one of the most common tasks in a processor. In Thumb code, the instruction mnemonic for moving data is MOV. There are several types of MOV instructions, based on the operand type and opcode suffix.

Instruction	MOV
Function	Move register into register
Syntax (UAL)	MOV <Rd>, <Rm>
Syntax (Thumb)	MOV <Rd>, <Rm>
	CPY <Rd>, <Rm>
Note	Rm and Rn can be high or low registers
	CPY is a pre-UAL synonym for MOV (register)

If we want to copy a register value to another and update the APSR at the same time, we could use MOVS/ADDS.

Instruction	MOVS/ADDS
Function	Move register into register
Syntax (UAL)	MOVS <Rd>, <Rm>
	ADDS <Rd>, <Rm>, #0
Syntax (Thumb)	MOVS <Rd>, <Rm>
Note	Rm and Rn are both low registers
	APSR.Z, APSR.N, and APSR.C (for ADDS) update

We can also load an immediate data element into a register using the MOV instruction.

Instruction	MOV
Function	Move immediate data (sign extended) into register
Syntax (UAL)	MOVS <Rd>, #immed8
Syntax (Thumb)	MOV <Rd>, #immed8
Note	Immediate data range 0 to +255
	APSR.Z and APSR.N update

If we want to load an immediate data element into a register that is out of the 8-bit value range, we need to store the data into a program memory space and then use a memory access instruction to read the data into the register. This can be written using a pseudo instruction LDR, which the assembler converts into a real instruction. This process will be covered later in this chapter.

The MOV instructions can cause a branch to happen if the destination register is R15 (PC). However, generally the BX instruction is used for this purpose.

Another type of data transfer in the Cortex-M0 processor is Special Registers accesses. To access the Special Registers (CONTROL, PRIMASK, xPSR, etc.), the MRS and MSR instructions are needed. These two instructions cannot be generated in C language. However, they can be created using inline assembler or Embedded Assembler,[3] or another C compiler–specific feature like the named register variables feature in ARM RVDS or Keil MDK.

Instruction	MRS
Function	Move Special Register into register
Syntax	MRS <Rd>, <SpecialReg>
Note	Example:
	MRS R0, CONTROL; Read CONTROL register into R0
	MRS R9, PRIMASK; Read PRIMASK register into R9
	MRS R3, xPSR; Read xPSR register into R3

Table 5.5 shows the complete list of special register symbols that are available on the Cortex-M0 processor when MSR and MRS instructions are used.

Table 5.5: Special Register Symbols for MRS and MSR Instructions

Symbol	Register	Access Type
APSR	Application Program Status Register (PSR)	Read/Write
EPSR	Execution PSR	Read only
IPSR	Interrupt PSR	Read only
IAPSR	Composition of IPSR and APSR	Read only
EAPSR	Composition of EPSR and APSR	Read only
IEPSR	Composition of IPSR and EPSR	Read only
XPSR	Composition of APSR, EPSR, and IPSR	Read only
MSP	Main stack pointer	Read/Write
PSP	Process stack pointer	Read/Write
PRIMASK	Primary exception mask register	Read/Write
CONTROL	CONTROL register	Read/Write in Thread mode
		Read only in Handler mode

Instruction	MSR
Function	Move register into Special Register
Syntax	MSR <SpecialReg>, <Rd>
Note	Example:
	MSR CONTROL, R0; Write R0 into CONTROL register
	MSR PRIMASK, R9; Write R9 into PRIMASK register

Memory Accesses

The Cortex-M0 processor supports a number of memory access instructions, which support various data transfer sizes and addressing modes. The supported data transfer sizes are Word, Half Word and Byte. In addition, there are separate instructions to support signed and unsigned data. Table 5.6 summarizes the memory address instruction mnemonics.

Most of these instructions also support multiple addressing modes. When the instruction is used with different operands, the assembler will generate different instruction encoding.

Table 5.6: Memory Access Instructions for Various Transfer Sizes

Transfer Size	Unsigned Load	Signed Load	Signed/Unsigned Store
Word	LDR	LDR	STR
Half word	LDRH	LDRSH	STRH
Byte	LDRB	LDRSB	STRB

> **Important**
>
> It is important to make sure the memory address accessed is aligned. For example, a word size access can only be carried out on address locations when address bits[1:0] are set to zero, and a half word size access can only be carried out on address locations when an address bit[0] is set to zero. The Cortex-M0 processor does not support unaligned transfers. Any attempt at unaligned memory access results in a hard fault exception. Byte-size transfers are always aligned on the Cortex-M0 processor.

For memory read operations, the instruction to carry out single accesses is LDR (load):

Instruction	LDR/LDRH/LDRB
Function	Read single memory data into register
Syntax	LDR <Rt>, [<Rn>, <Rm>] ; Word read
	LDRH <Rt>, [<Rn>, <Rm>] ; Half Word read
	LDRB <Rt>, [<Rn>, <Rm>] ; Byte read
Note	Rt = memory[Rn + Rm]
	Rt, Rn and Rm are low registers

The Cortex-M0 processor also supports immediate offset addressing modes:

Instruction	LDR/LDRH/LDRB
Function	Read single memory data into register
Syntax	LDR <Rt>, [<Rn>, #immed5] ; Word read
	LDRH <Rt>, [<Rn>, #immed5] ; Half Word read
	LDRB <Rt>, [<Rn>, #immed5] ; Byte read
Note	Rt = memory[Rn + ZeroExtend (#immed5 << 2)] ; Word
	Rt = memory[Rn + ZeroExtend(#immed5 << 1)] ; Half word
	Rt = memory[Rn + ZeroExtend(#immed5)] ; Byte
	Rt and Rn are low registers

The Cortex-M0 processor supports a useful PC relative load instruction for allowing efficient literal data accesses. This instruction can be generated when we use the LDR pseudo instruction for putting an immediate data value into a register. These data are stored alongside the instructions, called literal pools.

Instruction	LDR
Function	Read single memory data word into register
Syntax	LDR <Rt>, [PC, #immed8] ; Word read
Note	Rt = memory[WordAligned(PC+4) + ZeroExtend(#immed8 << 2)]
	Rt is a low register, and targeted address must be a word-aligned address, the reason for adding 4.

(Continued)

Instruction	LDR
Example: 　LDR　R0,=0x12345678 ; A pseudo instruction that uses literal load 　　　　　　　　　　　　　　; to put an immediate data into a register 　LDR　R0, [PC, #0x40]　; Load a data in current program address 　　　　　　　　　　　　　; with offset of 0x40 into R0 　LDR　R0, label　　　　; Load a data in current program 　　　　　　　　　　　　; referenced by label into R0	

There is also an SP-related load instruction, which supports a wider offset range. This instruction is useful for accessing local variables in C functions because often the local variables are stored on the stack.

Instruction	LDR
Function	Read single memory data word into register
Syntax	LDR　<Rt>, [SP, #immed8] ; Word read
Note	Rt = memory[SP + ZeroExtend(#immed8 << 2)] Rt is a low register

The Cortex-M0 processor can also sign extends the read data automatically using the LDRSB and LDRSH instructions. This is useful when a signed 8-bit/16-bit data type is used, which is common in C programs.

Instruction	LDRSH/LDRSB
Function	Read single signed memory data into register
Syntax	LDRSH <Rt>, [<Rn>, <Rm>] ; Half word read LDRSB <Rt>, [<Rn>, <Rm>] ; Byte read
Note	Rt = SignExtend(memory[Rn + Rm]) Rt, Rn and Rm are low registers

For single data memory writes, the instruction is STR (store):

Instruction	STR/STRH/STRB
Function	Write single register data into memory
Syntax	STR　　<Rt>, [<Rn>, <Rm>] ; Word write STRH　<Rt>, [<Rn>, <Rm>] ; Half Word write STRB　<Rt>, [<Rn>, <Rm>] ; Byte write
Note	memory[Rn + Rm] = Rt Rt, Rn and Rm are low registers

Like the load operation, the store operation supports an immediate offset addressing mode:

Instruction	STR/STRH/STRB
Function	Write single memory data into memory
Syntax	STR <Rt>, [<Rn>, #immed5] ; Word write
	STRH <Rt>, [<Rn>, #immed5] ; Half Word write
	STRB <Rt>, [<Rn>, #immed5] ; Byte write
Note	memory[Rn + ZeroExtend(#immed5 << 2)] = Rt ; Word
	memory[Rn + ZeroExtend(#immed5 << 1)] = Rt ; Half word
	memory[Rn + ZeroExtend(#immed5)] = Rt ; Byte
	Rt and Rn are low registers

An SP-relative store instruction, which supports a wider offset range, is also available. This instruction is useful for accessing local variables in C functions because often the local variables are stored on the stack.

Instruction	STR
Function	Write single memory data word into memory
Syntax	STR <Rt>, [SP, #immed8] ; Word write
Note	memory[SP + ZeroExtend(#immed8 << 2)] = Rt
	Rt is a low register

One of the important features in ARM processors is the ability to load or store multiple registers with one instruction. There is also an option to update the base address register to the next location. For load/store multiple instructions, the transfer size is always in word size.

Instruction	LDM (Load Multiple)
Function	Read multiple memory data word into registers, base address register update by memory read
Syntax	LDM <Rn>, {<Ra>, <Rb> ,....} ; Load multiple registers from memory
Note	Ra = memory[Rn],
	Rb = memory[Rn+4],
	...
	Rn, Ra, Rb .... are low registers. Rn is on the list of registers to be updated by memory read.
	For example,
	LDM R2, {R1, R2, R5 − R7} ; Read R1,R2,R5,R6 and R7 from memory.

Instruction	LDMIA (Load Multiple Increment After)/LDMFD − Base Address Register Update to Subsequence Address
Function	Read multiple memory data word into registers and update base register
Syntax	LDMIA <Rn>!, {<Ra>, <Rb> ,....} ; Load multiple registers from memory
	; and increment base register after completion
Note	Ra = memory[Rn],

(*Continued*)

Instruction	LDMIA (Load Multiple Increment After)/LDMFD — Base Address Register Update to Subsequence Address
	Rb = memory[Rn+4], ... and then update Rn to last read address plus 4 Rn, Ra, Rb .... are low registers. For example, LDMIA R0!, {R1, R2, R5 − R7} ; Read multiple registers, R0 update to address after last read operation. LDMFD is another name for the same instruction, which was used for restoring data from a Full Descending stack, in traditional ARM systems that use software managed stacks.

Instruction	STMIA (Store Multiple Increment After)/STMEA
Function	Write multiple register data into memory and update base register
Syntax	STMIA <Rn>!, {<Ra>, <Rb> ,....} ; Store multiple registers to memory ; and increment base register after completion
Note	memory[Rn] = Ra, memory[Rn+4] = Rb, ... and then update Rn to last store address plus 4 Rn, Ra, Rb .... are low registers. For example, STMIA R0!, {R1, R2, R5 − R7} ; Store R1, R2, R5, R6, and R7 to memory ; and update R0 to address after where R7 stored STMEA is another name for the same instruction, which was used for storing data to an Empty Ascending stack, in traditional ARM systems that use software managed stack. If <Rn> is in the register list, it must be the first register in the register list.

Stack Memory Accesses

Two memory access instructions are dedicated to stack memory accesses. The PUSH instruction is used to decrement the current stack pointer and store data to the stack. The POP instruction is used to read the data from the stack and increment the current stack pointer. Both PUSH and POP instructions allow multiple registers to be stored or restored. However, only low registers, LR (for PUSH operation) and PC (for POP operation), are supported.

Instruction	PUSH
Function	Write single or multiple registers (low register and LR) into memory and update base register (stack pointer)
Syntax	PUSH {<Ra>, <Rb> ,....} ; Store multiple registers to memory and ; decrement SP to the lowest pushed data address PUSH {<Ra>, <Rb>, ..., LR} ; Store multiple registers and LR to ; memory and decrement SP to the lowest pushed data address

(Continued)

Instruction	PUSH
Note	memory[SP-4] = Ra, memory[SP-8] = Rb, … and then update SP to last store address. For example, PUSH {R1, R2, R5 − R7, LR} ; Store R1, R2, R5, R6, R7, and LR to stack

Instruction	POP
Function	Read single or multiple registers (low register and PC) from memory and update base register (stack pointer)

Syntax	POP {<Ra>, <Rb> ,….} ; Load multiple registers from memory
	                ; and increment SP to the last emptied stack address plus 4
	POP {<Ra>, <Rb>, …., PC} ; Load multiple registers and PC from
	                ; memory and increment SP to the last emptied stack
	                ; address plus 4

Note	Ra = memory[SP], Rb = memory[SP+4], … and then update SP to last restored address plus 4. For example, POP {R1, R2, R5 − R7} ; Restore R1, R2, R5, R6, R7 from stack

By allowing the Link Register (LR) and Program Counter (PC) to be used with the PUSH and the POP instructions, a function call can combine the register restore and function return operations into a single instruction. For example,

```
my_function
  PUSH { R4, R5, R7, LR} ; Save R4, R5, R7 and LR (return address)
  ... ; function body
  POP  { R4, R5, R7, PC} ; Restore R4, R5, R7 and return
```

Arithmetic Operations

The Cortex-M0 processor supports a number of arithmetic operations. The most basic are add, subtract, twos complement, and multiply. For most of these instructions, the operation can be carried out between two registers, or between one register and an immediate constant.

Instruction	ADD
Function	Add two registers
Syntax (UAL)	ADDS <Rd>, <Rn>, <Rm>
Syntax (Thumb)	ADD <Rd>, <Rn>, <Rm>
Note	Rd = Rn + Rm, APSR update. Rd, Rn, Rm are low registers.

Instruction	ADD
Function	Add an immediate constant into a register
Syntax (UAL)	ADDS <Rd>, <Rn>, #immed3
	ADDS <Rd>, #immed8
Syntax (Thumb)	ADD <Rd>, <Rn>, #immed3
	ADD <Rd>, #immed8
Note	Rd = Rn + ZeroExtend(#immed3), APSR update, or
	Rd = Rd + ZeroExtend(#immed8), APSR update.
	Rd, Rn, Rm are low registers.

Instruction	ADD
Function	Add two registers without updating APSR
Syntax (UAL)	ADD <Rd>, <Rm>
Syntax (Thumb)	ADD <Rd>, <Rm>
Note	Rd = Rd + Rm.
	Rd, Rm can be high or low registers.

Instruction	ADD
Function	Add stack pointer to a register without updating APSR
Syntax (UAL)	ADD <Rd>, SP, <Rd>
Syntax (Thumb)	ADD <Rd>, SP
Note	Rd = Rd + SP.
	Rd can be high or low register.

Instruction	ADD
Function	Add stack pointer to a register without updating APSR
Syntax (UAL)	ADD SP, <Rm>
Syntax (Thumb)	ADD SP, <Rm>
Note	SP = SP + Rm.
	Rm can be high or low register.

Instruction	ADD
Function	Add stack pointer to a register without updating APSR
Syntax (UAL)	ADD <Rd>, SP, #immed8
Syntax (Thumb)	ADD <Rd>, SP, #immed8
Note	Rd = SP + ZeroExtend(#immed8 <<2).
	Rd is a low register.

Instruction	ADD
Function	Add an immediate constant to stack pointer
Syntax (UAL)	ADD SP, SP, #immed7
Syntax (Thumb)	ADD SP, #immed7
Note	SP = SP + ZeroExtend(#immed7 <<2).
	This instruction is useful for C functions to adjust the SP for local variables.

Instruction	ADR (ADD)
Function	Add an immediate constant with PC to a register without updating APSR
Syntax (UAL)	ADR <Rd>, <label> (normal syntax)
	ADD <Rd>, PC, #immed8 (alternate syntax)
Syntax (Thumb)	ADR <Rd>, (normal syntax)
	ADD <Rd>, PC, #immed8 (alternate syntax)
Note	Rd = (PC[31:2]<<2) + ZeroExtend(#immed8 <<2).
	This instruction is useful for locating a data address within the program memory near to the current instruction. The result address must be word aligned.
	Rd is a low register.

Instruction	ADC
Function	Add with carry and update APSR
Syntax (UAL)	ADCS <Rd>, <Rm>
Syntax (Thumb)	ADC <Rd>, <Rm>
Note	Rd = Rd + Rm + Carry
	Rd and Rm are low registers.

Instruction	SUB
Function	Subtract two registers
Syntax (UAL)	SUBS <Rd>, <Rn>, <Rm>
Syntax (Thumb)	SUB <Rd>, <Rn>, <Rm>
Note	Rd = Rn − Rm, APSR update.
	Rd, Rn, Rm are low registers.

Instruction	SUB
Function	Subtract a register with an immediate constant
Syntax (UAL)	SUBS <Rd>, <Rn>, #immed3
	SUBS <Rd>, #immed8
Syntax (Thumb)	SUB <Rd>, <Rn>, #immed3
	SUB <Rd>, #immed8
Note	Rd = Rn − ZeroExtend(#immed3), APSR update, or
	Rd = Rd − ZeroExtend(#immed8), APSR update.
	Rd, Rn are low registers.

Instruction	SUB
Function	Subtract SP by an immediate constant
Syntax (UAL)	SUB SP, SP, #immed7
Syntax (Thumb)	SUB SP, #immed7
Note	SP = SP - ZeroExtend(#immed7 <<2). This instruction is useful for C functions to adjust the SP for local variables.

Instruction	SBC
Function	Subtract with carry (borrow)
Syntax (UAL)	SBCS <Rd>, <Rd>, <Rm>
Syntax (Thumb)	SBC <Rd>, <Rm>
Note	Rd = Rd − Rm − Borrow, APSR update. Rd and Rm are low registers.

Instruction	RSB
Function	Reverse Subtract (negative)
Syntax (UAL)	RSBS <Rd>, <Rn>, #0
Syntax (Thumb)	NEG <Rd>, <Rn>
Note	Rd = 0 − Rm, APSR update. Rd and Rm are low registers.

Instruction	MUL
Function	Multiply
Syntax (UAL)	MULS <Rd>, <Rm>, <Rd>
Syntax (Thumb)	MUL <Rd>, <Rm>
Note	Rd = Rd * Rm, APSR.N, and APSR.Z update. Rd and Rm are low registers.

There are also a few compare instructions that compare (using subtract) values and update flags (APSR), but the result of the comparison is not stored.

Instruction	CMP
Function	Compare
Syntax (UAL)	CMP <Rn>, <Rm>
Syntax (Thumb)	CMP <Rn>, <Rm>
Note	Calculate Rn − Rm, APSR update but subtract result is not stored.

Instruction	CMP
Function	Compare
Syntax (UAL)	CMP <Rn>, #immed8
Syntax (Thumb)	CMP <Rn>, #immed8
Note	Calculate Rd − ZeroExtended(#immed8), APSR update but subtract result is not stored. Rn is a low register.

Instruction	CMN
Function	Compare negative
Syntax (UAL)	CMN <Rn>, <Rm>
Syntax (Thumb)	CMN <Rn>, <Rm>
Note	Calculate Rn − NEG(Rm), APSR update but subtract result is not stored. Effectively the operation is an ADD.

Logic Operations

Another set of essential operations in most processors is made up of logic operations. For logical operations, the Cortex-M0 processor has a number of instructions available, including basic features like AND, OR, and the like. In addition, it has a number of instructions for compare and testing.

Instruction	AND
Function	Logical AND
Syntax (UAL)	ANDS <Rd>, <Rd>, <Rm>
Syntax (Thumb)	AND <Rd>, <Rm>
Note	Rd = AND(Rd, Rm), APSR.N, and APSR.Z update. Rd and Rm are low registers.

Instruction	ORR
Function	Logical OR
Syntax (UAL)	ORRS <Rd>, <Rd>, <Rm>
Syntax (Thumb)	ORR <Rd>, <Rm>
Note	Rd = OR(Rd, Rm), APSR.N, and APSR.Z update. Rd and Rm are low registers.

Instruction	EOR
Function	Logical Exclusive OR
Syntax (UAL)	EORS <Rd>, <Rd>, <Rm>
Syntax (Thumb)	EOR <Rd>, <Rm>
Note	Rd = XOR(Rd, Rm), APSR.N, and APSR.Z update. Rd and Rm are low registers.

Instruction	BIC
Function	Logical Bitwise Clear
Syntax (UAL)	BICS <Rd>, <Rd>, <Rm>
Syntax (Thumb)	BIC <Rd>, <Rm>
Note	Rd = AND(Rd, NOT(Rm)), APSR.N, and APSR.Z update. Rd and Rm are low registers.

Instruction	MVN
Function	Logical Bitwise NOT
Syntax (UAL)	MVNS <Rd>, <Rm>
Syntax (Thumb)	MVN <Rd>, <Rm>
Note	Rd = NOT(Rm), APSR.N, and APSR.Z update. Rd and Rm are low registers.

Instruction	TST
Function	Test (bitwise AND)
Syntax (UAL)	TST <Rn>, <Rm>
Syntax (Thumb)	TST <Rn>, <Rm>
Note	Calculate AND(Rn, Rm), APSR.N, and APSR.Z update, but the AND result is not stored. Rd and Rm are low registers.

Shift and Rotate Operations

The Cortex-M0 also supports shift and rotate instructions. It supports both arithmetic shift operations (the datum is a signed integer value where MSB needs to be reserved) as well as logical shift.

Instruction	ASR
Function	Arithmetic Shift Right
Syntax (UAL)	ASRS <Rd>, <Rd>, <Rm>
Syntax (Thumb)	ASR <Rd>, <Rm>
Note	Rd = Rd >> Rm, last bit shift out is copy to APSR.C, APSR.N and APSR.Z are also updated. Rd and Rm are low registers.

Instruction	ASR
Function	Arithmetic Shift Right
Syntax (UAL)	ASRS <Rd>, <Rm>, #immed5

(Continued)

Instruction	ASR

Syntax (Thumb)
Note
	ASR <Rd>, <Rm>, #immed5
	Rd = Rm >> immed5, last bit shifted out is copied to
	APSR.C, APSR.N and APSR.Z are also updated.
	Rd and Rm are low registers.

When ASR is used, the MSB of the result is unchanged, and the Carry flag is updated using the last bit shifted out (Figure 5.3).

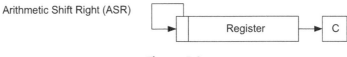

Figure 5.3:
Arithmetic Shift Right.

For logical shift operations, the instructions are LSL (Figure 5.4) and LSR (Figure 5.5).

Instruction	LSL
Function Syntax (UAL) Syntax (Thumb) Note	Logical Shift Left LSLS <Rd>, <Rd>, <Rm> LSL <Rd>, <Rm> Rd = Rd << Rm, last bit shifted out is copied to APSR.C, APSR.N and APSR.Z are also updated. Rd and Rm are low registers.

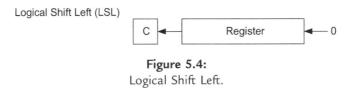

Figure 5.4:
Logical Shift Left.

Figure 5.5:
Logical Shift Right.

Instruction	LSL
Function	Logical Shift Left
Syntax (UAL)	LSLS <Rd>, <Rm>, #immed5
Syntax (Thumb)	LSL <Rd>, <Rm>, #immed5
Note	Rd = Rm << #immed5, last bit shifted out is copied to APSR.C, APSR.N and APSR.Z are also updated. Rd and Rm are low registers.

Instruction	LSR
Function	Logical Shift Right
Syntax (UAL)	LSRS <Rd>, <Rd>, <Rm>
Syntax (Thumb)	LSR <Rd>, <Rm>
Note	Rd = Rd >> Rm, last bit shifted out is copied to APSR.C, APSR.N and APSR.Z are also updated. Rd and Rm are low registers.

Instruction	LSR
Function	Logical Shift Right
Syntax (UAL)	LSRS <Rd>, <Rm>, #immed5
Syntax (Thumb)	LSR <Rd>, <Rm>, #immed5
Note	Rd = Rm >> #immed5, last bit shifted out is copied to APSR.C, APSR.N and APSR.Z are also updated. Rd and Rm are low registers.

There is only one rotate instruction, ROR (Figure 5.6).

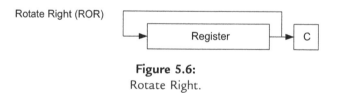

Figure 5.6:
Rotate Right.

Instruction	ROR
Function	Rotate Right
Syntax (UAL)	RORS <Rd>, <Rd>, <Rm>
Syntax (Thumb)	ROR <Rd>, <Rm>
Note	Rd = Rd rotate right by Rm bits, last bit shifted out is copied to APSR.C, APSR.N and APSR.Z are also updated. Rd and Rm are low registers.

If a rotate left operation is needed, this can be done using a ROR with a different offset:

$$\text{Rotate\_Left(Data, offset)} == \text{Rotate\_Right(Data, (32 - offset))}$$

Extend and Reverse Ordering Operations

The Cortex-M0 processor supports a number of instructions that can perform data reordering or extraction (Figures 5.7, 5.8, and 5.9).

Instruction	REV (Byte-Reverse Word)
Function	Byte Order Reverse
Syntax	REV <Rd>, <Rm>
Note	Rd = {Rm[7:0] , Rm[15:8], Rm[23:16], Rm[31:24]}
	Rd and Rm are low registers.

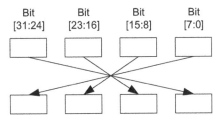

Figure 5.7:
REV operation.

Instruction	REV16 (Byte-Reverse Packed Half Word)
Function	Byte Order Reverse within half word
Syntax	REV16 <Rd>, <Rm>
Note	Rd = {Rm[23:16], Rm[31:24], Rm[7:0] , Rm[15:8]}
	Rd and Rm are low registers.

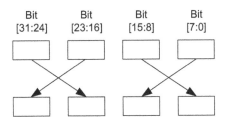

Figure 5.8:
REV16 operation.

Instruction	REVSH (Byte-Reverse Signed Half Word)
Function Syntax Note	Byte order reverse within lower half word, then sign extend result REVSH <Rd>, <Rm> Rd = SignExtend({Rm[7:0] , Rm[15:8]}) Rd and Rm are low registers.

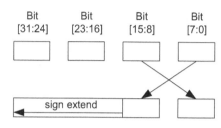

Figure 5.9:
REVSH operation.

These reverse instructions are usually used for converting data between little endian and big endian systems.

The SXTB, SXTH, UXT, and UXTH instructions are used for extending a byte or half word data into a word. They are usually used for data type conversions.

Instruction	SXTB (Signed Extended Byte)
Function Syntax Note	SignExtend lowest byte in a word of data SXTB <Rd>, <Rm> Rd = SignExtend(Rm[7:0]) Rd and Rm are low registers.

Instruction	SXTH (Signed Extended Half Word)
Function Syntax Note	SignExtend lower half word in a word of data SXTH <Rd>, <Rm> Rd = SignExtend(Rm[15:0]) Rd and Rm are low registers.

Instruction	UXTB (Unsigned Extended Byte)
Function Syntax Note	Extend lowest byte in a word of data UXTB <Rd>, <Rm> Rd = ZeroExtend(Rm[7:0]) Rd and Rm are low registers.

Instruction	UXTH (Unsign Extended Half Word)
Function	Extend lower half word in a word of data
Syntax	UXTH <Rd>, <Rm>
Note	Rd = ZeroExtend(Rm[15:0]) Rd and Rm are low registers.

With SXTB or SXTH, the data are extended using bit[7] or bit[15] of the input data, whereas for UXTB and UXTH, the data are extended using zeros. For example, if R0 is 0x55AA8765, the result of these extended instructions is

```
SXTB   R1, R0    ; R1 = 0x00000065
SXTH   R1, R0    ; R1 = 0xFFFF8765
UXTB   R1, R0    ; R1 = 0x00000065
UXTH   R1, R0    ; R1 = 0x00008765
```

Program Flow Control

There are five branch instructions in the Cortex-M0 processor. They are essential for program flow control like looping and conditional execution, and they allow program code to be partitioned into functions and subroutines.

Instruction	B (Branch)
Function	Branch to an address (unconditional)
Syntax	B <label>
Note	Branch range is +/− 2046 bytes of current program counter

Instruction	B<cond> (Conditional Branch)
Function	Depending of APSR, branch to an address
Syntax	B<cond> <label>

Note	Branch range is +/− 254 bytes of current program counter.
	For example,
	    CMP R0, #0x1 ; Compare R0 with 0x1
	    BEQ process1 ; Branch to process1 if R0 equal 1

The <cond> is one of the 14 possible condition suffixes (Table 5.7).

For example, a simple loop that runs three times could be

```
     MOVS    R0, #3    ; Loop counter starting value is 3
loop                   ; "loop" is an address label
     SUBS    R0, #1    ; Decrement by 1 and update flag
     BGT     loop      ; branch to loop if R0 is Greater Than (GT) 1
```

Table 5.7: Condition Suffixes for Conditional Branch

Suffix	Branch Condition	Flags (APSR)
EQ	Equal	Z flag is set
NE	Not equal	Z flag is cleared
CS/HS	Carry set / unsigned higher or same	C flag is set
CC/LO	Carry clear / unsigned lower	C flag is cleared
MI	Minus / negative	N flag is set (minus)
PL	Plus / positive or zero	N flag is cleared
VS	Overflow	V flag is set
VC	No overflow	V flag is cleared
HI	Unsigned higher	C flag is set and Z is cleared
LS	Unsigned lower or same	C flag is cleared or Z is set
GE	Signed greater than or equal	N flag is set and V flag is set, or N flag is cleared and V flag is cleared (N == V)
LT	Signed less than	N flag is set and V flag is cleared, or N flag is cleared and V flag is set (N != V)
GT	Signed greater then	Z flag is cleared, and either both N flag and V flag are set, or both N flag and V flag are cleared (Z == 0 and N == V)
LE	Signed less than or equal	Z flag is set, or either N flag set with V flag cleared, or N flag cleared and V flag set (Z == 1 or N != V)

The loop will execute three times. The third time, R0 is 1 before the SUBS instruction. After the SUBS instruction, the zero flag is set, so the condition for the branch failed and the program continues execution after the BGT instruction.

Instruction	BL (Branch and Link)
Function	Branch to an address and store return address to LR. Usually use for function calls, and can be used for long-range branch that is beyond the branch range of branch instruction (B <label>).
Syntax	BL <label>
Note	Branch range is +/− 16MB of current program counter. For example, BL functionA ; call a function called functionA

Instruction	BX (Branch and Exchange)
Function	Branch to an address specified by a register, and change processor state depending on bit[0] of the register.
Syntax	BX <Rm>
Note	Because the Cortex-M0 processor only supports Thumb code, bit[0] of the register content (Rm) must be set to 1, otherwise it means it is trying to switch to the ARM state and this will generate a fault exception.

BL is commonly used for calling a subroutine or function. When it is executed, the address of the next instruction will be stored to the Link Register (LR), with the LSB set to 1. When the subroutine or function completes the required task, it can then return to the calling program by executing a "BX LR" instruction (Figure 5.10).

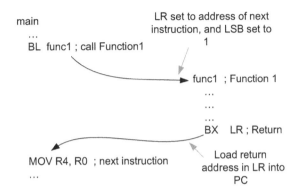

Figure 5.10:
Function call and return using BL and BX instructions.

BX can also be used to branch to an address that has an offset that is more than the normal branch instruction. Because the target is specified by a 32-bit register, it can branch to any address in the memory map.

Instruction	BLX (Branch and Link with Exchange)
Function	Branch to an address specified by a register, save return address to LR, and change processor state depending on bit[0] of the register.
Syntax	BLX <Rm>
Note	Because the Cortex-M0 processor only supports Thumb code, the bit [0] of the register content (Rm) must be set to 1, otherwise it means it is trying to switch to the ARM state and this will create a fault exception.

BLX is used when a function call is required but the address of the function is held inside a register (e.g., when working with function pointers).

Memory Barrier Instructions

Memory barrier instructions are often needed when the memory system is complex. In some cases, if the memory barrier instruction is not used, race conditions could occur and cause system failures. For example, in some ARM processors that support simultaneous bus transfers (as a processor can have multiple memory interfaces), the transfer sequence of these transfers might overlap. If the software code relies on strict ordering of memory access sequences, it could result in software errors in corner cases. The memory barrier instructions allow the processor to stop executing the next instruction, or stop starting a new transfer, until the current memory access has completed.

Because the Cortex-M0 processor only has a single memory interface to the memory system and does not have a write buffer in the system bus interface, the memory barrier instruction is rarely needed. However, memory barriers may be necessary on other ARM processors that have more complex memory systems. If the software needs to be portable to other ARM processors, then the uses of memory barrier instructions could be essential. Therefore, the memory barrier instructions are supported on the Cortex-M0 to provide better compatibility between the Cortex-M0 processor and other ARM processors.

There are three memory barrier instructions that support on the Cortex-M0 processor:

- DMB
- DSB
- ISB

Instruction	DMB
Function	Data Memory Barrier
Syntax	DMB
Note	Ensures that all memory accesses are completed before new memory access is committed

Instruction	DSB
Function	Data Synchronization Barrier
Syntax	DSB
Note	Ensures that all memory accesses are completed before the next instruction is executed

Instruction	ISB
Function	Instruction Synchronization Barrier
Syntax	ISB
Note	Flushes the pipeline and ensures that all previous instructions are completed before executing new instructions

Architecturally, there are various cases where these instructions are needed. Although in practice omitting the memory barrier instruction might not cause any issue on the Cortex-M0, it could be an issue when the same software is used on another ARM processor. For example, after changing the CONTROL register with MSR instruction, architecturally an ISB should be used after writing to the CONTROL register to ensure subsequent instructions use the updated settings. Although the Cortex-M0 omits the ISB instruction in this case, the omission does not cause an issue.

Another example is memory remap control. In some microcontrollers, a hardware register can change the memory map. After writing to the memory map switching register, you need to use the DSB instruction to ensure the write has been completed and memory configuration has been updated before carrying out the next step. Otherwise, if the memory switching is delayed, possibly because of a write buffer in the system bus interface (e.g., the Cortex-M3 has a write buffer in the system bus interface to allow higher performance), and the processor starts to access the switched memory region immediately, the access could be using the old memory mapping, or the transfer could become corrupted by the memory map switching.

Memory barrier instruction is also needed when the program contains self-modifying code. For example, if an application changes its own program code, the instruction execution that follows should use the updated program code. However, if the processor is pipelined or has a fetch buffer, the processor may have already fetched an old copy of the modified instruction. In this case, the program should use a DSB operation to ensure the write to the memory is completed; then it should use an ISB instruction to ensure the instruction fetch buffer is updated with the new instructions.

More details about memory barriers can be found in the ARMv6-M Architecture Reference manual (reference 3).

Exception-Related Instructions

The Cortex-M0 processor provides an instruction called supervisor call (SVC). This instruction causes the SVC exception to take place immediately if the exception priority level of SVC is higher than current level.

Instruction	SVC
Function	Supervisor call
Syntax	SVC #<immed8>
	SVC <immed8>
Note	Trigger the SVC exception. For example,
	SVC #3 ; SVC instruction, with parameter, equals 3.
	Alternative syntax without the "#" is also allowed. For example,
	SVC 3 ; this is the same as SVC #3.

An 8-bit immediate data element is used with SVC instruction. This parameter does not affect the SVC exception directly, but it can be extracted by the SVC handler and be used as an input to the SVC function. Typically the SVC can be used to provide access to system service or the application programming interface (API), and this parameter can be used to indicate which system service is required.

If the SVC instruction is used in an exception handler that has the same or a higher priority than the SVC, this will cause a fault exception. As a result, the SVC cannot be used in the hard fault handler, the NMI handler, or the SVC handler itself.

Besides using MSR instruction, the PRIMASK special register can also be changed using an instruction called CPS:

Instruction	CPS
Function	Change processor state: enable or disable interrupt
Syntax	CPSIE I ; Enable Interrupt (Clearing PRIMASK)
	CPSID I ; Disable Interrupt (Setting PRIMASK)
Note	PRIMASK only block external interrupts, SVC, PendSV, SysTick. But it does not block NMI and the hard fault handler.

The switching of PRIMASK to disable and enable the interrupt is commonly used for timing critical code.

Sleep Mode Feature—Related Instructions

The Cortex-M0 processor can enter sleep mode by executing the Wait-for-Interrupt (WFI) and Wait-for-Event (WFE) instructions. Note that for the Cortex-M1 processor, as the design is implemented in a FPGA design, which does not have sleep mode, these two instructions execute as NOP and will not cause the processor to stop.

Instruction	WFI
Function	Wait for Interrupt
Syntax	WFI
Note	Stops program execution until an interrupt arrives or until the processor enters a debug state.

WFE is just like WFI, except that it can also be awakened by events. An event can be an interrupt, the execution of an SEV instruction (see next page), or the entering of a debug state. A previous event also affects a WFE instruction: Inside the Cortex-M0 processor, there is an event register that records whether an event has occurred (exceptions, external events, or the execution of an SEV instruction). If the event register is not set when the WFE is executed, the WFE instruction execution will cause the processor to enter sleep mode. If the event register is set when WFE is executed, it will cause the event register to be cleared and the processor proceeds to the next instruction.

Instruction	WFE
Function	Wait for Event
Syntax	WFE
Note	If the internal event register is set, it clears the internal event register and continues execution. Otherwise, stop program execution until an event (e.g., an interrupt) arrives or until the processor enters a debug state.

WFE can also be awakened by an external event input signal, which is normally used in a multiprocessing environment.

The Send Event (SEV) instruction is normally used in multiprocessor systems to wake up other processors that are in sleep mode by means of the WFE instruction. For single-processor systems, where the processor does not have a multiprocessor communication interface or the multiprocessor communication interface is not used, the SEV can only affect the local event register inside the processor itself.

Instruction	SEV
Function	Send event to all processors in multiprocessing environment (including itself)
Syntax	SEV
Note	Set local event register and send out an event pulse to other microprocessor in a multiple processor system

Other Instructions

The Cortex-M0 processor supports an NOP instruction. This instruction can be used to produce instruction alignment or to introduce delay.

Instruction	NOP
Function	No operation
Syntax	NOP
Note	The NOP instruction takes one cycle minimum on Cortex-M0. In general, delay timing produced by NOP instruction is not guaranteed and can vary among different systems (e.g., memory wait states, processor type). If the timing delay needs to be accurate, a hardware timer should be used.

The breakpoint instruction is used to provide a breakpoint function during debug. Usually a debugger, replacing the original instruction, inserts this instruction. When the breakpoint is hit, the processor would be halted, and the user can then carry out the debug tasks through the debugger. The Cortex-M0 processor also has a hardware breakpoint unit. This is limited to four breakpoints. Because many microcontrollers use flash memory, which can be reprogrammed a number of times, using software breakpoint instruction allows more breakpoints to be set at no extra cost. The breakpoint instruction has an 8-bit immediate data field. This immediate value does not affect the breakpoint operation directly, but the debugger can extract this value and use it for debug operation.

Instruction	BKPT
Function	Breakpoint
Syntax	BKPT #<immed8>
	BKPT <immed8>
Note	BKPT instruction can have an 8-bit immediate data field. The debugger can use this as an identifier for the BKPT. For example,
	BKPT #0 ; breakpoint, with immediate field equal zero
	Alternative syntax without the "#" is also allowed. For example,
	BKPT 0 ; This is the same as BKPT #0.

The YIELD instruction is a hint instruction targeted for embedded operating systems. This is not implemented in the current releases of the Cortex-M0 processor and executes as NOP.

When used in multithread systems, YIELD can indicate that the current thread is delayed (e.g., waiting for hardware) and can be swapped out. In this case, the processor does not have to spend too much time on an idle task and can switch to other tasks earlier to get better system throughput. On the Cortex-M0 processor, this instruction is executed as an NOP (no operation) because it does not have special support for multithreading. This instruction is included for better software compatibility with other ARM processors.

Instruction	YIELD
Function	Indicate task is stalled
Syntax	YIELD
Note	Execute as NOP on the Cortex-M0 processor

Pseudo Instructions

Apart from the instructions listed in the previous section, a few pseudo instructions are also available. The pseudo instructions are provided by the assembler tools, which convert them into one or more real instructions.

The most commonly used pseudo instruction is the LDR. This allows a 32-bit immediate data item to be loaded into a register.

Pseudo Instruction	LDR
Function	Load a 32-bit immediate data into register Rd
Syntax	LDR <Rd>, =immed32
Note	This is translated to a PC-related load from a literal pool. For example,
	LDR R0, =0x12345678 ; Set R0 to hexadecimal value 0x12345678
	LDR R1, =10 ; Set R1 to decimal value 10
	LDR R2, ='A' ; Set R2 to character 'A'

Pseudo Instruction	LDR
Function	Load a data in specified address (label) into register
Syntax	LDR <Rd>, label
Note	The address of label must be word aligned and should be closed to the current program counter. For example, you can put a data item in program ROM using DCD and then access this data item using LDR.
	LDR R0, CONST_NUM ; Load CONST_NUM (0x17) in R0
	...
	ALIGN 4 ; make sure next data are word aligned
	CONST_NUM DCD 0x17 ; Put a data item in program code

Other pseudo instructions depend on the tool chain being used. For more information, please refer to the tools documentation for details.

## CHAPTER 6

Instruction Usage Examples

Overview

In the previous chapter we looked at the instruction set of the Cortex-M0 processor. In this chapter we will see how these instructions are used to carry out various operations. The examples in this chapter are useful for understanding the instruction set. Because most embedded programmers write their program in C, there is no need for most application to write code in assembly, as illustrated in these examples.

The following examples are written based on ARM assembly syntax. For the GNU assembler, the syntax is different in a number of ways, as highlighted in the previous chapter.

Program Control

If-Then-Else

One the most important functions of the instruction set is to handle conditional branches. For example, if we need to carry out the task

```
if (counter > 10) then
    counter = 0
else
    counter = counter + 1
```

Assume the R0 is used as a "counter" variable; the preceding operation can be implemented as

```
    CMP  R0, #10       ; compare to 10
    BLE  incr_counter  ; if less or equal, then branch
    MOVS R0, #0        ; counter = 0
    B    counter_done  ; branch to counter_done
incr_counter
    ADDS R0, R0, #1    ; counter = counter +1
counter_done
    ...
```

The program code first carries out a compare and then executes a conditional branch. The program then carries out a required task and finishes at the program address labeled as "counter_done."

The Definitive Guide to the ARM Cortex-M0. DOI: 10.1016/B978-0-12-385477-3.10006-0

Loop

Another important program control operation is looping. For example,

```
Total = 0;
for (i=0;i<5;i=i+1)
  Total = Total + i;
```

Assume "Total" is R0 and "i" is R1; the program can be implemented as

```
     MOVS R0, #0      ; Total = 0
     MOVS R1, #0      ; i = 0
loop
    ADDS R0, R0, R1  ; Total = Total + i
    ADDS R1, R1, #1  ; i = i + 1
    CMP  R1, #5      ; compare i to 5
    BLT  loop        ; if less than then branch to loop
```

More on the Branch Instructions

As Table 6.1 illustrates, there are various branch instructions.

Table 6.1: Various Branch Instructions

Branch Type	Examples
Normal branch. Branch always carries out.	B label (Branch to address marked as "label".)
Conditional branch. Branch depends on the current status of APSR and the condition specified in the instruction.	BEQ label (Branch if Z flag is set, which results from a equal comparison or ALU operation with result of zero.)
Branch and link. Branch always carries out and updates the Link Register (LR, R14) with the instruction address following the executed BL instruction.	BL label (Branch to address "label", and Link Register updated to the instruction after this BL instruction.)
Branch and exchange state. Branch to address stored in a register. The LSB of the register should be set to 1 to indicate the Thumb state. (Cortex-M0 does not support ARM instruction, so the Thumb state must be used.)	BX LR (Branch to address stored in the Link Register. This instruction is often used for function return.)
Branch and link with exchange state. Branch to address stored in a register, with the Link Register (LR/R14) updated to the instruction address following the executed BLX instruction. The LSB of the register should be set to 1 to indicate the Thumb state. (Cortex-M0 does not support ARM instruction, so can use this Thumb state must be used.)	BLX R4 (Branch to address stored in the R4, and LR is updated to the instruction following the BLX instruction. This instruction is often used for calling functions addressed by function pointers.)

The BL instruction (Branch and Link) is usually used for calling functions. It can also be used for normal branch operations when a long branch range is required. If the branch target offset is more than 16MB, we can use the BX instruction instead. An example is illustrated in Table 6.2.

Table 6.2: Instruction for Branch Range

Branch Range	Available Instruction
Under +/− 254 bytes	B label B<cond> label
Under +/− 2KB	B label
Under +/− 16MB	BL label
Over +/− 16MB	LDR R0,=label; Load the address value of label in R0 BX R0; Branch to address pointed to by R0, or BLX R0; Branch to address pointed to by R0 and update LR

Typical Usages of Branch Conditions

A number of conditions are available for the conditional branch. They allow the result of signed and unsigned data operations or compare operations to be used for branch control. For example, if we need to carry out a conditional branch after a compare operation "CMP R0, R1," we can use one of the conditional branch instructions shown in Table 6.3.

Table 6.3: Conditional Branch Instructions for Value Comparison

Required Branch Control	Unsigned Data	Signed Data
If (R0 equal R1) then branch	BEQ label	BEQ label
If (R0 not equal R1) then branch	BNE label	BNE label
If (R0 > R1) then branch	BHI label	BGT label
If (R0 >= R1) then branch	BCS label / BHS label	BGE label
If (R0 < R1) then branch	BCC label / BLO label	BLT label
If (R0 <= R1) then branch	BLS label	BLE label

To detect value overflow in add or subtract operations, we can use the instructions shown in Table 6.4.

Table 6.4: Conditional Branch Instructions for Overflow Detection

Required Branch Control	Unsigned Data	Signed Data
If (overflow (R0 + R1)) then branch	BCS label	BVS label
If (no_overflow (R0 + R1)) then branch	BCC label	BVC label
If (overflow (R0 − R1)) then branch	BCC label	BVS label
If (no_overflow (R0 − R1)) then branch	BCS label	BVC label

To detect whether an operation result is a positive value or negative value (signed data), the "PL" and "MI" suffixes can be used for the conditional branch (Table 6.5).

Table 6.5: Conditional Branch Instructions for Positive or Negative Value Detection

Required Branch Control	Unsigned Data	Signed Data
If (result >= 0) then branch	Not applicable	BPL label
If (result < 0) then branch	Not applicable	BMI label

Apart from using the compare (CMP) instruction, conditional branches can also be controlled by the result of arithmetic operations and logical operations, or instructions like compare negative (CMN) and test (TST). For example, a simple loop that executes five times can be written as

```
     MOVS R0, #5          ; Loop counter
loop
     SUBS R0, R0, #1      ; Decrement loop counter
     BNE  loop            ; if result is not 0 then branch to loop
```

A polling loop that waits until a status register bit 3 to be set can be written as

```
     LDR  R0, =Status     ; Load address of status register in R0
     MOVS R2, #0x8        ; Bit 3 is set
loop
     LDR  R1, [R0]        ; Read the status register
     TST  R1, R2          ; Compute "R1 AND 0x8"
     BEQ  loop            ; if result is 0 then try again
```

Function Calls and Function Returns

When carrying out function call (or subroutine call), we need to save the return address, which is the address of the instruction following the call instruction, so that we can resume the execution of the current instruction sequence. There are two instructions that can be used for the function call, as shown in Table 6.6.

Table 6.6: Instructions for Function or Subroutine Calls

Instruction Example	Scenarios
BL function	Target function address is fixed, and the offset is within +/− 16MB.
LDR R0, =function; (other registers could also be used) BLX R0	Target function address can be changed during run time, or the offset is over +/− 16MB.

After executing the BL/BLX instructions, the return address is stored in the Link Register (LR/R14) for function return when the function completed. In the simple cases, the function executed will be terminated using "BX LR" (Figure 6.1).

If the value of LR can be changed during "FunctionA," we will need to save the return address to prevent it from being lost. This happens when the BL or BLX instruction is executed within FunctionA, for example, when a nested function call is required. For illustration, Figure 6.2 shows when FunctionA calls another function, called FunctionB.

In the Cortex-M0 processor, you can push multiple low registers (R0 to R7) and the return address in LR onto the stack with just one instruction. Similarly, you can carry out the pop

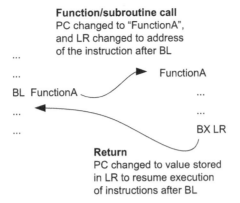

Figure 6.1:
Simple function call and function return.

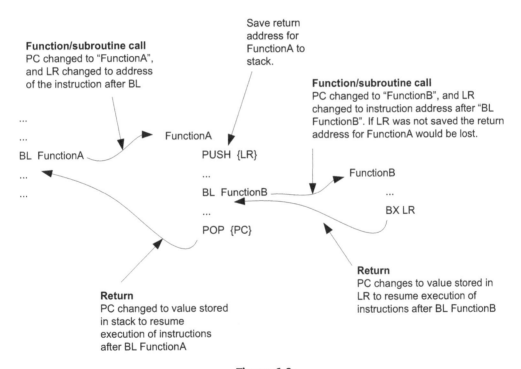

Figure 6.2:
Nested function call and function return.

operation to low registers and the Program Counter (PC) in one instruction. This allows you to combine register values restore and return with a single instruction. For example, if the registers R4 to R6 are being modified in "FunctionA," and needed to be saved to the stack, we can write "FunctionA," as shown in Figure 6.3.

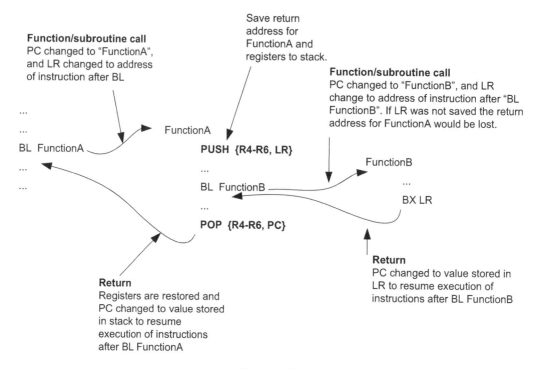

Figure 6.3:
Using push and pop of multiple registers in functions.

Branch Table

In C programming, sometime we use the "switch" statement to allow a program to branch to multiple possible address locations based on an input. In assembly programming, we can handle the same operation by creating a table of branch destination addresses, issue a load (LDR) to the table with offset computed from the input, and then use BX to carry out the branch. In the following example, we have a selection input of 0 to 3 in R0, which allows the program to branch to Dest0 to Dest3. If the input value is larger than 3, it will cause a branch to the default case:

```
        CMP   R0, #3          ; Compare input to maximum valid choice
        BHI   default_case    ; Branch to default case if higher than 3
        MOVS  R2, #4          ; Multiply branch table offset by 4
        MULS  R0, R2, R0      ; (size of each entry)
        LDR   R1,=BranchTable ; Get base address of branch table
        LDR   R2,[R1,R0]      ; Get the actual branch destination
        BX    R2              ; Branch to destination
        ALIGN 4               ; Alignment control. The table has
              ; to be word aligned to prevent unaligned read
BranchTable   ; table of each destination addresses
```

```
        DCD Dest0
        DCD Dest1
        DCD Dest2
        DCD Dest3
default_case
        ... ; Instructions for default case
Dest0
        ... ; Instructions for case '0'
Dest1
        ... ; Instructions for case '1'
Dest2
        ... ; Instructions for case '2'
Dest3
        ... ; Instructions for case '3'
```

Additional examples of complex branch conditional handling are presented in Chapter 16.

Data Accesses

Data accesses are vital to embedded applications. The Cortex-M0 processor provides a number of load (memory read) and store (memory write) instructions with various address modes. Here we will go through a number of typical examples of how these instructions can be applied.

Simple Data Accesses

Normally the memory locations (physical addresses) of software variables are defined by the linker. However, we can write the software code to access to the variables as long as we know the symbol of the variables. For example, if we need to calculate the sum of an integer array "DataIn" with 10 elements (32-bit each) and put the result in another variable called "Sum" (also 32-bit), we can use the following assembly code:

```
        LDR   r0,=DataIn; Get the address of variable 'DataIn'
        MOVS r1, #10    ; loop counter
        MOVS r2, #0     ; Result - starting from 0
add_loop
        LDM   r0!,{r3}   ; Load result and increment address
        ADDS r2, r3     ; add to result
        SUBS r1, #1     ; increment loop counter
        BNE   add_loop
        LDR   r0,=Sum    ; Get the address of variable 'Sum'
        STR   r2,[r0]    ; Save result to Sum
```

In the preceding example, we use the LDM instruction rather than a normal LDR instruction. This allows us to read the memory and increment the address to the next array element at the same time.

When using assembly to access data, we need to pay attention to a few things:

• Use correct instruction for corresponding data size. Different instructions are available for different data sizes.

• Make sure that the access is aligned. If an access is unaligned, it will trigger a fault exception. This can happen if an instruction of incorrect data size is used to access a data.

- Various addressing modes are available and can simplify your assembly codes. For example, when programming/accessing a peripheral, you can set a register to its base address value and then use an immediate offset addressing mode for accessing each register. In this way, you do not have to set up the register address every time a different register is accessed.

Example of Using Memory Access Instruction

To demonstrate how different memory access instructions can be used, this section presents several simple examples of memory copying functions. The most basic approach is to copy the data byte by byte, thus allowing any number of bytes to be copied, and this approach does not have memory alignment issues:

```
          LDR  r0,=0x00000000 ; Source address
          LDR  r1,=0x20000000 ; Destination address
          LDR  r2,=100     ; number of bytes to copy
copy_loop
          LDRB r3, [r0]    ; read 1 byte
          ADDS r0,  r0, #1 ; increment source pointer
          STRB r3, [r1]    ; write 1 byte
          ADDS r1,  r1, #1 ; increment destination pointer
          SUBS r2,  r2, #1 ; decrement loop counter
          BNE  copy_loop   ; loop until all data copied
```

The program code uses a number of add and subtract instructions in the loop, which reduce the performance. We could modify the code to reduce the program size using a register offset address mode:

```
          LDR  r0,=0x00000000 ; Source address
          LDR  r1,=0x20000000 ; Destination address
          LDR  r2,=100          ; number of bytes to copy, also
copy_loop                       ; acts as loop counter
          SUBS r2, r2, #1       ; decrement offset and loop counter
          LDRB r4,[r0, r2]      ; read 1 byte
          STRB r4,[r1, r2]      ; write 1 byte
          BNE  copy_loop        ; loop until all data copied
```

By using the loop counter as a memory offset, we have reduced the code size and improved execution speed. The only side effect is that the copying operation will start from the end of the memory block and finish at the start of the memory block.

For copying large amounts of data, we can use multiple load and store instructions to increase the performance. Because the load store multiple instructions can only be used with word accesses, we usually use them in memory-copying functions only when we know that the size of the memory being copied is large and the data are word aligned:

```
          LDR  r0,=0x00000000 ; Source address
          LDR  r1,=0x20000000 ; Destination address
          LDR  r2,=128          ; number of bytes to copy, also
```

```
copy_loop                       ;    acts as loop counter
        LDMIA r0!,{r4-r7}   ; Read  4 words and increment r0
        STMIA r1!,{r4-r7}   ; Store 4 words and increment r1
        LDMIA r0!,{r4-r7}   ; Read  4 words and increment r0
        STMIA r1!,{r4-r7}   ; Store 4 words and increment r1
        LDMIA r0!,{r4-r7}   ; Read  4 words and increment r0
        STMIA r1!,{r4-r7}   ; Store 4 words and increment r1
        LDMIA r0!,{r4-r7}   ; Read  4 words and increment r0
        STMIA r1!,{r4-r7}   ; Store 4 words and increment r1
        SUBS  r2, r2, #64    ; Each time 64 bytes are copied
        BNE   copy_loop  ; loop until all data copied
```

In the preceding code, each loop iteration copies 64 bytes. This greatly increases the performance of data transfer.

Another type of useful memory access instruction is the load and store instruction with stack pointer-related addressing. This is commonly used for local variables, as C compilers often store simple local variables on the stack memory. For example, let's say we need to create two local variables in a function called "function1." The code can be written as follows:

```
function1
        SUB  SP, SP, #0x8 ; Reserve 2 words of stack
                           ; (8 bytes) for local variables
        ;Data processing in function
        MOVS r0, #0x12 ; set a dummy value
        STR  r0, [sp, #0]  ; Store 0x12 in 1st local variable
        STR  r0, [sp, #4]  ; Store 0x12 in 2nd local variable
        LDR  r1, [sp, #0]  ; Read from 1st local variable
        LDR  r2, [sp, #4]  ; Read from 2nd local variable
        ADD  SP, SP, #0x8; Restore SP to original position
        BX   LX
```

In the beginning of the function, a stack pointer adjustment is carried out so that the data reserved will not be overwritten by further stack push operations. During the execution of the function, SP-related addressing with immediate offset allows the local variables to be accessed efficiently. The value of SP can also be copied to another register if further stack operations are required or if the some of the local variables are in byte or half-word size (in ARMv6-M, SP-related addressing mode only supports word-size data). In such cases, load/store instructions accessing the local variables would use the copied version of SP (Figure 6.4).

At the end of the function, the local variables can be discarded and we restore the SP value to the position as when the function started using an ADD instruction.

Data Type Conversion

The Cortex-M0 processor supports a number of instructions for converting data among different data types.

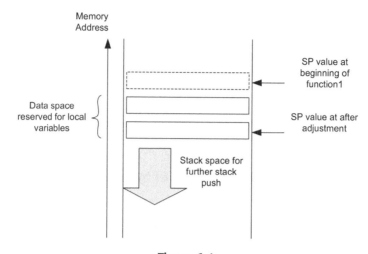

Figure 6.4:
Reserving two words of stack space for local variables.

Conversion of Data Size

On ARM compilers, different data types have different sizes. Table 6.7 shows a number of commonly used data types and their corresponding sizes on ARM compilers.

When converting a data value from one type to another type with a larger size, we need to sign-extend or zero-extend it. A number of instructions are available to handle this conversion (Table 6.8).

Table 6.7: Size of Commonly Used Data Types in C Compilers for ARM

C Data Type	Number of Bits
"char," "unsigned char"	8
"enum"	8/16/32 (smallest is chosen)
"short," "unsigned short"	16
"int," "unsigned int"	32
"long," "unsigned long"	32

Table 6.8: Instructions for Signed-Extend and Zero-Extend Data Value Operations

Conversion Operation	Instruction
Converting an 8-bit signed data value to 32-bit or 16-bit signed data value	SXTB (signed-extend byte)
Converting a 16-bit signed data value to a 32-bit signed data value	SXTH (signed-extend half word)
Converting an 8-bit unsigned data value to a 32-bit or 16-bit data value	UXTB (zero extend byte)
Converting a 16-bit unsigned data value to a 32-bit data value	UXTH (zero extend half word)

If the data are in the memory, we can read the data and carry out the zero-extend or signed-extend operation in a single instruction (Table 6.9).

Table 6.9: Memory Read Instructions with Signed-Extend and Zero Extend Data Value Operations

Conversion Operation	Instruction
Read an 8-bit signed data value from memory and convert it to a 16-bit or 32-bit signed value	LDRSB
Read a 16-bit signed data value from memory and convert it to a 32-bit signed value	LDRSH
Read an 8-bit unsigned data value from memory and convert it to a 16-bit or 32-bit value	LDRB
Read a 16-bit unsigned data value from memory and convert it to a 32-bit value	LDRH

Endian Conversion

The memory system of a Cortex-M0 microcontroller can be in either little endian configuration or big endian configuration. It is defined in hardware and cannot be changed by programming. Occasionally we might need to convert data between little endian and big endian format. Table 6.10 presents several instructions to handle this situation.

Table 6.10: Instructions for Conversion between Big Endian and Little Endian Data

Conversion Operation	Instruction
Convert a little endian 32-bit value to big endian, or vice versa	REV
Convert a little endian 16-bit unsigned value to big endian, or vice versa	REV16
Convert a little endian 16-bit signed value to big endian, or vice versa	REVSH

Data Processing

Most of the data processing operations can be carried out in a simple instruction sequence. However, there are situations when more steps are required. Here we will look at a number of examples.

64-Bit/128-Bit Add

Adding two 64-bit values together is fairly straightforward. Assume that you have two 64-bit values (X and Y) stored in four registers. You can add them together using ADDS followed up by ADCS instruction:

```
LDR   r0,=0xFFFFFFFF ; X_Low  (X = 0x3333FFFFFFFFFFFF)
LDR   r1,=0x3333FFFF ; X_High
LDR   r2,=0x00000001 ; Y_Low  (Y = 0x3333000000000001)
LDR   r3,=0x33330000 ; Y_High
ADDS  r0,r0,r2 ; lower 32-bit
ADCS  r1,r1,r3 ; upper 32-bit
```

In this example, the result is in R1, R0, which is 0x66670000 and 0x00000000. The operation can be extended to 96-bit values, 128-bit values, or more by increasing the number of ADCS instructions in the sequence (Figure 6.5).

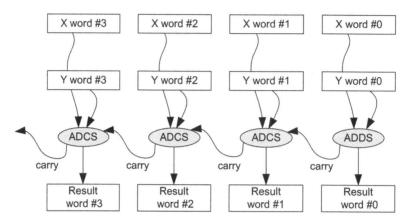

Figure 6.5:
Adding of two 128-bit numbers.

64-Bit/128-Bit Sub

The operation of 64-bit subtract is similar to the one for 64-bit add. Assume that you have two 64-bit values (X and Y) in four registers. You can subtract them (X − Y) using SUBS followed by SBCS instruction:

```
LDR   r0,=0x00000001 ; X_Low(X = 0x0000000100000001)
LDR   r1,=0x00000001 ; X_High
LDR   r2,=0x00000003 ; Y_Low(Y = 0x0000000000000003)
LDR   r3,=0x00000000 ; Y_High
SUBS  r0,r0,r2 ; lower 32-bit
SBCS  r1,r1,r3 ; upper 32-bit
```

In this example, the result is in R1, R0, which is 0x00000000 and 0xFFFFFFFE. The operation can be extended to 96-bit values, 128-bit values, or more by increasing the number of SBCS instructions in the sequence (Figure 6.6).

Integer Divide

Unlike the Cortex-M3/M4 processor, the Cortex-M0 processor does not have integer divide instructions. For users who program their application in C language, the C compiler automatically inserts the required C library function that handles integer divide if required. Users who prefer to write their application entirely in assembly language can create an assembly function like that shown in Figure 6.7, which handles unsigned integer divide.

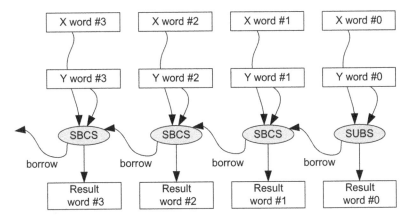

Figure 6.6:
Subtracting two 128-bit values.

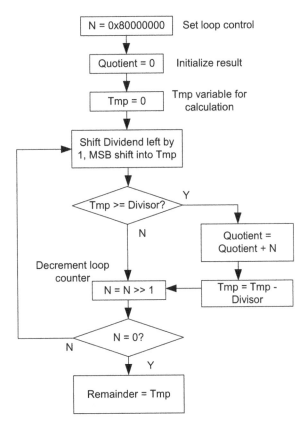

Figure 6.7:
Simple unsigned integer divide function.

The divide function contains a loop that iterates 32 times and computes 1 bit of the result each time. Instead of using an integer loop counter, the loop control is done by a value N, which has 1 bit set (one hot), and shifts right by 1 bit each time the loop is executed. The corresponding assembly code can be written as follows:

```
simple_divide
    ; Inputs
    ;   R0 = dividend
    ;   R1 = divider
    ; Outputs
    ;   R0 = quotient
    ;   R1 = remainder
    PUSH {R2-R4}  ; Save registers to stack
    MOV   R2, R0  ; Save dividend to R2 as R0 will be changed
    MOVS  R3, #0x1 ; loop control
    LSLS  R3, R3, #31 ; N = 0x80000000
    MOVS  R0, #0   ; initial Quotient
    MOVS  R4, #0   ; initial Tmp
simple_divide_loop
    LSLS  R2, R2, #1 ; Shift dividend left by 1 bit, MSB go into carry
    ADCS  R4, R4, R4 ; Shift Tmp left by 1 bit, carry move into LSB
    CMP   R4, R1
    BCC   simple_divide_lessthan
    ADDS  R0, R0, R3 ; Increment quotient
    SUBS  R4, R4, R1
simple_divide_lessthan
    LSRS  R3, R3, #1 ; N = N >> 1
    BNE   simple_divide_loop
    MOV   R1, R4  ; Put remainder in R1, Quotient is already in R0
    POP   {R2-R4} ; Restore used register
    BX    LR      ; Return
```

This simple example does not handle signed data and there is no special handling for divide-by-zero cases. If you need to handle signed data division, you can create a wrapper to convert the dividend and divisor into unsigned data first, and then run the unsigned divide and convert the result back to the signed value afterward.

Unsigned Integer Square Root

Another mathematical calculation that is occasionally needed in embedded systems is the square root. Because the square root can only deal with positive numbers (unless complex numbers are used), the following example only handles unsigned integers (Figure 6.8). For the following implementation, the result is rounded to the next lower integer.

The corresponding assembly code can be written as follows:

```
simple_sqrt
    ; Input  : R0
    ; Output : R0 (square root result)
    PUSH    {R1-R3}    ; Save registers to stack
    MOVS    R1, #0x1   ; Set loop control register
```

```
    LSLS   R1, R1, #15 ; R1 = 0x00008000
    MOVS   R2, #0      ; Initialize result
simple_sqrt_loop
    ADDS   R2, R2, R1  ; M = (M + N)
    MOVS   R3, R2      ; Copy (M + N) to R3
    MULS   R3, R3, R3  ; R3 = (M + N)^2
    CMP    R3, R0
    BLS    simple_sqrt_lessequal
    SUBS   R2, R2, R1  ; M = (M − N)
simple_sqrt_lessequal
    LSRS   R1, R1, #1  ; N = N >> 1
    BNE    simple_sqrt_loop
    MOV    R0, R2      ; Copy to R0 and return
    POP    {R1-R3}     ;
    BX     LR          ; Return
```

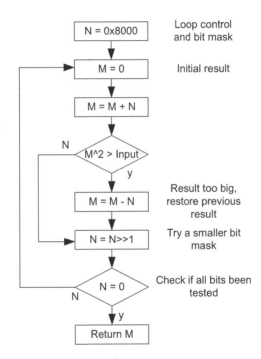

Figure 6.8:
Simple unsigned integer square root function.

Bit and Bit Field Computations

Bit data processing is common in microcontroller applications. From the previous divide example code, we have already seen some basic bit computation on the Cortex-M0 processor. Here we will cover a few more examples of bit and bit field processing.

To extract a bit from a value stored in a register, we first need to determine how the result will be used. If the result is to be used for controlling a conditional branch, the best solution is to use shift or rotate instructions to copy the required bit in the Carry flag in the APSR, and then carry out the conditional branch using a BCC or BCS instruction. For example,

```
LSRS  R0, R0,   #<n+1> ; Shift bit "n" into carry flag in APSR
BCS   <label>         ; branch if carry is set
```

If the result is going to be used for other processing, then we could extract the bit by a logic shift operation. For example, if we need to extract bit 4 in the register R0, this can be carried out as follows:

```
LSLS R0, R0,   #27 ; Remove un-needed top bits
LSRS R0, R0,   #31 ; Move required bit into bit 0
```

This extraction method can be generalized to support the extraction of bit fields. For example, if we need to extract a bit field in a data value that is "W" bits wide, starting with bit position "P" (LSB of the bit field), we can extract the bit field using the following instruction:

```
LSLS R0, R0,   #(32-W-P) ; Remove un-needed top bits
LSRS R0, R0,   #(32-W)   ; Align required bits to bit 0
```

For example, if we need to extract an 8-bit-width bit field from bit 4 to bit 11 (Figure 6.9), we can use this instruction sequence:

```
LSLS R0, R0, #(32-8-4) ; Remove un-needed top bits
LSRS R0, R0, #(32-8)   ; Align required bits to bit 0
```

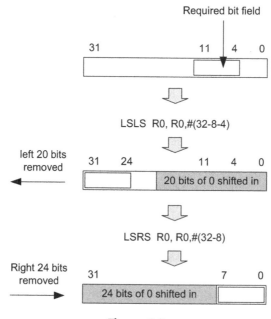

Figure 6.9:
Bit field extract operation.

In a similar way, we can clear the bit field in a register by a few shift and rotate instructions (Figure 6.10):

```
RORS  R0, R0,  #4        ; Shift unneeded bit to bit 0
LSRS  R0, R0,  #8        ; Align required bits to bit 0
RORS  R0, R0,  #(32-8-4) ; store value to original position
```

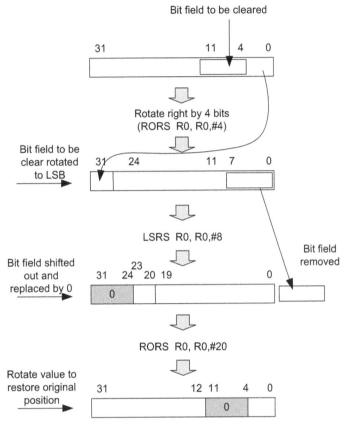

Figure 6.10:
Bit field clear operation.

For masking other bit patterns, we can use the Bit Clear (BICS) instruction. For example,

```
LDR  R1, =Bit_Mask ; Bit to clear
BICS R0, R0, R1       ; Clear bits that are not required
```

The "Bit_Mask" is a value that reflects the bit pattern you want to clear. The BICS instruction does not have any limitation of the bit pattern to be cleared, but it might require a slightly larger program size, as the program might need to store the value of the "Bit_Mask" pattern as a word-size constant.

## CHAPTER 7

Memory System

Overview

In this chapter we will look into the memory architecture of the Cortex-M0 processor and how it affects software development.

The Cortex-M0 processor has a 32-bit system bus interface with 32-bit address lines (4 GB of address space). The system bus is based on a bus protocol called AHB-Lite (Advanced High-performance Bus), which is a protocol defined in the Advanced Microcontroller Bus Architecture (AMBA) standard. The AMBA standard is developed by ARM, and is widely used in the semiconductor industry.

Although the AHB-Lite protocol provides high-performance accesses to the memory system, very often a secondary bus segment can also be found for slower devices including peripherals. In ARM microcontrollers, the peripheral bus system is normally based on the Advanced Peripheral Bus (APB) protocol. The APB is connected to the AHB-Lite via a bus bridge and may run at a different clock speed compared to the AHB system bus. The data path on the APB is also 32-bit, but the address lines are often less than 32-bit as the peripheral address space is relatively small (Figure 7.1).

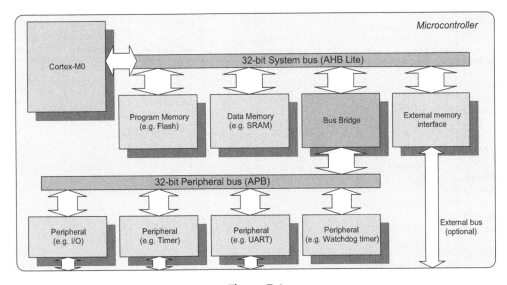

Figure 7.1:
Separation of system and peripheral bus in typical 32-bit microcontrollers.

The Definitive Guide to the ARM Cortex-M0. DOI: 10.1016/B978-0-12-385477-3.10007-2

Because of the separation of main system bus and peripheral bus, and in some cases with separated clock frequency controls, an application might need to initialize some clock control hardware in the microcontroller before accessing the peripherals. In some cases, there can be multiple peripheral bus segments in a microcontroller running at different clock frequencies. Besides allowing some part of the system to run at a slower speed, the separation of bus segments also provides the possibility of power reduction by allowing the clock to a peripheral system to be stopped.

Depending on the microcontroller design, some high-speed peripherals might be connected to the AHB-Lite system bus instead of the APB. This is because the AHB-Lite protocol requires fewer clock cycles for each transfer when compared to the APB. The bus protocol behavior affects the system operation and the programmer's view on the memory system in a number of ways. This subject will be covered in various places in this chapter.

Memory Map

The 4GB memory space of the Cortex-M0 processor is architecturally divided into a number of regions (Figure 7.2). Each region has its recommended usage, and the memory access behavior could depend on which memory region you are accessing to. This memory region

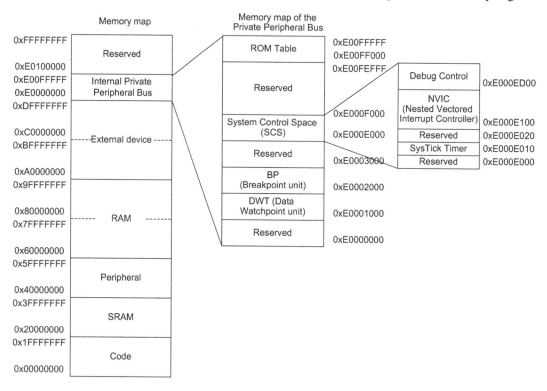

Figure 7.2:
Architecturally defined memory map of the Cortex-M0 processor.

definition helps software porting between different ARM Cortex microcontrollers, as they all have similar arrangements.

Despite having an architecturally defined memory map, the actual usage of the memory map is very flexible. There are only a few limitations—for example, a few memory regions that are allocated for peripherals do not allow program code execution, and a number of internal components have a fixed memory address to ensure software portability.

Next we will look into the usage of each region.

Code Region (0x00000000–0x1FFFFFFF)

The size of the code region is 512 MB. It is primarily used to store program code, including the exception vector table, which is a part of the program image. It can also be used for data memory (connection to RAM).

SRAM Region (0x20000000–0x3FFFFFFF)

The SRAM region is the located in the next 512 MB of the memory map. It is primarily used to store data, including stack. It can also be used to store program code. For example, in some cases you might want to copy program code from slow external memory to the SRAM and execute it from there. Despite the name given to this region (it is called "SRAM"), the actual memory devices being used could be SRAM, SDRAM, or some other type.

Peripheral Region (0x40000000–0x5FFFFFFF)

The peripheral region also has the size of 512 MB. It is primarily used for peripherals and can also be used for data storage. However, program execution is not allowed in the peripheral region. The peripherals connected to this memory region can be either the AHB-Lite peripheral or APB peripherals (via a bus bridge).

RAM Region (0x60000000–0x9FFFFFFF)

The RAM region consists of two 512 MB blocks, which results in total of 1 GB of space. Both 512 MB memory blocks are primarily used to stored data, and in most cases the RAM region can be used as a 1GB continuous memory space. The RAM region can also be used for program code execution. The only differences between the two halves of the RAM region is the memory attributes, which might cause differences in cache behavior if a system-level cache (level-2 cache) is used. Memory attributes will be covered in more detail later in this chapter.

Device Region (0xA0000000–0xDFFFFFFF)

The external device region consists of two 512 MB memory blocks, which results in a total of 1 GB of space. Both 512 MB memory blocks are primarily used for peripherals and I/O usage. The device region does not allow program execution, but it can be used for general data storage. Similar to the RAM region, the two halves of the device region have different memory attributes.

Internal Private Peripheral Bus (PPB) (0xE0000000–0xE00FFFFF)

The internal PPB memory space is allocated for peripherals inside the processor, such as the interrupt controller NVIC, as well as the debug components. The internal PPB memory space is 1 MB in size, and program execution is not allowed in this memory range.

Within the PPB memory range, a special range of memory is defined as the System Control Space (SCS). The SCS address is from 0xE000E000 to 0xE000EFFF. It contains the interrupt control registers, system control registers, debug control registers, and the like. The NVIC registers are part of the SCS memory space. The SCS also contains an optional timer called the SysTick. This will be covered in Chapter 10.

Reserved Memory Space (0xE0100000–0xFFFFFFFF)

The last section of the memory map is a 511 MB reserved memory space. This may be reserved in some microcontrollers for vendor-specific usages.

Although the Cortex-M0 processor has this fixed memory map, the memory usage is very flexible. For example, it can have multiple SRAM memory blocks placed in the SRAM region as well as the CODE region, and it can execute program code from external memory components located in RAM region. Microcontroller vendors can also add their own system-level memory features, such as system-level cache, if needed.

So how does the memory map of a typical real system look like? For a typical microcontroller developed with the Cortex-M0 processor, normally you can find the following elements:

- Flash memory (for program code)
- Internal SRAM (for data)
- Internal peripherals
- External memory interface (for external memories as well as external peripherals (optional))
- Interfaces for other external peripherals (optional)

After putting all these components together, an example microcontroller could be illustrated as shown in Figure 7.3.

Figure 7.3 shows how some memory regions can be used. However, in many low-cost microcontrollers, the system designs do not have any external memory interface or Secure Digital (SD) card interface. In these cases, some of the memory regions, like the external RAM or the external device regions, might not be used.

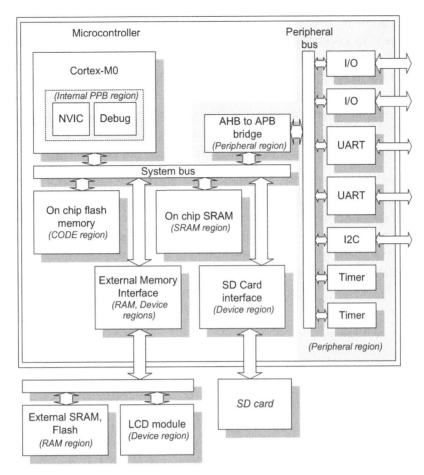

Figure 7.3:
Examples of various memory regions in a microcontroller design.

Program Memory, Boot Loader, and Memory Remapping

Usually the program memory of the Cortex-M0 is implemented with on chip flash memory. However, it is also possible that the program is stored externally or using other types of memory devices (e.g. EEPROM).

When the Cortex-M0 processor comes out of reset, it accesses the vector table in address zero for initial MSP value and reset vector value, and it then starts the program execution from the reset vector. To ensure that the system works correctly, a valid vector table and a valid program memory must be available in the system to prevent the processor from executing rogue program code.

Usually this is done by a flash memory starting from address zero. However, an off-the-shelf microcontroller product might not have any program in the flash memory before the user programs it. To allow the processor to start up correctly, some Cortex-M0 based micro-controllers come with a boot loader, a small program located on the microcontroller chip that executes after the processor powers up and branches to the user application in the flash memory only if the flash is programmed. The boot loader is preprogrammed by the chip manufacturer. Sometimes it is stored on the on-chip flash memory with a separate memory section from user applications (to allow the user to update the program without affecting the boot loader); other times is it stored on a nonvolatile memory that is separate from the user programmable flash memory (to prevent the users from accidentally erasing the boot loader).

When a boot loader is present, it is common for the microcontroller vendor to implement a memory map-switching feature called "remap" on the system bus. The switching of the memory map is controlled by a hardware register, which is programmed when the boot loader is executed. There are various types of remap arrangements. One common remap arrangement is to allow the boot loader to be mapped to the start of the memory during the power-up phase using address alias, as shown in Figure 7.4.

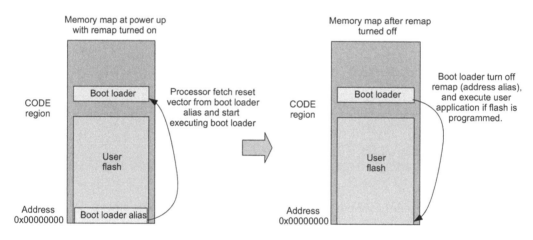

Figure 7.4:
An example of a memory-remap implementation with the boot loader.

The boot loader might also support additional features like hardware initialization (clock and PLL setup), supporting of multiple boot configurations, firmware protection, or even flash erase utilities. The memory remap feature is implemented on the system bus and is not a part of the Cortex-M0 processor, therefore different microcontrollers from different vendors have different implementations.

Another common type of remap feature implemented on some ARM microcontrollers allows an SRAM block to be remapped to address 0x0 (Figure 7.5). Normally nonvolatile memory used on microcontrollers like flash memory is slower than SRAM. When the microcontroller is running at a high clock rate, wait states would be required if the program is executed from the

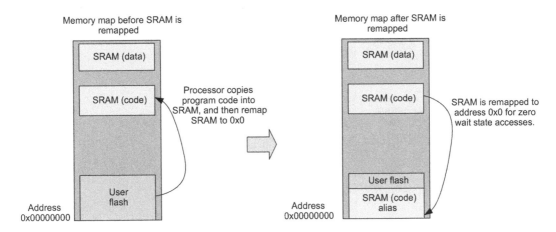

Figure 7.5:
A different example of memory-remap implementation—SRAM for fast program accesses.

flash memory. By allowing an SRAM memory block to be remapped to address 0x0, then the program can be copied to SRAM and executed at maximum speed. This also avoids wait states in vector table fetch, which affects interrupt latency.

Data Memory

The data memory in Cortex-M0 processor is used for software variables, stack memory, and, in some cases, heap memory. Sometimes local variables in C functions could be stored onto the stack memory. The heap memory is needed when the applications use C functions that require dynamically allocated memory space.

In most embedded applications without operating systems (OSs), only one stack is used (only the main stack pointer is required). In this case, the data memory can be arranged as shown in (Figure 7.6).

Because the stack operation is based on a full descending stack arrangement, and heap memory allocation is ascending, it is common to put the stack at the end of the memory block and heap memory just after normal memory to get the most efficient arrangement.

For embedded applications with embedded OS, each task might have its own stack memory range (see Figure 4.15 from Chapter 4). It is also possible for each task to have its own allocated memory block, with each memory block containing a memory layout consisting of stack, heap, and data.

Little Endian and Big Endian Support

The Cortex-M0 processor supports either the little endian or big endian memory format (Figures 7.7 and 7.8). The microcontroller vendor makes the choice when the system is

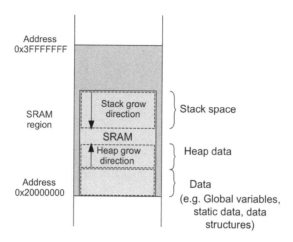

Figure 7.6:
An example of common SRAM usage.

designed, and embedded programmers cannot change it. Software developers must configure their development tools project options to match the endianness of the targeted microcontroller.

The big endian mode supported on the Cortex-M0 processor is called the Byte-Invariant big endian mode, or "BE8." It is one of the big endian modes in ARM architectures. Traditional ARM processors, like ARM7TDMI, use a different big endian mode called the Word-Invariant big endian mode, or "BE32." The difference between the two is on the hardware interface level and does not affect the programmer's view.

Bits	[31:24]	[23:16]	[15:8]	[7:0]
0x00000008	Byte 0xB	Byte 0xA	Byte 9	Byte 8
0x00000004	Byte 7	Byte 6	Byte 5	Byte 4
0x00000000	Byte 3	Byte 2	Byte 1	Byte 0

Figure 7.7:
Little endian 32-bit memory.

Most of the Cortex-M0 processor—based microcontrollers are using the little endian configuration. With the little endian arrangement, the lowest byte of a word-size data is stored in bit 0 to bit 7 (Figure 7.7).

In the big endian configuration, the lowest byte of a word-size data is stored in bit 24 to bit 31 (Figure 7.8)

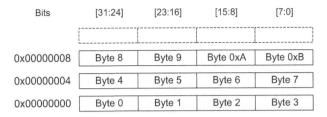

Figure 7.8:
Big endian 32-bit memory.

Both memory configurations support data handling of different sizes. The Cortex-M0 processor can generate byte, half-word, and word transfers. When the memory is accessed, the memory interface selects the data lanes base on the transfer size and the lowest two bits of the address. Figure 7.9 illustrates the data access for little endian systems.

Address	Size	Bits 31-24	Bits 23-16	Bits 15-8	Bits 7-0
0x00000000	Word	Data[31:24]	Data[23:16]	Data[15:8]	Data[7:0]
0x00000000	Half word			Data[15:8]	Data[7:0]
0x00000002	Half word	Data[15:8]	Data[7:0]		
0x00000000	Byte				Data[7:0]
0x00000001	Byte			Data[7:0]	
0x00000002	Byte		Data[7:0]		
0x00000003	Byte	Data[7:0]			

Figure 7.9:
Data access in little endian systems.

Similarly, a big endian system supports data access of different sizes (Figure 7.10).

Note that there are two exceptions in big endian configurations: (1) the instruction fetch is always in little endian, and (2) the accesses to Private Peripheral Bus (PPB) address space is always in little endian.

Data Type

The Cortex-M0 processor supports different data types by providing various memory access instructions for different transfer sizes and by providing a 32-bit AHB-LITE interface, which

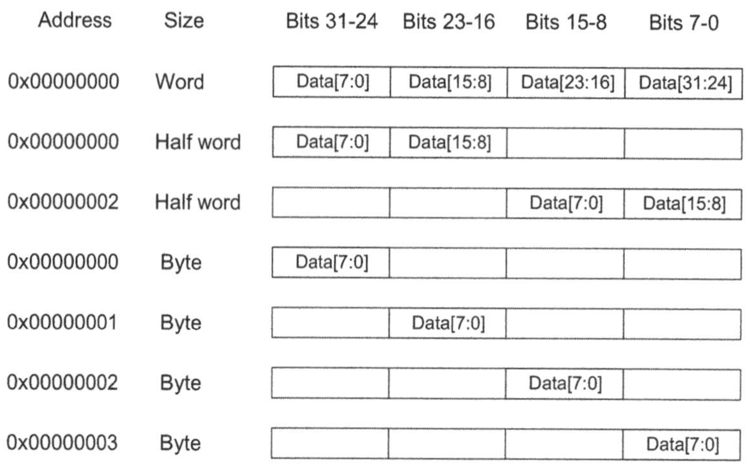

Address	Size	Bits 31-24	Bits 23-16	Bits 15-8	Bits 7-0
0x00000000	Word	Data[7:0]	Data[15:8]	Data[23:16]	Data[31:24]
0x00000000	Half word	Data[7:0]	Data[15:8]		
0x00000002	Half word			Data[7:0]	Data[15:8]
0x00000000	Byte	Data[7:0]			
0x00000001	Byte		Data[7:0]		
0x00000002	Byte			Data[7:0]	
0x00000003	Byte				Data[7:0]

Figure 7.10:
Data access in big endian system.

supports 32-bit, 16-bit, and 8-bit transfers. For example, in C language development, the data types presented in Table 7.1 are commonly used.

Table 7.1: Commonly Used Data Types in C Language Development

Type	Number of Bits in ARM	Instructions
"char", "unsigned char"	8	LDRB, LDRSB, STRB
"enum"	8/16/32 (smallest is chosen)	LDRB, LDRH, LDR, STRB, STRH, STR
"short", "unsigned short"	16	LDRH, LDRSH, STRH
"int", "unsigned int"	32	LDR, STR
"long", "unsigned long"	32	LDR, STR

If "stdint.h" in C99 is used, the data types shown in Table 7.2 are available.

Table 7.2: Commonly Used Data Types Provided in "stdint.h" in C99

Type	Number of Bits in ARM	Instructions
"int8_t", "uint8_t"	8	LDRB, LDRSB, STRB
"int16_t", "uint16_t"	16	LDRH, LDRSH, STRH
"int32_t", "uint32_t"	32	LDR, STR

For other data types that require a larger size (e.g., int64_t, uint64_t), the C compilers automatically convert the data transfer into multiple memory access instructions.

Note that for peripheral register accesses, the data type being used should match the hardware register size. Otherwise the peripheral might ignore the transfer or not functioning as expected.

In most cases, peripherals connected to the peripheral bus (APB) should be accessed using word-size transfers. This is because APB protocol does not have transfer size signals, hence all the transfers are assumed to be word size. Therefore, peripheral registers accessed via the APB are normally declared to be "volatile unsigned integers."

Effect of Hardware Behavior to Programming

The design of the processor hardware and the behavior of the bus protocol affect the software in a number of ways. In a previous section, we mentioned that peripherals connected to the APB are usually accessed using word-size transfers because of the nature of the APB protocol. In this section, we will look into other similar aspects.

Data Alignment

The Thumb instruction set supported by the Cortex-M0 processor can only generate aligned transfers (Figure 7.11). It means that the transfer address must be a multiple of the transfer size. For example, a word size transfer can only access addresses like 0x0, 0x4, 0x8, 0xC, and so forth. Similarly, a half-word transfer can only access addresses like 0x0, 0x2, 0x4, and so forth. All byte data accesses are aligned.

If the program attempts to generate an unaligned transfer, this will result in a fault exception and cause the hard fault handler to be executed. In normal cases, C compilers do not generate

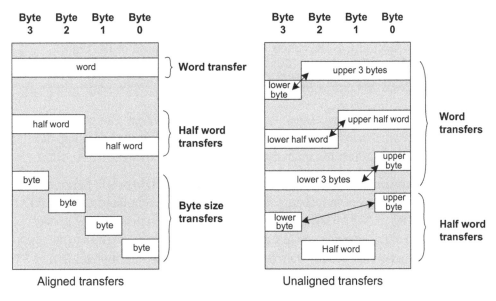

Figure 7.11:
Example of aligned and unaligned transfers (little endian memory).

any unaligned transfers, but an unaligned transfer can still be generated if a C program directly manipulated a pointer.

Unaligned transfers can also be generated accidentally when programming in assembly language—for example, when load store instructions of wrong transfer size is used. In the case of a half-word data type located in address 0x1002, which is an aligned data type, it can be accessed using LDRH, LDRSH, or STRH instructions without problems. But if the program code used LDR or STR instruction to access the data, an unaligned access fault would be triggered.

Access to Invalid Addresses

Unlike most 8-bit or 16-bit processors, a memory access to an invalid memory address generates a fault exception on ARM Cortex-M0 microcontrollers. This provides better program error detection and allows software bugs to be detected earlier.

In the AHB system connected to the Cortex-M0 processor, the address decoding logic detects the address being accessed and the bus system response with an error signal if the access is going to an invalid location. The bus error can be caused by either data access or instruction fetch.

One exception to this behavior is the branch shadow instruction fetch. Because of the pipeline nature of the Cortex-M0 processor, instructions are fetched in advanced. Therefore, if the program execution reaches the end of a valid memory region and a branch is executed, there might be chance that the addresses beyond the valid instruction memory region could have been fetched, resulting in a bus fault in the AHB system. However, in this case the bus fault would be ignored if the faulted instruction is not executed because of the branch.

Use of Multiple Load and Store Instructions

The multiple load and store instructions in the Cortex-M0 processor can greatly increase the system performance when used correctly. For example, it can be used to speed up data transfer processes or as a way to adjust the memory pointer automatically.

When handling peripheral accesses, we need to avoid the use of LDM or STM instructions. If the Cortex-M0 processor receives an interrupt request during the execution of LDM or STM instructions, the LDM or STM instruction will be abandoned and the interrupt service will be initiated. At the end of the interrupt service, the program execution will return to the interrupted LDM or STM instruction and restart from the *first* transfer of the interrupted LDM or STM.

As a result of this restart behavior, some of the transfers in this interrupt LDM or STM instruction could be carried out twice. It is not a problem for normal memory devices, but if the access is carried on a peripheral, then the repeating of the transfer could cause error. For example, if the LDM instruction is used for reading a data value in a First-In-First-Out (FIFO) buffer, then some of the data in the FIFO could be lost as the read operation is repeated.

As a precaution, we should avoid the use of LDM or STM instructions on peripheral accesses unless we are sure that the restart behavior does not cause incorrect operation to the peripheral.

Memory Attributes

The Cortex-M0 processor can be used with wide range of memory systems and devices. To make porting of software between different devices easier, a number of memory attribute settings are available for each region in the memory map. Memory attributes are characteristics of the memory accesses; they can affect data and instruction accesses to memory as well as accesses to peripherals.

In the ARMv6-M architecture, which is used by the Cortex-M0 processor, a number of memory access attributes are defined for different memory regions:

- *Executable*. The shareable attribute defines whether program execution is allowed in that memory region. If a memory region is define as nonexecutable, in ARM documentation it is marked as eXecute Never (XN).
- *Bufferable*. When a data write is carried out to a bufferable memory region, the write transfer can be buffered, which means the processor can continue to execute the next instruction without waiting for the current write transfer to complete.
- *Cacheable*. If a cache device is present on the system, it can keep a local copy of the data during a data transfer and reuse it the next time the same memory location is accessed to speed up the system. The cache device can be a cache memory unit or a small buffer in a memory controller.
- *Shareable*. The shareable attribute defines whether more than one processor can access a shareable memory region. If a memory region is shareable, the memory system needs to ensure coherency between memory accesses by multiple processors in this region.

For most users of the Cortex-M0 products, only the XN attribute is relevant because it defines which regions can be used for program execution. The other attributes are used only if cache unit or multiple processors are used. Because the Cortex-M0 processor does not have an internal cache unit, in most cases these memory attributes are not used. If a system-level cache is used or when the memory controller has a built-in cache, then these memory attributes, exported by the processor via the AHB interface, could be used.

Based on the memory attributes, various memory types are architecturally defined and are used to define what type of devices could be used in each memory region:

- *Normal memory*. Normal memories can be shareable or nonshareable and cacheable or noncacheable. For cacheable memories, the caching behavior can be further be divided into Write Through (WT) or Write Back Write Allocate (WBWT).

- *Device memory.* Device memories are noncacheable. They can be shareable or nonshareable.
- *Strongly-ordered (SO) memory.* A memory region that is nonbufferable, noncacheable, and transfers to/from a strongly ordered region takes effect immediately. Also, the orders of strongly ordered transfers on the memory interface must be identical to the orders of the corresponding memory access instructions (no access reordering for speed optimization; the Cortex-M0 does not have any access reordering feature). Strongly ordered memory regions are always shareable.

The memory attribute for each memory region in the Cortex-M0 processor is defined using these memory type definitions (Table 7.3). During the memory accesses, the memory attributes are exported from the processor to the AHB system.

The PPB memory region is defined as strongly ordered (SO). This means the memory region is nonbufferable and noncacheable. In the Cortex-M0, operations following an access to a strongly order region do not begin until the access has been completed. This behavior is important for changing registers in the System Control Space (SCS), where we often expect the operation of changing a control register to take place immediately, before next instruction is executed.

In some other ARM processors like the Cortex-M3, there can also be default memory access permission for each region. Because the Cortex-M0 processor does not have a separated

Table 7.3: Memory Attribute Map

Address	Region	Memory Type	Cache	XN	Descriptions
0x00000000– 0x1FFFFFFF	CODE	Normal	WT	—	Memory for program code including vector table
0x20000000– 0x3FFFFFFF	SRAM	Normal	WBWA	—	SRAM, typically used for data and stack memory
0x40000000– 0x5FFFFFFF	Peripheral	Device	—	XN	Typically used for on-chip devices
0x60000000– 0x7FFFFFFF	RAM	Normal	WBWA	—	Normal memory with Write Back, Write Allocate cache attributes
0x80000000– 0x9FFFFFFF	RAM	Normal	WT	—	Normal memory with Write Through cache attributes
0xA0000000– 0xBFFFFFFF	Device	Device, shareable	—	XN	Shareable device memory
0xC0000000– 0xDFFFFFFF	Device	Device	—	XN	Nonshareable device memory
0xE0000000– 0xE00FFFFF	PPB	Strongly ordered, shareable	—	XN	Internal private peripheral bus
0xE0100000– 0xFFFFFFFF	Reserved	Reserved	—	—	Reserved (vendor-specific usage)

privileged and nonprivileged (user) access level, the processor is in the privilege access level all the time and therefore does not have a memory map for default memory access permission.

In practice, most of the memory attributes and memory type definitions are unimportant (apart from the XN attribute) to users of Cortex-M0 microcontrollers. However, if the software code has to be reused on high-end processors, especially on systems with multiple processors and cache memories, these details can be important.

## CHAPTER 8

Exceptions and Interrupts

What Are Exceptions and Interrupts?

Exceptions are events that cause changes in program flow control outside a normal code sequence. When it happens, the program that is currently executing is suspended, and a piece of code associated with the event (the exception handler) is executed. The events could either be external or internal. When an event is from an external source, it is commonly known as an interrupt or interrupt request (IRQ). Exceptions and interrupts are supported in almost all modern processors. In microcontrollers, the interrupts can also be generated using on-chip peripherals or by software.

The software code that is executed when an exception occurs is called exception handler. If the exception handler is associated with an interrupt event, then it can also be called an interrupt handler, or interrupt service routine (ISR). The exception handlers are part of the program code in the compiled program image.

When the exception handler has finished processing the exception, it will return to the interrupted program and resume the original task. As a result, the exception handling sequence requires some way to store the status of the interrupted program and allow this information to be restored after the exception handler has completed its task. This can be done by a hardware mechanism or by a combination of hardware and software operations. In the next couple of chapters, we will see how the Cortex-M0 processor handles this process.

It is common to divide exceptions into multiple levels of priority, and while running an exception handler of a low priority exception, a higher priority exception can be triggered and serviced. This is commonly known as a nested exception. The priority level of an exception could be programmable or fixed. Apart from priority settings, some exceptions (including most interrupts) can also be disabled or enabled by software.

Exception Types on the Cortex-M0 Processor

The Cortex-M0 processor contains a built-in interrupt controller, which supports up to 32 interrupt request (IRQ) inputs, and a nonmaskable interrupt (NMI) input. Depending on the design of the microcontroller product, the IRQ and the NMI can be generated either from on-chip peripherals or external sources. In addition, the Cortex-M0 processor also supports a number of internal exceptions.

The Definitive Guide to the ARM Cortex-M0. DOI: 10.1016/B978-0-12-385477-3.10008-4

Each exception source in the Cortex-M0 processor has a unique exception number. The exception number for NMI is 2, and the exception numbers for the on-chip peripherals and external interrupt sources are from 16 to 47. The other exception numbers, from 1 to 15, are for system exceptions generated inside the processor, although some of the exception numbers in this range are not used.

Each exception type also has an associated priority. The priority levels of some exceptions are fixed and some are programmable. Table 8.1 shows the exception types, exception numbers, and priority levels.

Table 8.1: List of Exceptions in the Cortex-M0 Processor

Exception Number	Exception Type	Priority	Descriptions
1	Reset	−3 (Highest)	Reset
2	NMI	−2	Nonmaskable interrupt
3	Hard fault	−1	Fault handling exception
4−10	Reserved	NA	—
11	SVC	Programmable	Supervisor call via SVC instruction
12−13	Reserved	NA	—
14	PendSV	Programmable	Pendable request for system service
15	SysTick	Programmable	System tick timer
16	Interrupt #0	Programmable	External interrupt #0
17	Interrupt #1	Programmable	External interrupt #1
...	...	...	...
47	Interrupt #31	Programmable	External interrupt #31

Nonmaskable Interrupt (NMI)

The NMI is similar to IRQ, but it cannot be disabled and has the highest priority apart from the reset. It is very useful for safety critical systems like industrial control or automotive. Depending on the design of the microcontroller, the NMI could be used for power failure handling, or it can be connected to a watchdog unit to restart a system if the system stopped responding. Because the NMI cannot be disabled by control registers, the responsiveness is guaranteed.

Hard Fault

Hard fault is an exception type dedicated to handling fault conditions during program execution. These fault conditions can be trying to execute an unknown opcode, a fault on a bus interface or memory system, or illegal operations like trying to switch to ARM state.

SVCall (SuperVisor Call)

SVC exception takes place when the SVC instruction is executed. SVC is usually used in systems with an operating system (OS), allowing applications to have access to system services.

PendSV (Pendable Service Call)

PendSV is another exception for applications with OS. Unlike the SVC exception, which must start immediately after the SVC instruction has been executed, PendSV can be delayed. The OS commonly uses PendSV to schedule system operations to be carried out only when high-priority tasks are completed.

SysTick

The System Tick Timer inside the NVIC is another feature for OS application. Almost all operating systems need a timer to generate regular interrupt for system maintenance work like context switching. By integrating a simple timer in the Cortex-M0 processor, porting an OS from one device to another is much easier. The SysTick timer and its exception are optional in the Cortex-M0 microcontroller implementation.

Interrupts

The Cortex-M0 microcontroller could support from 1 to 32 interrupts. The interrupt signals could be connected from on-chip peripherals or from an external source via the I/O port. In some cases (depending on the microcontroller design), the external interrupt number might not match the interrupt number on the Cortex-M0 processor.

External interrupts need to be enabled before being used. If an interrupt is not enabled, or if the processor is already running another exception handler with same or higher priority, the interrupt request will be stored in a pending status register. The pended interrupt request can be triggered when the priority level allows—for example, when a higher-priority interrupt handler has been completed and returned. The NVIC can accept interrupt request signals in the form of a high logic level, as well as an interrupt pulse (with a minimum of one clock cycle). Note that in the external interface of a microcontroller, the external interrupt signals can be active high or active low, or they can have programmable configurations.

Exception Priority Definition

In the Cortex-M0 processor, each exception has a priority level. The priority level affects whether the exception will be carried out or if it will wait until later (stay in a pending state). The Cortex-M0 processor supports three fixed highest priority levels and four programmable levels. For exceptions with programmable priority levels, the priority level configuration registers are 8 bits wide, but only the two MSBs are implemented (Figure 8.1).

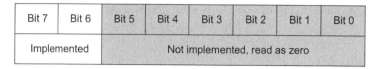

Figure 8.1:
A priority-level register with 2 bits implemented.

Because bit 5 to bit 0 are not implemented, they are always read as zero, and write to these bits are ignored. With this setup, we have possible priority levels of 0x00 (high priority), 0x40, 0x80, and 0xC0 (low priority). This is similar to the Cortex-M3 processor, except that on the Cortex-M3 processor it has at least 3 bits implemented, and therefore the Cortex-M3 processor has at least eight programmable priority levels, whereas the Cortex-M0 processor has only four. When combine with the three fixed priority levels, the Cortex-M0 processor has total of seven priority levels (Figure 8.2).

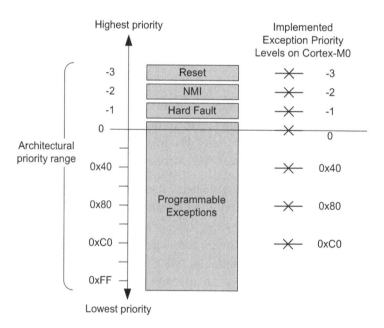

Figure 8.2:
Available priority levels in the Cortex-M0 processor.

The reason for removing the LSB of the priority level register instead of the MSB is to make it easier to port software from one Cortex-M0/M3 device to another. In this way, a program written for devices with wider priority width registers is likely to be able run on devices with narrower priority widths. If the MSB is removed instead of the LSB, you might get an inversion of priority level arrangement among several exceptions during porting of the application. This might result in an exception that is expected to have a lower exception priority preempting another exception that was expected to have a higher priority.

If an enabled exception event occurs (e.g., interrupt, SysTick timer, etc.) while no other exception handler is running, and the exception is not blocked because of PRIMASK (the interrupt masking register, see descriptions in Chapter 3), then the processor will accept it and the exception handler will be executed. The process of switching from a current running task to an exception handler is called preemption.

If the processor is already running another exception handler but the new exception has higher priority level than the current level, then preemption will also take place. The running exception handler will be suspended, and the new exception handler will be executed. This is commonly known as nested interrupt or nested exception. After the new exception handler is completed, the previous exception handler can resume execution and return to thread when it is completed.

However, if the processor is already running another exception handler that has the same or a higher priority level, the new exception will have to wait by entering a pending state. A pending exception can wait until the current exception level changes—for example, after the running exception handler has been completed and returned, and the current priority level has been lowered to be below the priority level of the pending exception. The pending status of exceptions can be accessed via memory-mapped registers in the NVIC. It is possible to clear the pending status of an exception by writing to an NVIC register in software. If the pending status of an exception is cleared, it will not be executed.

If two exceptions happen at the same time and they have the same programmed priority level, the exception with the lower exception type number will be processed first. For example, if both IRQ #0 and IRQ #1 are enabled, both have the same priority level, and both are asserted at the same time, IRQ #0 will be handled first. This rule only applies when the processor is accepting the exceptions, but not when one of these exceptions is already being processed.

The interrupt nesting support in the Cortex-M0 processor does not require any software intervention. This is different from traditional ARM7TDMI, as well as some 8-bit and 16-bit microcontroller architectures where interrupts are disabled automatically during interrupt services and require additional software processing to enable nest interrupt supports.

Vector Table

Traditionally, in processors like the ARM7TDMI, the starting address of the exception handler is fixed. Because the ARM7TDMI has only one IRQ input, multiple IRQs have to share the same IRQ handler starting point, and the handler has to access the status of the interrupt controller to determine which interrupt to be serviced. In the Cortex-M0 processor, the built-in interrupt controller NVIC supports vectored interrupts, which means the exception vectors for different interrupts are separated and the starting-of-interrupt-service routines are located automatically without software intervention.

When the Cortex-M0 processor starts to process an interrupt request, it needs to locate the starting address of the exception handler. This information is stored in the beginning of the memory space, called the vector table (Figure 8.3). The vector table contains the exception vectors for available exceptions in the system, as well as the starting value of the main stack pointer (MSP).

The order of exception vector being stored is the same order of the exception number. Because each vector is one word (four bytes), the address of the exception vector is the exception number times four. Each exception vector is the starting address of the exception handler, with the LSB set to 1 to indicate that the exception handler is in Thumb code.

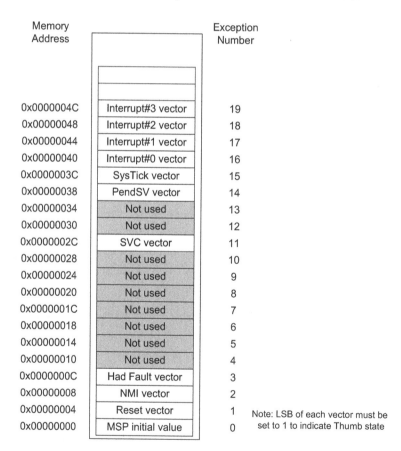

Figure 8.3:
Vector table.

Some of the spaces in the vector table are not used because the Cortex-M0 only has a few system exceptions. Some of the unused exceptions are used on other ARM processors like the Cortex-M3/M4 processor, as these processors have additional system exceptions.

Exception Sequence Overview

Acceptance of Exception Request

The processor accepts an exception if the following conditions are satisfied:

- For interrupt and SysTick interrupt requests, the interrupt has to be enabled
- The processor is not running an exception handler of the same or a higher priority
- The exception is not blocked by the PRIMASK interrupt masking register

Note that for SVC exception, if the SVC instruction is accidentally used in an exception handler that has the same or a higher priority than the SVC exception itself, it will cause the hard fault exception handler to execute.

Stacking and Unstacking

To allow an interrupted program to be resumed correctly, some parts of the current state of the processor must be saved before the program execution switches to the exception handler that services the occurred exception. Different processor architectures have different ways to do this. In the Cortex-M0 processor, the architecture uses a mixture of automatic hardware arrangement and, only if necessary, additional software steps for saving and restoring processor status.

When an exception is accepted on the Cortex-M0 processor, some of the registers in the register banks (R0 to R3, R12, R14), the return address (PC), and the Program Status Register (xPSR) are pushed to the current active stack memory automatically. The Link Register (LR/R14) is then updated to a special value to be used during exception return (EXC_RE-TURN, to be introduced later in this chapter), and then the exception vector is automatically located and the exception handler starts to execute.

At the end of the exception handling process, the exception handler executes a return using the special value (EXC_RETURN, previously generated in LR) to trigger the exception return mechanism. The processor checks to determine if there is any other exception to be serviced. If not, the register values previously stored on the stack memory are restored and the interrupted program is resumed.

The actions of automatically saving and restoring of the register contents are called "stacking" and "unstacking" (Figure 8.4). These mechanisms allow exception handlers to be implemented as normal C functions, thereby reducing the software overhead of exception

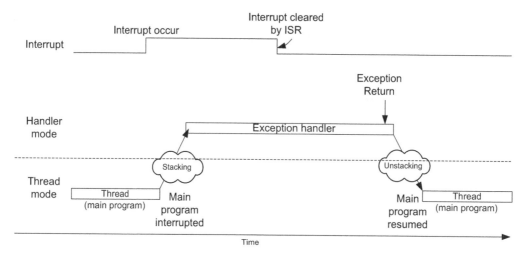

Figure 8.4:
Stacking and unstacking of registers at exception entry and exit.

handling as well as the circuit size (no need to have extra banked registers), and hence lowering the power consumption of the design.

The registers not saved by the automatic stacking process will have to be saved and restored by software in the exception handler, if the exception handler had modified them. However, this does not affect the use of normal C functions as exception handlers, because it is a requirement for C compilers to save and restore the other registers (R4-R11) if they will be modified during the C function execution.

Exception Return Instruction

Unlike some other processors, there is no special return instruction for exception handlers. Instead, a normal return instruction is used and the value load into PC is used to trigger the exception return. This allows exception handlers to be implemented as a normal C function.

Two different instructions can be used for exception return:

```
BX <Reg>; Load a register value into PC (e.g., "BX LR")
```

and

```
POP { <Reg1>,< Reg2>,...., PC} ; POP instruction with PC being one of the registers
                          being updated
```

When one of these instructions is executed with a special value called EXC_RETURN being loaded into the program counter (PC), the exception return mechanism will be triggered. If the value being loaded into the PC does not match the EXC_RETURN pattern, then it will be executed as a normal BX or POP instruction.

Tail Chaining

If an exception is in a pending state when another exception handler has been completed, instead of returning to the interrupted program and then entering the exception sequence again, a tail-chain scenario will occur. When tail chain occurs, the processor will not have to restore all register values from the stack and push them back to the stack again. The tail chaining of exceptions allows lower exception processing overhead and hence better energy efficiency (Figure 8.5).

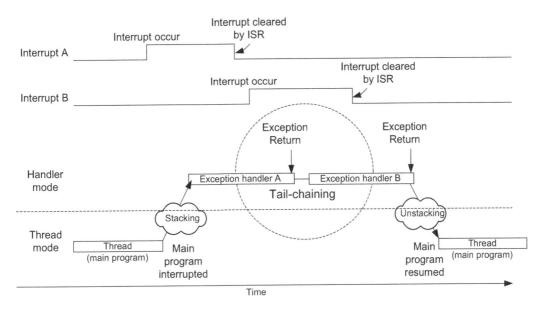

Figure 8.5:
Tail chaining of an interrupt service routine.

Late Arrival

Late arrival is an optimization mechanism in the Cortex-M0 to speed up the processing of higher-priority exceptions. If a higher priority exception occurs during the stacking process of a lower-priority exception, the processor switches to handle the higher-priority exception first (Figure 8.6).

Because processing of either interrupt requires the same stacking operation, the stacking process continues as normal when the late-arriving, higher-priority interrupt occurs. At the end of the stacking process, the vector for the higher-priority exception is fetched instead of the lower-priority one.

Without the late arrival optimization, a processor will have to preempt and enter the exception entry sequence again at the beginning of the lower-priority exception handler. This results in longer latency as well as larger stack memory usage.

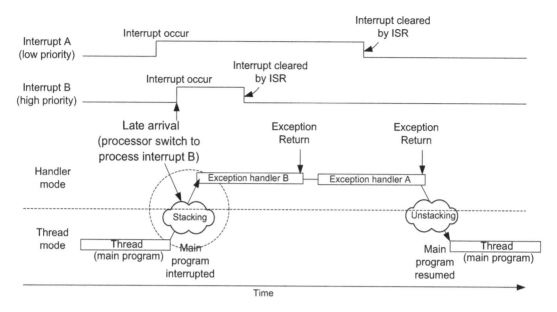

Figure 8.6:
Late arrival optimization.

EXC_RETURN

The EXC_RETURN is a special architecturally defined value for triggering and helping the exception return mechanism. This value is generated automatically when an exception is accepted and is stored into the Link Register (LR, or R14) after stacking. The EXC_RETURN is a 32-bit value; the upper 28 bits are all set to 1, and bits 0 to 3 are used to provide information for the exception return mechanism (Table 8.2).

Bit 0 of EXC_RETURN on the Cortex-M0 processor is reserved and must be 1.
Bit 2 of EXC_RETURN indicates whether the unstacking should restore registers from the main stack (using MSP) or process stack (using PSP).
Bit 3 of EXC_RETURN indicates whether the processor is returning to Thread mode or Handler mode.

Table 8.2: Bit Fields in EXC_RETURN Value

Bits	31:28	27:4	3	2	1	0
Descriptions	EXC_RETURN indicator	Reserved	Return mode	Return stack	Reserved	Reserved
Value	0xF	0xFFFFFF	1 (thread) or 0 (handler)	0 (main stack) or 1 (process stack)	0	1

Table 8.3 shows the valid EXC_RETURN values for the Cortex-M0 processor.

Table 8.3: Valid EXC_RETURN Value

EXC_RETURN	Condition
0xFFFFFFF1	Return to Handler mode (nested exception case)
0xFFFFFFF9	Return to Thread mode and use the main stack for return
0xFFFFFFFD	Return to Thread mode and use the process stack for return

Because the EXC_RETURN value is loaded into LR automatically at the exception entry, the exception handler treats it as a normal return address. If the return address does not need to be saved onto the stack, the exception handler can trigger the exception return and return to the interrupt program by executing "BX LR," just like a normal function. Alternatively, if the exception handler needs to execute function calls, it will need to push the LR to the stack. At the end of the exception handler, the stacked EXC_RETURN value can be loaded into PC directly by a POP instruction, thus triggering the exception return sequence and the return to the interrupt program.

The following diagrams show situations in which different EXC_RETURN values are generated and used.

If the thread is using main stack (CONTROL register bit 1 is zero), the value of the LR will be set to 0xFFFFFFF9 when it enters an exception and 0xFFFFFFF1 when a nested exception is entered, as shown in Figure 8.7.

If the thread is using process stack (CONTROL register bit 1 is set to 1), the value of LR will be 0xFFFFFFFD when entering the first exception and 0xFFFFFFF1 when entering a nested exception, as shown in Figure 8.8.

As a result of EXC_RETURN format, a normal return instruction cannot return to an address in the range of 0xFFFFFFFX, because this will be treated as an exception return rather than a normal one. However, because the address range 0xFXXXXXXX is reserved and should not contain program code, it is not a problem.

Details of Exception Entry Sequence

When an exception takes place, a number of things happen:

- The stack pointer stacks and updates
- The processor fetches the vector and updates it to the PC
- The registers update (LR, IPSR, NVIC registers)

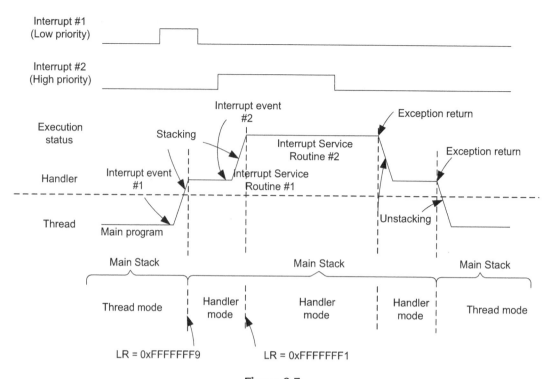

Figure 8.7:
LR set to EXC_RETURN values at exceptions (main stack is used in Thread mode).

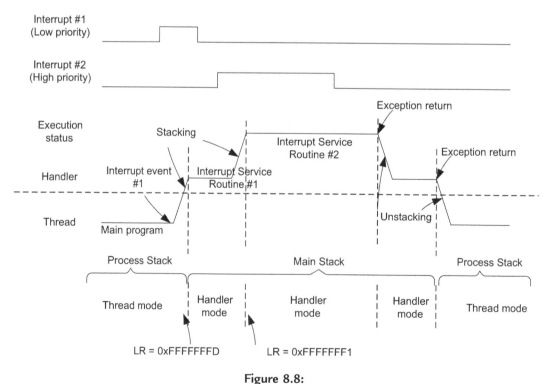

Figure 8.8:
LR set to EXC_RETURN values at exceptions (process stack is used in Thread mode).

Stacking

When an exception takes place, eight registers are pushed to the stack automatically (Figure 8.9). These registers are R0 to R3, R12, R14 (the Link Register), the return address (address of the next instruction, or Program Counter), and the program status register (xPSR). The stack being used for stacking is the current active stack. If the processor was in Thread mode when the exception happened, the stacking could be done using either the process stack or the main stack, depending on the setting in the CONTROL register bit 1. If CONTROL[1] was 0, the main stack would be used for the stacking.

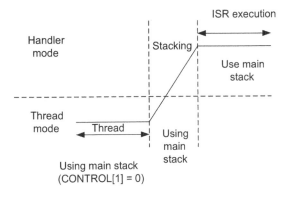

Figure 8.9:
Exception stacking in nested interrupt with the main stack used in Thread mode.

If the processor was in Thread mode and CONTROL[1] was set to 1 when the exception occurred, the stacking will be done using the process stack (Figure 8.10).

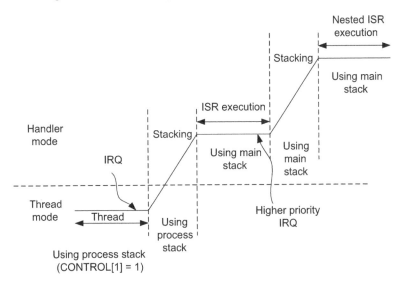

Figure 8.10:
Exception stacking in nested interrupt with process stack used in Thread mode.

For nested exceptions, the stacking always uses the main stack because the processor is already in Handler mode, which can only use the main stack.

The reason for the registers R0–R3, R12, PC, LR, and xPSR to be saved to stack is that these are called "caller saved registers." According to the AAPCS (ARM Architecture Procedure Call Standard, reference 4), a C function does not have to retain the values of these registers. To allow exception handlers to be implemented as normal C functions, these registers have to be saved and restored by hardware, so that when the interrupt program resumes, all these registers will be the same as they were before the exception occurred.

The grouping of the register contents that are pushed onto the stack during stacking is called a "stack frame" (Figure 8.11). In the Cortex-M0 processor, a stack frame is always double word aligned. This ensures that the stack implementation conforms to the AAPCS standard (reference 4). If the position of the last pushed data could be in an address that is not double word aligned, the stacking mechanism automatically adjusts the stacking position to the next double-word-aligned location and sets a flag (bit 9) in the stacked xPSR to indicate that the double word stack adjustment has been made.

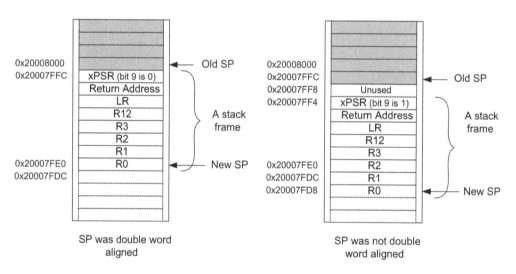

Figure 8.11:
Stack frame and double word stack alignment.

During unstacking, the processor checks the flag in the stacked xPSR and adjusts the stack pointer accordingly.

The stacking of registers is carried out in the order shown in Figure 8.12.

When the stacking has been completed, the stack pointer is updated, and the main stack pointer is selected as the current stack pointer (handlers always use the main stack), then the exception vector will be fetched.

Figure 8.12:
Order of register stacking during the exception sequence in the Cortex-M0 processor.

Vector Fetch and Update PC

After the stacking is done, the processor then fetches the exception vector from the vector table. The vector is then updated to the PC, and the instruction fetch of the exception handler execution starts from this address.

Registers Update

As the exception handler starts to execute, the value of LR is updated to the corresponding EXC_RETURN value. This value is to be used for exception return. In addition, the IPSR is also updated to the exception number of the currently serving exception.

In addition, a number of NVIC registers might also be updated. This includes the status registers for external interrupts if the exception taken is an interrupt, or the Interrupt Control and Status Register if it is a system exception.

Details of Exception Exit Sequence

When an exception return instruction is executed (loading of EXC_RETURN into PC by POP or BX instruction), the exception exit sequence begins. This includes the following steps:

- Unstacking of registers
- Fetching and executing from the restored return address

Unstacking of Registers

To restore the registers to the state they were in before the exception was taken, the register values that were stored on to the stack during stacking are read (POP) and restored back to the registers (Figure 8.13). Because the stack frame can be stored either on the main stack or on the processor stack, the processor first checks the value of the EXC_RETURN being used. If bit 2 of EXC_RETURN is 0, it starts the unstacking from the main stack. If this bit is 1, it starts the unstacking from the process stack.

After the unstacking is done, the stack pointer needs to be adjusted. During stacking, a 4-byte space might have been included in the stack memory so as to ensure the stack frame is double word aligned. If this is the case, the bit 9 of the stack xPSR would be set to 1, and the value of SP could be adjusted accordingly to remove the 4-byte padding space.

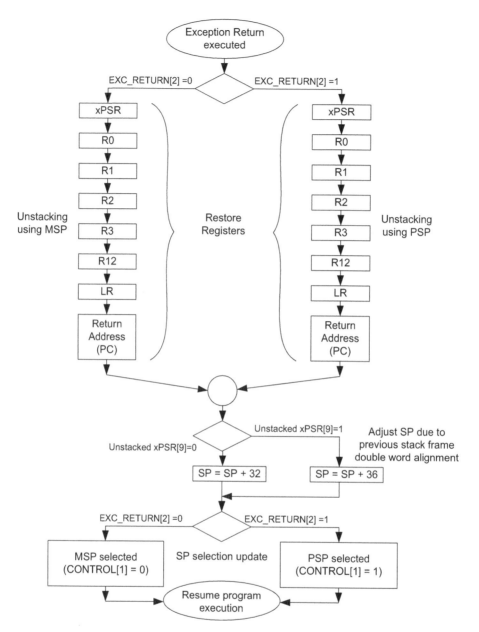

Figure 8.13:
Unstacking at the exception exit.

In addition, the current SP selection may be switched back to the process stack if bit 2 of EXC_RETURN was set to 1 and bit 3 of the EXC_RETURN was set, indicating the exception exit is returning to Thread mode.

Fetch and Execute from Return Address

After the exception return process is completed, the processor can then fetch instructions from the restored return address in the program counter and resume execution of the interrupted program.

## CHAPTER 9

Interrupt Control and System Control

Overview of the NVIC and System Control Block Features

The Nested Vectored Interrupt Controller (NVIC) is an integrated part of the Cortex-M0 processor. It is closely linked to the processor core logic and provides the functions of interrupt control and system exception support.

Apart from the NVIC, there is also a System Control Block (SCB), which shares the same System Control Space (SCS) memory address range. The SCB contains features for operating system support, like an internal timer for the SysTick exception. The OS-related features will be covered in Chapter 10.

The NVIC features were explored in Chapter 3. Those features include the following:

- Flexible interrupt management include enable/disable, priority configurations
- Hardware nested interrupt support
- Vectored exception entry
- Interrupt masking

The NVIC in the Cortex-M0 processor supports up to 32 external interrupts and one nonmaskable interrupt (NMI). The interrupt input requests can be level triggered, or they can be pulsed with a minimum of one clock cycle. Each external interrupt can be enabled or disabled independently, and its pending status can also be set or clear manually.

The NVIC control registers are memory mapped and can be easily accessed in C language. The location of the NVIC registers starts from address 0xE000E100. For the Cortex-M0 processor, the accesses to the NVIC register must be in word size. (Note that in the Cortex-M3 processor, access to the NVIC can be in the form of word, half-word, or byte transfers.)

Similar to the NVIC registers, the SCB registers are also word accessible, and the address starts from 0xE000E010. The SCB registers handle features like the SysTick timer operations, system exception management and priority control, and sleep mode control.

Interrupt Enable and Clear Enable

The Interrupt Enable control register is a programmable register that is used to control the enable/disable of the interrupt requests (exception 16 and above). The width of this register depends on how many interrupts are supported; the maximum size is 32 bit and the minimum

The Definitive Guide to the ARM Cortex-M0. DOI: 10.1016/B978-0-12-385477-3.10009-6

size is 1 bit. This register is programmed via two separate addresses. To enable an interrupt, the SETENA address is used, and to disable an interrupt, the CLRENA address is used (Table 9.1).

Table 9.1: Interrupt Enable Set and Clear Register

Address	Name	Type	Reset Value	Descriptions
0xE000E100	SETENA	R/W	0x00000000	Set enable for interrupt 0 to 31. Write 1 to set bit to 1, write 0 has no effect. Bit[0] for Interrupt #0 (exception #16) Bit[1] for Interrupt #1 (exception #17) … Bit[31] for Interrupt #31 (exception #47) Read value indicates the current enable status
0xE000E180	CLRENA	R/W	0x00000000	Clear enable for interrupt 0 to 31. Write 1 to clear bit to 0, write 0 has no effect. Bit[0] for Interrupt #0 (exception #16) … Bit[31] for Interrupt #31 (exception #47) Read value indicates the current enable status

Separating the set and clear operations in two different addresses has various advantages. First, it reduces the steps needed for enabling an interrupt, thus getting small code and shorter execution time. For example, to enable interrupt #2, we only need to program the NVIC with one access:

```
*((volatile unsigned long *)(0xE000E100))=0x4;//Enable interrupt #2
```

or in assembly,

```
LDR    R0,=0xE000E100 ; Setup address in R0
MOVS   R1,#0x4        ; interrupt #2
STR    R1,[R0]        ; write to set interrupt enable
```

The second advantage is that this arrangement prevents race conditions between multiple application processes that can result in the loss of programmed control information. For example, if the enable control is implemented using a simple read/write register, a read-modify-write process is required for enabling an interrupt (e.g., interrupt #2 in this case), and if between the read operation and write operation, an interrupt occurred and the ISR changed another bit in the interrupt-enabled register, the change done by the ISR could be overwritten when the interrupted program resumed.

An interrupt enable can be cleared with the use of similar code, only the address is different. For example, to disable interrupt #2, we use the following code:

```
*((volatile unsigned long *)(0xE000E180))=0x4;//Disable interrupt #2
```

Or in assembly

```
LDR   R0,=0xE000E180; Setup address in R0
MOVS  R1,#0x4         ; interrupt #2
STR   R1,[ R0]        ; write to clear interrupt enable
```

In normal application development, it is best to use the NVIC control functions provided in the CMSIS-compliant device driver library to enable or disable interrupts. This gives your application code the best software portability. CMSIS is part of the device driver library from your microcontroller vendor and is covered in Chapter 4. To enable or disable interrupt using CMSIS, the functions provided are as follows:

```
void NVIC_EnableIRQ(IRQn_Type IRQn);    // Enable Interrupt –
                                        // IRQn value of 0 refer to Interrupt #0
void NVIC_DisableIRQ(IRQn_Type IRQn);   // Disable Interrupt –
                                        // IRQn value of 0 refer to Interrupt #0
```

Interrupt Pending and Clear Pending

If an interrupt takes place but cannot be processed immediately—for example, if the processor is serving a higher-priority interrupt—the interrupt request will be pended. The pending status is held in a register and will remain valid until the current priority of the processor is lowered so that the pending request is accepted or until the application clears the pending status manually.

The interrupt pending status can be accessed or modified, through the Interrupt Set Pending (SETPEND) and Interrupt Clear Pending (CLRPEND) register addresses (Table 9.2). Similar to the Interrupt Enable control register, the Interrupt Pending status register is physically one register, but it uses two addresses to handle the set and clear the bits. This allows each bit to be

Table 9.2: Interrupt Pending Set and Clear Register

Address	Name	Type	Reset Value	Descriptions
0xE000E200	SETPEND	R/W	0x00000000	Set pending for interrupt 0 to 31. Write 1 to set bit to 1, write 0 has no effect. Bit[0] for Interrupt #0 (exception #16) Bit[1] for Interrupt #1 (exception #17) … Bit[31] for Interrupt #31 (exception #47) Read value indicates the current pending status
0xE000E280	CLRPEND	R/W	0x00000000	Clear pending for interrupt 0 to 31. Write 1 to clear bit to 0, write 0 has no effect. Bit[0] for Interrupt #0 (exception #16) … Bit[31] for Interrupt #31 (exception #47) Read value indicates the current pending status

modified independently, without the risk of losing information because of race conditions between two application processes.

The Interrupt Pending status register allows an interrupt to be triggered by software. If the interrupt is already enabled, no higher-priority exception handler is running, and no interrupt masking is set, then the interrupt service routine will be carried out almost immediately. For example, if we want to trigger interrupt #2, we can use the following code:

```
*((volatile unsigned long *)(0xE000E100))=0x4; //Enable interrupt #2
*((volatile unsigned long *)(0xE000E200))=0x4; //Pend interrupt #2
```

or in assembly,

```
    LDR    R0,=0xE000E100 ; Setup address in R0
    MOVS   R1,#0x4        ; interrupt #2
    STR    R1,[R0]        ; write to set interrupt enable
    LDR    R0,=0xE000E200 ; Setup address in R0
    STR    R1,[R0]        ; write to set pending status
```

In some cases we might need to clear the pending status of an interrupt. For example, when an interrupt-generating peripheral is being reprogrammed, we can disable the interrupt for this peripheral, reprogram its control registers, and clear the interrupt pending status before re-enabling the peripheral (in case unwanted interrupt requests might be generated during reprogramming). For example, to clear the pending status of interrupt 2, we can use the following code:

```
*((volatile unsigned long *)(0xE000E280))=0x4;// Clear interrupt #2
                                              // pending status
```

or in assembly,

```
    LDR    R0,=0xE000E280 ; Setup address in R0
    MOVS   R1,#0x4        ; interrupt #2
    STR    R1,[R0]        ; write to clear pending status
```

In the CMSIS-compliant device driver libraries, three functions are provided for accessing the pending status registers:

```
void NVIC_SetPendingIRQ(IRQn_Type IRQn); // Set pending status of a interrupt
void NVIC_ClearPendingIRQ(IRQn_Type IRQn); // Clear pending status of a interrupt
uint32_t NVIC_GetPendingIRQ(IRQn_Type IRQn); // Return true if the interrupt pending
                                             //status is 1
```

Interrupt Priority Level

Each external interrupt has an associated priority-level register. Each of them is 2 bits wide, occupying the two MSBs of the Interrupt Priority Level Registers. Each Interrupt Priority Level Register occupies 1 byte (8 bits). NVIC registers in the Cortex-M0 processor can only be

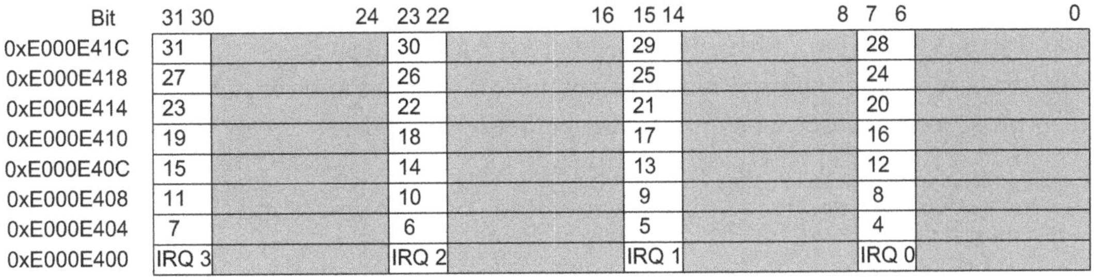

Bit	31 30	24	23 22	16	15 14	8	7 6	0
0xE000E41C	31		30		29		28	
0xE000E418	27		26		25		24	
0xE000E414	23		22		21		20	
0xE000E410	19		18		17		16	
0xE000E40C	15		14		13		12	
0xE000E408	11		10		9		8	
0xE000E404	7		6		5		4	
0xE000E400	IRQ 3		IRQ 2		IRQ 1		IRQ 0	

Figure 9.1:
Interrupt Priority Level Registers for each interrupt.

accessed using word-size transfers, so for each access, four Interrupt Priority Level Registers are accessed at the same time (Figure 9.1).

The unimplemented bits are read as zero. Write values to those unimplemented bits are ignored, and read values of the unimplemented bits return zeros (Table 9.3).

Table 9.3: Interrupt Priority Level Registers (0xE000E400−0xE000E41C)

Address	Name	Type	Reset Value	Descriptions
0xE000E400	IPR0	R/W	0x00000000	Priority level for interrupt 0 to 3 [31:30] Interrupt priority 3 [23:22] Interrupt priority 2 [15:14] Interrupt priority 1 [7:6] Interrupt priority 0
0xE000E404	IPR1	R/W	0x00000000	Priority level for interrupt 4 to 7 [31:30] Interrupt priority 7 [23:22] Interrupt priority 6 [15:14] Interrupt priority 5 [7:6] Interrupt priority 4
0xE000E408	IPR2	R/W	0x00000000	Priority level for interrupt 8 to 11 [31:30] Interrupt priority 11 [23:22] Interrupt priority 10 [15:14] Interrupt priority 9 [7:6] Interrupt priority 8
0xE000E40C	IPR3	R/W	0x00000000	Priority level for interrupt 12 to 15 [31:30] Interrupt priority 15 [23:22] Interrupt priority 14 [15:14] Interrupt priority 13 [7:6] Interrupt priority 12
0xE000E410	IPR4	R/W	0x00000000	Priority level for interrupt 16 to 19
0xE000E414	IPR5	R/W	0x00000000	Priority level for interrupt 20 to 23
0xE000E418	IPR6	R/W	0x00000000	Priority level for interrupt 24 to 27
0xE000E41C	IPR7	R/W	0x00000000	Priority level for interrupt 28 to 31

Because each access to the priority level register will access four of them in one go, if we only want to change one of them, we need to read back the whole word, change 1 byte, and then write back the whole value. For example, if we want to set the priority level of interrupt #2 to 0xC0, we can do it by using the following code:

```
unsigned long temp; // a temporary variable
temp = * ((volatile unsigned long *) (0xE000E400)); // Get IPR0
temp = temp & (0xFF00FFFF) | (0xC0 << 16); // Change Priority level
* ((volatile unsigned long *) (0xE000E400)) = temp; // Set IPR0
```

Or in assembly, we use this code:

```
LDR    R0,=0xE000E400 ; Setup address in R0
LDR    R1,[R0]         ; Get PRIORITY0
MOVS   R2, #0xFF       ; Byte mask
LSLS   R2, R2, #16     ; Shift mask to interrupt #2's position
BICS   R1, R1, R2      ; R1 = R1 AND (NOT(0x00FF0000))
MOVS   R2, #0xC0       ; New value for priority level
LSLS   R2, R2, #16     ; Shift left by 16 bits
ORRS   R1, R1, R2      ; Put new priority level
STR    R1,[R0]         ; write back value
```

Alternatively, if the mask value and new value are fixed in the application code, we can set the mask value and new priority level values using LDR instructions to shorten the code:

```
LDR    R0,=0xE000E400 ; Setup address in R0
LDR    R1,[R0]         ; Get PRIORITY0
LDR    R2,=0x00FF0000 ; Mask for interrupt #2's priority
BICS   R1, R1, R2      ; R1 = R1 AND (NOT(0x00FF0000))
LDR    R2,=0x00C00000 ; New value for interrupt #2's priority
ORRS   R1, R1, R2      ; Put new priority level
STR    R1,[R0]         ; write back value
```

With CMSIS-compliant device driver libraries, the interrupt priority level can be accessed by two functions:

```
void NVIC_SetPriority(IRQn_Type IRQn, uint32_t priority); // Set the priority
                   // level of an interrupt or a system exception
uint32_t NVIC_GetPriority(IRQn_Type IRQn); // return the priority
                   // level of an interrupt or a system exception
```

Note that these two functions automatically shift the priority level values to the implemented bits of the priority level registers. Therefore, when we want to set the priority value of interrupt #2 to 0xC0, we should use this code:

```
NVIC_SetPriority(2, 0x3); // priority value 0x3 is shifted to become 0xC0
```

The Interrupt Priority Level Registers should be programmed before the interrupt is enabled. Usually this is done at the beginning of the program. Changing of interrupt priority when the interrupt is already enabled should be avoided, as this is architecturally unpredictable in the

ARMv6-M architecture and is not supported in Cortex-M0 processor. The case is different for the Cortex-M3/M4 processor. The Cortex-M3/M4 processor supports the dynamic switching of interrupt priority levels. Another difference between the Cortex-M3 processor and Cortex-M0 processor is that the interrupt priority registers can be accessed using byte or half word transfers in the Cortex-M3, so that you can access an individual priority level setting with byte-size accesses. More details of the differences between various Cortex-M processors are covered in Chapter 21.

Generic Assembly Code for Interrupt Control

For users who are programming the Cortex-M0 processor using assembly language, it could be handy to have a set of generic functions for handling interrupt control in the NVIC. For C language users, a function library is already provided in the Cortex Microcontroller Software Interface Standard (CMSIS). The CMSIS is included in the device driver libraries from all major microcontroller vendors and is openly accessible. More about the CMSIS is covered in Chapter 4.

Enable and Disable Interrupts

The enable and disable of interrupts are quite simple. The following functions—"nvic_set_enable" and "nvic_clr_enable"—require the interrupt number as input, which is stored in R0 before the function call:

```
;-------------------
; Enable IRQ
; - input R0 : IRQ number. E.g., IRQ#0 = 0
    ALIGN
nvic_set_enable FUNCTION
    PUSH    { R1, R2}
    LDR     R1,=0xE000E100 ; NVIC SETENA
    MOVS    R2, #1
    LSLS    R2, R2, R0
    STR     R2, [ R1]
    POP     { R1, R2}
    BX      LR    ; Return
    ENDFUNC
;-------------------
; Disable IRQ
; - input R0 : IRQ number. E.g., IRQ#0 = 0
    ALIGN
nvic_clr_enable FUNCTION
    PUSH    { R1, R2}
    LDR     R1,=0xE000E180 ; NVIC CLRENA
    MOVS    R2, #1
    LSLS    R2, R2, R0
    STR     R2, [ R1]
    POP     { R1, R2}
    BX      LR    ; Return
    ENDFUNC
;-------------------
```

To use the functions, just put the interrupt number in R0, and call the function. For example,

```
MOVS    R0,  #3 ; Enable Interrupt #3
BL      nvic_set_enable
```

The FUNCTION and ENDFUNC keywords are used to identify the start and end of a function in the ARM assembler (including the Keil MDK). This is optional. The "ALIGN" keyword ensures correct alignment of the starting of the function.

Set and Clear Interrupt Pending Status

The assembly functions for setting and clearing interrupt pending status are similar to the ones used for enabling and disabling interrupts. The only changes are labels and NVIC register address values:

```
;--------------------
; Set IRQ Pending status
; - input R0 : IRQ number. E.g., IRQ#0 = 0
ALIGN
nvic_set_pending FUNCTION
    PUSH    { R1, R2}
    LDR     R1,=0xE000E200 ; NVIC SETPEND
    MOVS    R2, #1
    LSLS    R2, R2, R0
    STR     R2, [ R1]
    POP     { R1, R2}
    BX      LR   ; Return
    ENDFUNC
;--------------------
; Clear IRQ Pending
; - input R0 : IRQ number. E.g., IRQ#0 = 0
ALIGN
nvic_clr_pending FUNCTION
    PUSH    {R1, R2}
    LDR     R1,=0xE000E280 ; NVIC CLRPEND
    MOVS    R2, #1
    LSLS    R2, R2, R0
    STR     R2, [ R1]
    POP     {R1, R2}
    BX      LR   ; Return
    ENDFUNC
;--------------------
```

Note that sometimes clearing the pending status of an interrupt might not be enough to stop the interrupt from happening. If the interrupt source generates an interrupt request continuously (level output), then the pending status could remain high, even if you try to clear it at the NVIC.

Setting up Interrupt Priority Level

The assembly function to set up the priority level for an interrupt is a bit more complex. First, it requires two input parameters: the interrupt number and the new priority level. Second, the calculation of priority level register address has to be modified, as there are up to eight priority registers. Finally, it needs to perform a read-modify-write operation to the correct byte inside the 32-bit priority level register, as the priority level registers are word access only:

```
;--------------------
; Set interrupt priority
; - input R0 : IRQ number. E.g., IRQ#0 = 0
; - input R1 : Priority level
     ALIGN
nvic_set_priority FUNCTION
     PUSH    {R2-R5}
     LDR     R2,=0xE000E400 ; NVIC Interrupt Priority #0
     MOV     R3, R0   ; Make a copy of IRQ number
     MOVS    R4, #3   ; clear lowest two bit of IRQ number
     BICS    R3, R4
     ADDS    R2, R3   ; address of priority register in R2
     ANDS    R4, R0   ; byte number (0 to 3) in priority register
     LSLS    R4, R4, #3 ; Number of bits to shift for priority & mask
     MOVS    R5, #0xFF  ; byte mask
     LSLS    R5, R5, R4 ; byte mask shift to right location
     MOVS    R3, R1
     LSLS    R3, R3, R4 ; Priority shift to right location
     LDR     R4, [R2]   ; Read existing priority level
     BICS    R4, R5     ; Clear existing priority value
     ORRS    R4, R3     ; Set new level
     STR     R4, [R2]   ; Write back
     POP     {R2-R5}
     BX      LR         ; Return
     ENDFUNC
;--------------------
```

In most applications, however, you can use a much simpler code to set up priority levels of multiple interrupts in one go at the beginning of the program. For example, you can predefine the priority levels in a table of constant values and then copy it to the NVIC priority level registers using a short instruction sequence:

```
     LDR     R0,=PrioritySettings ; address of priority setting table
     LDR     R1,=0xE000E400 ; address of interrupt priority registers
     LDMIA   R0!,{R2-R5} ; Read Interrupt Priority 0-15
     STMIA   R1!,{R2-R5} ; Write Interrupt Priority 0-15
     LDMIA   R0!,{R2-R5} ; Read Interrupt Priority 16-31
     STMIA   R1!,{R2-R5} ; Write Interrupt Priority 16-31
     ...
     ALIGN 4 ; Ensure that the table is word aligned
 PrioritySettings ; Table of priority level values (example values)
     DCD     0xC0804000 ; IRQ 3- 2- 1- 0
     DCD     0x80808080 ; IRQ  7- 6- 5- 4
     DCD     0xC0C0C0C0 ; IRQ 11-10- 9- 8
     DCD     0x40404040 ; IRQ 15-14-13-12
     DCD     0x40404080 ; IRQ 19-18-17-16
     DCD     0x404040C0 ; IRQ 23-22-21-20
     DCD     0x4040C0C0 ; IRQ 27-26-25-24
     DCD     0x004080C0 ; IRQ 31-30-29-28
```

Exception Masking Register (PRIMASK)

In some applications, it is necessary to disable all interrupts for a short period of time for time-critical processes. Instead of disabling all interrupts and restoring them using the interrupt enable/disable control register, the Cortex-M0 processor provides a separate feature for this usage. One of the special registers, called PRIMASK (introduced in Chapter 3), can be used to mask all interrupts and system exceptions, apart from the NMI and hard fault exceptions.

The PRIMASK is a single bit register and is set to 0 at reset. When set to 0, interrupts and system exceptions are allowed. When set to 1, only NMI and hard fault exceptions are allowed. Effectively, when it is set to 1, it changes the current priority level to 0 (the highest programmable level).

There are various ways to program the PRIMASK register. In assembly language, you can set or clear the PRIMASK register using the MSR instruction. For example, you can use the following code to set PRIMASK (disable interrupt):

```
MOVS    R0, #1      ; New value for PRIMASK
MSR     PRIMASK, R0 ; Transfer R0 value to PRIMASK
```

You can enable the interrupt in the same way, by just changing the R0 value to 0.

Alternatively, you can use the CPS instructions to set or clear PRIMASK:

```
CPSIE  i  ; Clear PRIMASK (Enable interrupt)
CPSID  i  ; Set PRIMASK (Disable interrupt)
```

In C language, users of CMSIS device drivers can use the following function to set and clear PRIMASK. Even if CMSIS is not used, most C compilers for ARM processors handle these two functions automatically as intrinsic functions:

```
void __enable_irq(void); // Clear PRIMASK
void __disable_irq(void); // Set PRIMASK
```

These two functions get compiled into the CPS instructions.

It is important to clear the PRIMASK after the time-critical routine is finished.
Otherwise the processor will stop accepting new interrupt requests. This applies even if the __disable_irq() function (or setting of PRIMASK) is used inside an interrupt handler. This behavior differs from that of the ARM7TDMI; in the ARM7TDMI processor, the I-bit can be reset (to enable interrupts) at exception return because of the restoration of the CPSR. When in the Cortex-M processors, PRIMASK and xPSR are separated and therefore the interrupt masking is not affected by exception return.

Interrupt Inputs and Pending Behavior

The Cortex-M0 processor supports interrupt requests in the form of a level trigger as well as pulse input. This feature involves a number of pending status registers associated with interrupt inputs, including the NMI input. For each interrupt input, there is a 1-bit register called the pending status register that holds the interrupt request even if the interrupt request is de-asserted (e.g., an interrupt pulse generated from external hardware connected via the I/O port). When the exception starts to be served by the processor, the hardware clears the pending status automatically.

In the case of NMI it is almost the same, apart from the fact that the NMI request is usually served almost immediately because it is the highest priority interrupt type. Otherwise it is quite similar to the IRQs: the pending status register for NMI allows software to trigger NMI and allows new NMI to be held in pending state if the processor is still serving the previous NMI request.

Simple Interrupt Process

Most peripherals developed for ARM processors use level trigger interrupt output. When an interrupt event take place, the interrupt signal connect from the peripheral to the NVIC will be asserted. The signal will remain high until the processor clears the interrupt request at the peripheral during the interrupt service routine. Inside the NVIC, the pending status register of the interrupt is set when the interrupt is detected and is cleared as the processor accepts and starts the interrupt service routine execution (Figure 9.2).

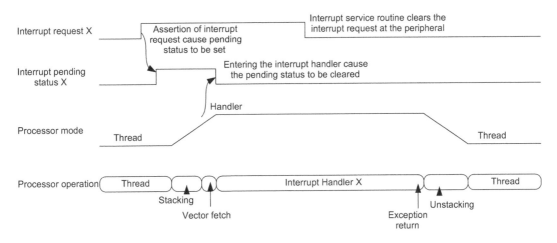

Figure 9.2:
A simple case of interrupt activation and pending status behavior.

Some interrupt sources might generate interrupt requests in the form of a pulse (for at least one clock cycle). In this case, the pending status register will hold the request until the interrupt is being served (Figure 9.3).

If the interrupt request is not carried out immediately and is de-asserted, and the pending status is cleared by software, then the interrupt request will be ignored and the processor will not execute the interrupt handler (Figure 9.4). The pending status can be cleared by writing to the

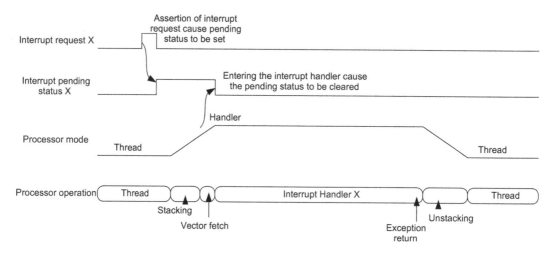

Figure 9.3:
A simple case of pulsed interrupt activation and pending status behavior.

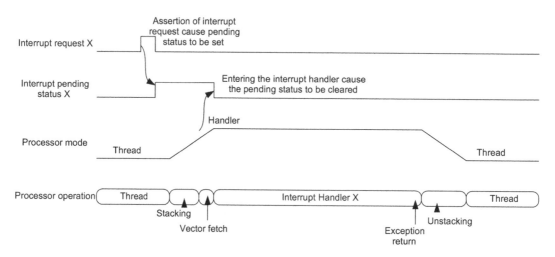

Figure 9.4:
Interrupt pending status is cleared and is not taken by the processor.

NVIC CLRPEND register. This is often necessary when setting up a peripheral, and the peripheral might have generated a spurious interrupt request previously.

If the interrupt request is still asserted by the peripheral when the software clears the pending status, the pending status will be asserted again immediately (Figure 9.5).

Now let us go back to the normal interrupt processing scenarios. If the interrupt request from the peripheral is not cleared during the execution of the exception handler, the pending status will be activated again at the exception return and will cause the exception handler to be

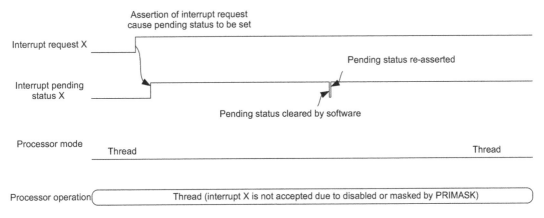

Figure 9.5:
Interrupt pending status is cleared and then reasserted.

executed again. This might happen if the peripheral got more data to be processed (for example, a data receiver might want to hold the interrupt request high, as long as data remain in its received data FIFO) (Figure 9.6).

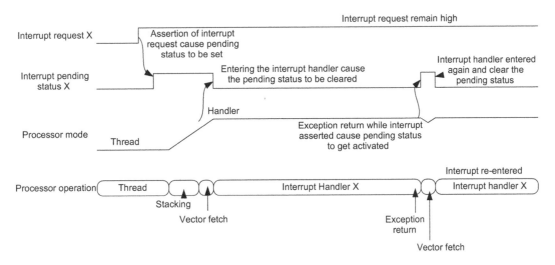

Figure 9.6:
Interrupt request remain high at interrupt exit cause reentering of the same interrupt handler.

For pulsed interrupts, if the interrupt request is pulsed several times before the processor starts the interrupt service routine (for example, the processor could be handling another interrupt request), then the multiple interrupt pulses will be treated as just one interrupt request (Figure 9.7).

If the pulsed interrupt request is triggered again during the execution of the interrupt service routine, it will be processed as a new interrupt request and will cause the interrupt service routine to be entered again after the interrupt exit (Figure 9.8).

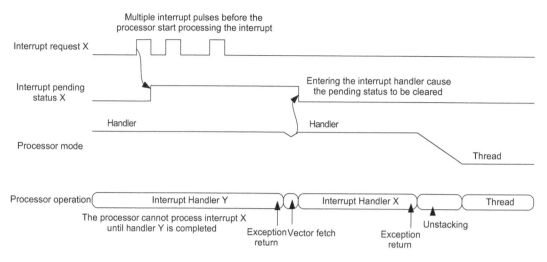

Figure 9.7:
Multiple interrupt request pulses can be merged into one request.

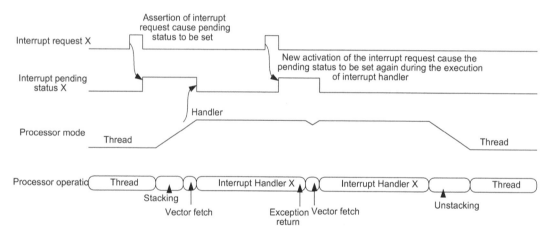

Figure 9.8:
Interrupt pending status can be asserted by a new interrupt request even during its handler execution.

The second interrupt request does not cause the interrupt to be serviced immediately because it is at the same priority level as the current execution priority. Once the processor exits the handler, then the current priority level is lowered, thus allowing the pending interrupt request to be serviced.

The pending status of an interrupt can be activated even when the interrupt is disabled. Therefore, when reprogramming a peripheral and setting up its interrupt and if the previous state of the peripheral is unknown, you might need to clear its interrupt pending status in the NVIC before reenabling the interrupt in the NVIC. This can be done by writing to the Interrupt Clear Pending register in address 0xE000E280.

Interrupt Latency

Under normal situations, the interrupt latency of the Cortex-M0 processor is 16 cycles. The interrupt latency is defined as from the processor clock cycle the interrupt is asserted, to the start of the execution of the interrupt handler. This interrupt latency assumes the following:

- The interrupt is enabled and is not masked by PRIMASK or other executing exception handlers.
- The memory system does not have any wait state. If the memory system has wait state, the interrupt could be delayed by wait states that occur at the last bus transfer before interrupt processing, stacking, vector fetch, or instruction fetch at the start of the interrupt handler.

There are some situations that can result in different interrupt latency:

- *Tail chaining of interrupt.* If the interrupt request occurs just as another exception handler returns, the unstacking and stacking process can be skipped, thus reducing the interrupt latency.
- *Late arrival.* If the interrupt request occurs during the stacking process of another lower-priority interrupt, the late arrival mechanism allows the new high-priority intercept to take place first. This can result in lower latency for the higher-priority interrupt.

These two behaviors allow interrupt latency to be reduced to a minimum. However, in some embedded applications, zero jitter interrupt response is required. Fortunately, the Cortex-M0 processor is equipped with a zero jitter feature.

On the interface of the Cortex-M0 processor, there is an 8-bit signal called IRQLATENCY, which is connected to the NVIC. This signal can be used to control the interrupt latency behavior. If this signal is connected to a 0, then the Cortex-M0 processor will start to process the interrupt request as soon as possible. If the signal is set to a specific value depending on the timing of the memory system, then it can enable the zero jitter behavior to force the interrupt latency to a higher number of cycles, but it is guaranteed to have zero jitter. The IRQLA-TENCY signal is normally controlled by configurable registers developed by microcontroller vendors and is not visible on the microcontroller interface.

Control Registers for System Exceptions

Besides the external interrupts, some of the system exceptions can also have programmable priority levels and pending status registers. First, we will look at the priority level registers for system exceptions. On the Cortex-M0 processor, only the three OS-related system exceptions have programmable priority levels. These include SVC, PendSV, and SysTick. Other system exceptions, like NMI and hard fault, have fixed priority levels (Figure 9.9).

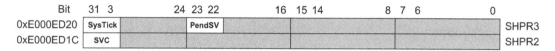

Figure 9.9:
Priority Level Registers for programmable system exceptions.

The unimplemented bits are read as zero. Write to those unimplemented bits are ignored. On the Cortex-M0 processor, only the System Handler Priority Register 2 (SHPR2) and SHPR3 are implemented (Table 9.4). SHPR1 is not available on the Cortex-M0 processor (it is available on the ARMv7-M architecture—for example, the Cortex-M3 processor).

Table 9.4: System Handler Priority Level Registers (0xE000ED1C−0xE000ED20)

Address	Name	Type	Reset Value	Descriptions
0xE000ED1C	SHPR2	R/W	0x00000000	System Handler Priority Register 2 [31:30] SVC priority
0xE000ED20	SHPR3	R/W	0x00000000	System Handler Priority Register 3 [31:30] SysTick priority [23:22] PendSV priority

Users of CMSIS-compliant device drivers can access to the SHPR2 and SHPR3 registers using the register names shown in Table 9.5.

Table 9.5: CMSIS Register Names for System Handler Priority Level Registers

Register	CMSIS Register Name	Descriptions
SHPR2	SCB->SHP[0]	System Handler Priority Register 2
SHPR3	SCB->SHP[1]	System Handler Priority Register 3

Another NVIC register useful for system exception handling is the Interrupt Control State Register (ISCR) (Table 9.6). This register allows the NMI exception to be pended by software, and it accesses the pending status of PendSV and SysTick exceptions. This register also provides information useful for the debugger, such as the current active exception number and whether or not any exception is currently pended. Because the SysTick implementation is optional, the SysTick exception pending set/clear bits are only available when the SysTick option is presented. As a result, bits 26 and 25 of this register might not be available.

Users of the CMSIS-compliant device driver library can access ICSR using the register name "SCB->ICSR."

Table 9.6: Interrupt Control State Register (0xE000ED04)

Bits	Field	Type	Reset Value	Descriptions
31	NMIPENDSET	R/W	0	Write 1 to pend NMI, write 0 has no effect. On reads return pending state of NMI.
30:29	Reserved	—	—	Reserved
28	PENDSVSET	R/W	0	Write 1 to set PendSV, write 0 has no effect. On reads return the pending state of PendSV.
27	PENDSVCLR	R/W	0	Write 1 to clear PendSV, write 0 has no effect. On reads return the pending state of PendSV.
26	PENDSTSET	R/W	0	Write 1 to pend SysTick, write 0 has no effect. On reads return the pending state of SysTick.
25	PENDSTCLR	R/W	0	Write 1 to clear SysTick pending, write 0 has no effect. On reads return the pending state of SysTick.
24	Reserved	—	—	Reserved.
23	ISRPREEMPT	RO	—	During debugging, this bit indicates that an exception will be served in the next running cycle, unless it is suppressed by debugger using the C_MASKINTS in Debug Control and Status Register.
22	ISRPENDING	RO	—	During debugging, this bit indicates that an exception is pended.
21:18	Reserved	—	—	Reserved.
17:12	VECTPENDING	RO	—	Indicates the exception number of the highest priority pending exception. If it is read as 0, it means no exception is currently pended.
11:6	Reserved	—	—	Reserved.
5:0	VECTACTIVE	RO	—	Current active exception number, same as IPSR. If the processor is not serving an exception (Thread mode), this field reads as 0.

Some of the fields (e.g., the ISRPREEMPT and ISRPENDING fields) in the ICSR are used by the debug system only. In most cases, application code only uses the ICSR to control or check the system exception pending status.

System Control Registers

The NVIC address range (from 0xE000E000 to 0xE000EFFF) also covers a number of system control registers. Therefore, the whole memory range for NVIC is referred to as the System Control Space (SCS). A few of these registers are for the SysTick timer; they will be introduced in the next chapter, where the OS supporting features will be covered. Here the rest of the system control registers are introduced.

CPU ID Base Register

The CPU ID Base register is a read-only register containing the processor ID (Figure. 9.10). It allows application software as well the debugger to determine the processor core type and version information.

Figure 9.10:
CPU ID Base Register.

The current release of the Cortex-M0 processor (r0p0) has CPU ID values of 0x410CC200. The variant (bit[23:20]) or revision numbers (bit[3:0]) advance for each new release of the core. The CPUID register can be accessed with CMSIS-compliant device drivers such as "SCB->CPUID" (Table. 9.7).

Table 9.7: CPU ID Base Register (0xE000ED00)

Bits	Field	Type	Reset Value	Descriptions
31:0	CPU ID	RO	0x410CC200 (r0p0)	CPU ID value. Used by debugger as well as application code to determine processor type and revision.

Software can also use this register to determine the CPU type. Bit[7:4] of the CPUID is "0" for the Cortex-M0, "1" for Cortex-M1, "3" for Cortex-M3, and "4" for Cortex-M4.

Application Interrupt and Reset Control Register

The Application Interrupt and Reset Control Register (AIRCR) has several functions (Table. 9.8). It allows an application to request a system reset, determine the endianness of the system, and clear all exception active statuses (this can be done by the debugger only). It can be accessed in CMSIS-compliant device drivers such as "SCB->AIRCR."

The VECTKEY field is used to prevent accidental write to this register from resetting the system or clearing of the exception status.

The application can use the ENDIANNESS bit as well as the debugger to determine the endianness of the system. This endianness of a Cortex-M0 processor system cannot be changed, as the setup is defined by the microcontroller vendor.

The SYSRESETREQ bit is used to request a system reset. When a value of 1 is written to this bit with a valid key, it causes a signal called SYSRESETREQ on the processor to be

Table 9.8: Application Interrupt and Reset Control Register (0xE000ED0C)

Bits	Field	Type	Reset Value	Descriptions
31:16	VECTKEY (during write operation)	WO	—	Register access key. When writing to this register, the VECTKEY field needs to be set to 0x05FA; otherwise the write operation would be ignored.
31:16	VECTKEYSTAT (during read operation)	RO	0xFA05	Read as 0xFA05.
15	ENDIANNESS	RO	0 or 1	1 indicates the system is big endian. 0 indicates the system is little endian.
14:3	Reserved	—	—	Reserved.
2	SYSRESETREQ	WO	—	Write 1 to this bit causes the external signal SYSRESETREQ to be asserted.
1	VECTCLRACTIVE	WO	—	Write 1 to this bit causes —exception active status to be cleared —processor to return to Thread mode —IPSR to be cleared This bit can be only be used by a debugger.
0	Reserved	—	—	Reserved.

asserted. The actual reset timing of the system depends on how this signal is connected. There can be a small delay from the time this bit is written to the actual reset, depending on the design of the system reset control. In typical microcontroller design, the SYSRESETREQ generates system reset for the processor and most parts of the system, but it should not affect the debug system of the microcontroller. This allows the debug operation to work correctly even when the software triggers a reset.

To reset the system using the AIRCR register, you can use the CMSIS function:

```
void NVIC_SystemReset(void);
```

Alternatively, the following code can be used:

```
__DSB(); // Data Synchronization Barrier (include for portability)
        // Ensure all memory accessed are completed
__disable_irq(); // Disable interrupts
*((volatile unsigned long *)(0xE000ED0C))=0x05FA0004;//System reset
while(1); // dead loop, waiting for reset
```

The "while" loop after the write prevents the processor from executing more instructions after the reset request has been issued. The disabling of interrupt is optional; if an interrupt is generated when the system reset request is set, and if the actual reset is delayed because of the reset controller design, there is a chance that the processor will enter the exception handler as the system reset starts. In most cases, it is not an issue, but we can prevent this from happening by setting the exception mask register PRIMASK to disable interrupts before setting the SYSRESETREQ bit.

The use of the DSB instruction allows the code to be used with other ARM processors that have write buffers in the memory interface. In these processors, a memory write operation might be delayed, and if the system reset and memory write happened at the same time, the memory could become corrupted. By inserting a DSB, we can make sure the reset will not happen until the last memory access is completed. Although this is not required in the Cortex-M0 (because there is no write buffer in the Cortex-M0), the DSB is included for better software portability.

The same reset request code can be written in assembly. In the following example code, the step to setting up PRIMASK is optional:

```
      DSB                     ; Data Synchronization Barrier
      CPSID   i               ; Set PRIMASK
      LDR     R0,=0xE000ED0C  ; AIRCR register address
      LDR     R1,=0x05FA0004  ; Set System reset request
      STR     R1,[R0]         ; write back value
Loop
      B       Loop            ; dead loop, waiting for reset
```

The debugger uses the VECTCLRACTIVE bit to clear exception status—for example, when the debugger tries to rerun a program without resetting the processor. Application code running on the processor should not use this feature.

Configuration and Control Register

The Configuration and Control Register (CCR) in the Cortex-M0 processor is a read-only register (Table. 9.9). It determines the double word stack alignment behavior and the trapping of unaligned access. On the ARMv6-M architecture, such as the Cortex-M0 processor, these behaviors are fixed and not configurable. This register is included to make it compatible to ARMv7-M architecture such as the Cortex-M3 processor. On the Cortex-M3 processor, these two behaviors are controllable.

Table 9.9: Configuration and Control Register (0xE000ED14)

Bits	Field	Type	Reset Value	Descriptions
31:10	Reserved	—	—	Reserved.
9	STKALIGN	RO	1	Double word exception stacking alignment behavior is always used.
8:4	Reserved	—	—	Reserved.
3	UNALIGN_TRP	RO	1	Instruction trying to carry out an unaligned access always causes a fault exception.
2:0	Reserved	—	—	Reserved.

The STKALIGN bit is set to 1, indicating that when exception stacking occurs, the stack frame is always automatically aligned to a double-word-aligned memory location. The UNALIGN_TRP bit is set to 1, indicating that when an instruction attempts to carry out an unaligned transfer, a fault exception will result. Users of CMSIS-compliant device drivers can access to the Configuration Control Register using the register name "SCB->CCR."

## CHAPTER 10

Operating System Support Features

Overview of the OS Support Features

The Cortex-M0 processor includes a number of features that target the embedded operating system (OS). These include the following:

- *A SysTick timer.* This 24-bit down counter can be used to generate a SysTick exception at regular intervals.
- *A second stack pointer called the process stack pointer.* This feature allows the stack of the applications and the OS kernel to be separated.
- *An SVC exception and SVC instruction.* Applications use the SVC to access OS services via the exception mechanism.
- *A PendSV exception.* The PendSV can be used by an OS, device drivers, or the application to generate service requests that can be deferred.

This chapter describes each of these features and provides some example usages.

Why Use an Embedded OS?

When the term "operating system" is mentioned, most people will first think of desktop operating systems like Windows and Linux. These desktop OSs require a powerful processor, a large amount of memory, and other hardware features to run. For embedded devices, the type of OSs being used are very different. Embedded operating systems can run on very low power microcontrollers with a small amount of memory (relative to desktop machines), and they run at a much lower clock frequency. For example, the Keil RTX, which will be covered later in this chapter, requires around 4 KB of program code space and around 0.5 KB of SRAM. Very often these embedded systems do not even have a display or keyboard, although you can add some display interfaces and input devices and access these input and output interfaces via application tasks running within the OS.

In embedded applications, an OS is normally used for managing multiple tasks. In this situation, the OS divides the processor time into a number of time slots and executes different tasks in each slot. At the end of each time slot, the OS task scheduler is executed, and then the execution might be switched to a different task at the beginning of the next time slot. This switching of tasks is commonly known as context switching (Figure 10.1).

The Definitive Guide to the ARM Cortex-M0. DOI: 10.1016/B978-0-12-385477-3.10010-2

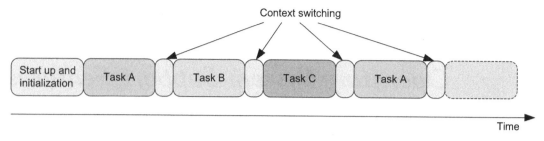

Figure 10.1:
Multitasking and context switching.

The length of each time slot depends on the hardware as well as the OS design. Some embedded OSs switch tasks several hundred times per second.

Some embedded OSs also define priority levels for each task so that a high-priority task is executed before lower-priority tasks. If the task has a higher priority than others, an OS might execute the task for a number of time slots continuously until the task reaches an idle state. Note that the priority definition in an OS is completely separate from the exception priority (for example, the interrupt priority level). The definition of task priority is based on the OS design and varies among different OSs.

Besides supporting multitasking, embedded OSs can also provide the functions of resource management, memory management, power management, and an application programming interface (API) for accessing peripherals, hardware, and communication channels (Figure 10.2).

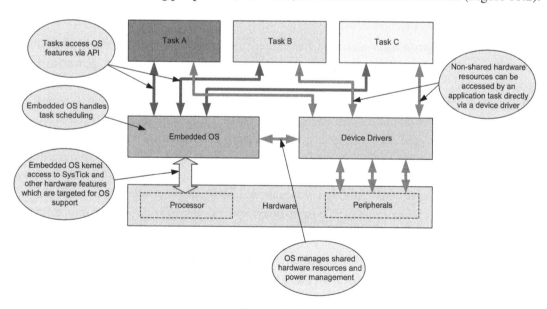

Figure 10.2:
Example roles of an embedded OS.

Use of an embedded OS is not always beneficial, because it requires extra program memory for the OS kernel and increases overhead in execution cycles. Most simple applications do not require an embedded OS. However, in complex embedded applications that demand the execution of tasks in parallel, using an OS can make the software design much easier and reduce the chance of a system design error.

A number of embedded OSs are already available for the Cortex-M0 processor. For example, the Keil Microcontroller Development Kit (MDK) provides a RTX kernel, which is easy to use and free of charge. In addition, embOS from SEGGER (www.segger.com), μC/OS-II and μC/OS-III from Micriμm (micrium.com), ThreadX from Express Logic (www.rtos.com), and μCLinux from the open source community are also supported on the Cortex-M0 processor. The list is growing rapidly, so by the time you read this book, some more OSs will be available for Cortex-M0 microcontrollers.

Because the Cortex-M0 processor does not have a memory management unit (MMU), it cannot run an embedded OS which requires virtual address capability like Windows CE or Symbian OS. The normal Linux OS also requires an MMU, so it cannot work on the Cortex-M0, but the μCLinux is a special version of Linux targeted at embedded devices without an MMU, and therefore the μCLinux can work on a Cortex-M0 processor.

In the next few sections, we will cover a number of features included on the Cortex-M0 to support efficient OS operations. The OS support features in the Cortex-M0 processor are also available in other ARM Cortex-M profile processors including the Cortex-M3 and the Cortex-M1. This consistency allows embedded OSs to be ported between these processors easily.

The SysTick Timer

To allow an OS to carry out context switching regularly to support multiple tasking, the program execution must be interrupted by a hardware source like a timer. When the timer interrupt is triggered, an exception handler that handles OS task scheduling is executed. The handler might also carry out other OS maintenance tasks. For the Cortex-M0 processor, a simple timer called the SysTick is included to generate this regular interrupt request.

The SysTick timer is a 24-bit down counter. It reloads automatically after reaching zero, and the reload value is programmable. When reaching zero, the timer can generate a SysTick exception (exception number 15). This exception event triggers the execution of the SysTick exception handler, which is a part of the OS.

For systems that do not required an OS, the SysTick timer can be used for other purposes like time keeping, timing measurement, or as a interrupt source for tasks that need to be executed regularly. The SysTick exception generation is controllable. If the exception generation is disabled, the SysTick timer can still be used by polling, for example, by checking the current value of the counter or polling of the counter flag.

SysTick Registers

The SysTick counter is controlled by four registers located in the System Control Space memory region (Figure 10.3 and Tables 10.1, 10.2, 10.3, and 10.4)

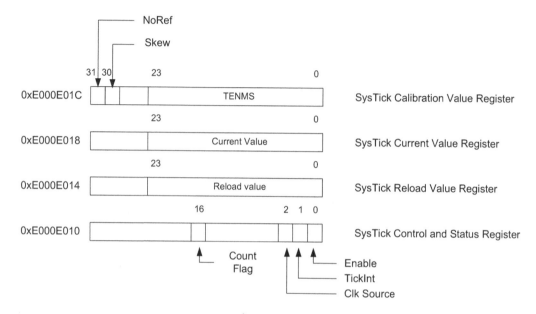

Figure 10.3:
SysTick registers.

Table 10.1: SysTick Control and Status Register (0xE000E010)

Bits	Field	Type	Reset Value	Descriptions
31:17	Reserved	—	—	Reserved.
16	COUNTFLAG	RO	0	Set to 1 when the SysTick timer reaches zero. Clear to 0 by reading of this register.
15:3	Reserved	—	—	Reserved.
2	CLKSOURCE	R/W	0	Value of 1 indicates that the core clock is used for the SysTick timer. Otherwise a reference clock frequency (depending on MCU design) would be used.
1	TICKINT	R/W	0	SysTick interrupt enabler. When this bit is set, the SysTick exception is generated when the SysTick timer counts down to 0.
0	ENABLE	R/W	0	When set to 1 the SysTick timer is enabled. Otherwise the counting is disabled.

Table 10.2: SysTick Reload Value Register (0xE000E014)

Bits	Field	Type	Reset Value	Descriptions
31:24	Reserved	—	—	Reserved.
23:0	RELOAD	R/W	Undefined	Specify the reload value of the SysTick timer.

Table 10.3: SysTick Current Value Register (0xE000E018)

Bits	Field	Type	Reset Value	Descriptions
31:24	Reserved	—	—	Reserved.
23:0	CURRENT	R/W	Undefined	On read returns the current value of the SysTick timer. Write to this register with any value to clear the register and the COUNTFLAG to 0. (This does not cause SysTick exception to generate.)

Table 10.4: SysTick Calibration Value Register (0xE000E01C)

Bits	Field	Type	Reset Value	Descriptions
31	NOREF	RO	—	If it is read as 1, it indicates SysTick always uses a core clock for counting, as no external reference clock is available. If it is 0, then an external reference clock is available and can be used. The value is MCU design dependent.
30	SKEW	RO	—	If set to 1, the TENMS bit field is not accurate. The value is MCU design dependent.
29:24	Reserved	—	—	Reserved.
23:0	TENMS	RO	—	Ten millisecond calibration value. The value is MCU design dependent.

For users of CMSIS-compliant device driver libraries, the SysTick registers can be accessed by the register definitions shown in Table 10.5 and included in CMSIS.

Table 10.5: SysTick Register Names in CMSIS

Name	Register	Address
SysTick->CTRL	SysTick Control and Status Register	0xE000E010
SysTick->LOAD	SysTick Reload Value Register	0xE000E014
SysTick->VAL	SysTick Current Value Register	0xE000E018
SysTick->CALIB	SysTick Calibration Value Register	0xE000E01C

Setting up SysTick

Because the reload value and current values of the SysTick timer are undefined at reset, the SysTick setup code needs to be in a certain sequence to prevent unexpected results (Figure 10.4).

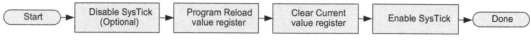

Figure 10.4:
Setup sequence for SysTick timer.

For users of CMSIS-compliant device driver libraries, a function called *SysTick_Config (uint32_t ticks)* is available that enables a SysTick exception to occur regularly. For example,

```
SysTick_Config (1000);   //Use CMSIS function to setup SysTick exception for every 1000 cpu
                         //cycles.
```

Alternatively, you can also program the SysTick by accessing the SysTick registers directly:

```
SysTick->CTRL = 0;      // Disable SysTick
SysTick->LOAD = 999;    // Count down from 999 to 0
SysTick->VAL  = 0;      // Clear current value to 0
SysTick->CTRL = 0x7;    // Enable SysTick, enable SysTick
                        // exception and use processor clock
```

The SysTick timer can be used with a polling method or by interrupt. Programs that use a polling method can read the SysTick Control and Status Registers to detect the COUNT-FLAG (bit 16). If the flag is set, the SysTick counter has counted down to 0.

For example, if we want to toggle a LED connected to an output port every 100 CPU cycles, we can develop a simple application that uses the SysTick timer with a polling loop. The polling loop reads the SysTick Control and Status Register and toggles the LED when a 1 is detected in the counter flag. Because the flag is cleared automatically when the SysTick Control and Status Register is read, there is no need to clear the counter flag (Figure 10.5).

You might wonder why the value of 99, and not 100, is written into the Reload value register. This is because the counter counts from 99 down to 0. To obtain a regular counter reload, or exception, from the SysTick timer, the reload value should be programmed to the interval value minus 1.

The SysTick Calibrate Value Register can be used to provide information for calculating the desired reload value for the SysTick. If a timing reference is available on the microcontroller, the TENMS field in the register may provide the tick count for 10 milliseconds. In some Cortex-M0 microcontrollers, a reference value is not available. In such cases, the NOREF bit is set to 1 to reflect this.

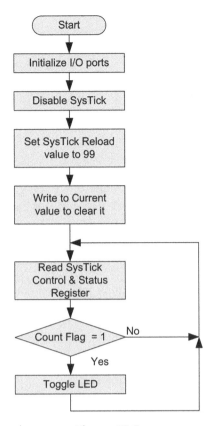

Figure 10.5:
A simple example of using SysTick with polling.

Users of CMSIS-compliant device driver libraries can also use a variable called *System-Frequency* (for CMSIS version 1.0 to version 1.2) or *SystemCoreClock* (for CMSIS version 1.3) for reload value calculation. This software variable can be linked to clock control functions in the device driver libraries to provide the actual processor clock frequency being used.

Using SysTick Timer for Timing Measurement

If neither the application code nor the OS uses the SysTick timer, it can be used as a simple solution for measuring the number of clock cycles required for a processing task. For example, the following setup code can be used to carry out timing measurements if the number of clock cycles is fewer than 16.7 million:

```
unsigned int START_TIME, STOP_TIME, DURATION;
SysTick->CTRL = 0;   // Disable SysTick
SysTick->LOAD = 0xFFFFFF; // Count down from maximum value
```

```
SysTick->VAL  = 0;   // Clear current value to 0
SysTick->CTRL = 0x5; // Enable SysTick, and use processor clock
while (SysTick->VAL==0);   // Wait until SysTick reloaded
START_TIME = SysTick->VAL; // Read start time value
processing();                 // Processing function being measured
STOP_TIME = SysTick->VAL;  // Read stop time value
SysTick->CTRL = 0;            // Disable SysTick
if ((SysTick->CTRL & 0x10000)==0) // if no overflow

  DURATION = START_TIME - STOP_TIME; // Calculate total cycles
else

  printf ("Timer overflowed\n");
```

Because the SysTick is a down counter, the value of START_TIME is larger than the value of STOP_TIME. The preceding example code assumes that the SysTick does not overflow during the execution of the processing task. If the duration is more than 16.7 million cycles ($2^{24} = 16777216$), a SysTick interrupt handler has to be used to count the number of times the timer overflows.

Besides generating regular interrupts and timing measurement, the SysTick timer can also be used to produce short delays. An example of a single-shot timer operation using the SysTick is covered in Chapter 15.

Process Stack and Process Stack Pointer

The Cortex-M0 processor has two stack pointers (SPs): the main stack pointer (MSP) and the process stack pointer (PSP). Both are 32-bit registers and can be referenced as R13, but only one is used at one time. The MSP is the default stack pointer and is initialized at reset by loading the value from the first word of the memory. For simple applications, we can use the MSP all the time. In this case, we only have one stack region. For systems with higher reliability requirements, usually with an embedded OS involved, we can define multiple stack regions: one for the OS kernel and exceptions and the others for different tasks (Figure 10.6).

The main reason for separating the kernel stack from the task's stacks is to enable easier context switching. During context switching, the stack pointer for the exiting application task will have to be saved and the stack pointer will change to point to the next task's stack. Very often the kernel code also requires a stack to operate. The kernel stack has to be separated to prevent data from being lost during stack pointer updates.

The separation of memory for tasks reduces the chance of a stack error by one task corrupting another task's stack or the OS kernel's stack. Although a rogue task can corrupt data in the RAM (e.g., stack overflow), an embedded OS can check the stack pointer value during context switching to detect stack errors. As a result, it can help improve the reliability of an embedded system.

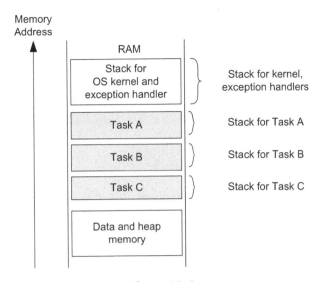

Figure 10.6:
Separate memory ranges for OS and application tasks.

In a typical OS environment, the MSP and PSP usages are as follows:

- MSP, for the OS kernel and exception handlers
- PSP, for application tasks

The OS kernel has to keep track of the stack pointer values for each task during context switching, and it will switch over the PSP value to allow each task to have its own stack (Figure 10.7).

As covered in Chapter 3, the selection of the pointer is determined by the current mode of the Cortex-M0 processor and the value of the CONTROL register. When the processor comes out of reset, it is in Thread mode, the CONTROL register's value is 0, and the MSP is selected as the default SP.

From the default state, the current stack pointer selection can be changed to use PSP by programming the CONTROL register. Note that an ISB instruction should be used (an ARMv6-M architectural recommendation) after programming the CONTROL register bit 1 to 1. You can also switch back to MSP by clearing bit 1 of the CONTROL register (Figure 10.8).

If an exception occurs, the processor will enter Handler mode and the MSP will be selected. The stacking process that pushes R0-R3, R12, LR, PC, and xPSR can be carried out using either MSP or PSP, depending on the value of the CONTROL register before the exception.

When an exception handler is completed, the PC is loaded with the EXC_RETURN value. Depending on the value of the lowest 4 bits of the EXC_RETURN, the processor can return to Thread mode with MSP selected, Thread mode with PSP selected, or Handler mode with MSP

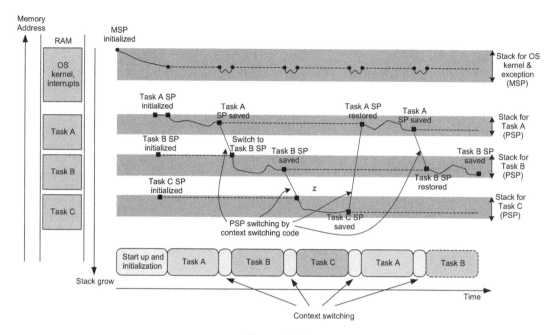

Figure 10.7:
MSP and PSP activities with a simple OS running three tasks.

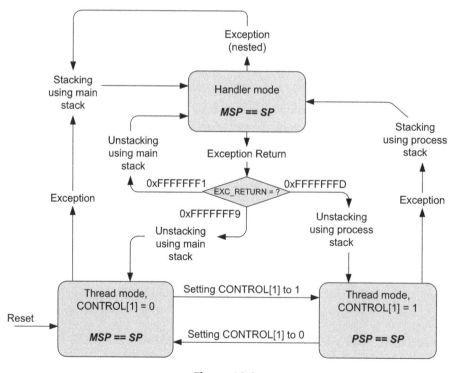

Figure 10.8:
Switching of stack pointer selection by software or exception entry/exit.

selected. The value of the CONTROL register is updated to match bit 2 of the EXC_RETURN value.

The value of MSP and PSP can be accessed using the MRS and MSR instructions. In general, changing the value of the currently selected stack pointer in C language is a bad idea because access to local variables and function parameters can be dependent on the stack pointer value. If it is changed, the values of these variables cannot be accessed.

If you are using CMSIS-compliant device driver libraries, you can access the value of the MSP and PSP with the functions presented in Table 10.6.

Table 10.6: CMSIS Functions for Accessing MSP and PSP

Functions	Description
uint32_t __get_MSP(void)	Read the current value of the main stack pointer
void __set_MSP(uint32_t topOfMainStack)	Set the value of the main stack pointer
uint32_t __get_PSP(void)	Read the current value of the process stack pointer
void __set_PSP(uint32_t topOfProcStack)	Set the value of the process stack pointer

To implement the context switching sequence as in Figure 10.7, you can use the procedures described in Figure 10.9 for OS initialization, and Figure 10.10 for context switching.

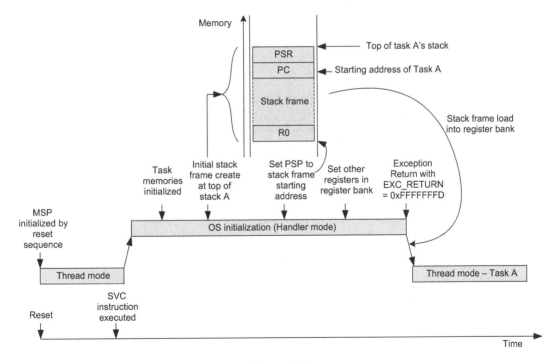

Figure 10.9:
Initialization of stack for a task in a simple OS.

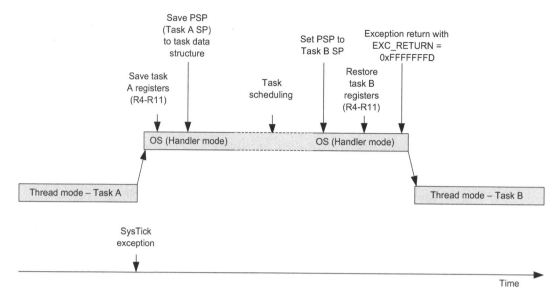

Figure 10.10:

Example of context switching from one task to another task in a simple OS.

There are various ways to implement an embedded OS; the illustration in Figure 10.9 and Figure 10.10 are only an example.

SVC

To build a complete OS, we need a few more features from the processor. The first one is a software interrupt mechanism to allow tasks to trigger a dedicated OS exception. In ARM processors this is called supervisor call (SVC). The SVC is an instruction as well as an exception type. When the SVC instruction is executed, the SVC exception is triggered and the processor will execute the SVC exception handler immediately, unless an exception with a higher or same priority is currently being handled.

An SVC can be used as a gateway for applications to access a system service provided by the OS (Figure 10.11). An application can pass parameters to the SVC handler inside the OS for different services.

In some development environments, SVCs can make the access to OS functions easier, as the accesses to OS functions do not require any address information. Therefore, the OS and the applications can be compiled and delivered separately. The application can interact with the OS by calling the correct OS service and providing the required parameters.

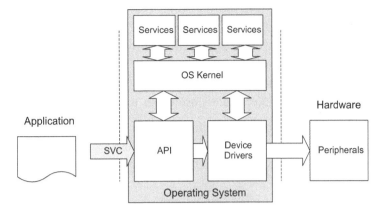

Figure 10.11:
SVC as a gateway to system services in the OS.

The SVC instruction contains an 8-bit immediate value. The SVC handler can extract this immediate value to determine which OS service is required. The syntax for SVC instruction in assembly is

```
SVC # 0×3 ;Call SVC service 3
```

Traditional ARM development tools support a slightly different syntax (without the "#" sign):

```
SVC 0×3 ;Call SVC service 3
```

This syntax can still be used.

In C language, there is no standard way to access SVC functions. In ARM development tools (including RealView Development Suite and Keil MDK), you can use the __svc keyword. This topic is covered in more depth in Chapter 18.

If you were a user of ARM7TDMI or similar classic ARM processors, you might notice that the SVC is similar to the SWI instruction on these processors. In fact, the binary encoding of SVC is identical to the SWI Thumb instruction. However, this instruction is renamed to SVC in newer architectures, and the SVC handler code is different from the SWI handler code for the ARM7TDMI.

Because of the interrupt priority behavior of the Cortex-M0 processor, the SVC can only be used in Thread mode or exception handlers that have a lower priority than the SVC itself. Otherwise, a hard fault exception is generated. As a result, you cannot use the SVC instruction inside another function accessed by an SVC, as it has the same priority level. You also cannot use an SVC inside an NMI handler or hard fault handler.

PendSV

The PendSV is an exception type that can be activated by setting a pending status bit in the NVIC. Unlike SVC, PendSV activation can be deferred. Therefore, you can set its pending status even when you are running an exception handler with a higher priority level than the PendSV exception. The PendSV exception is useful for the following functions:

- The context switching operation in an embedded OS
- Separating an interrupt processing task into two halves:
  - The first half must be executed quickly and is handled by a high-priority interrupt service routine
  - The second half is less timing critical and can be handled by a deferred PendSV handler with a lower priority; therefore, it allows other high-priority interrupt requests to be processed quickly

The second use of PendSV is fairly easy to understand, and more details of this usage are covered in Chapter 18 with a programming example. The use of PendSV for context switching is more complex. In a typical OS design, context switching can be triggered by the following:

- Task scheduling during a SysTick handler
- A task waiting for data/events calling an SVC service to swap in another task

Usually the SysTick exception is set up as a high-priority exception. As a result, the SysTick handler (part of the OS) can be invoked even if another interrupt handler is running. However, the actual context switching should not be carried out while an interrupt service routine is running. Otherwise, the interrupt service would be broken into multiple parts. Traditionally, if the OS detects that an interrupt service routine is running, it will not carry out the context switching and will wait until next OS tick (as shown in Figure 10.12).

By deferring the context switching to the next SysTick exception, the IRQ handler can complete the execution. However, if the IRQ is generated regularly and the IRQ rate coincides with the pattern of task switching activities, then some tasks might receive a larger share of processing time, or in some cases the context switching cannot be carried out for a long period, for example, if the IRQ occurs too frequently.

To solve this problem, the actual context switching process can be separated from the SysTick handler and implemented in a low-priority PendSV handler. By setting the priority of the PendSV exception to the lowest priority level, the PendSV handler can only be executed when no other interrupt service is running.

Take the activities in Figure 10.13 as an example. The SysTick exception periodically triggers the OS task scheduler for task scheduling. The OS task scheduler sets the pending state of the PendSV exception before exiting the exception. If no IRQ handler is running, the PendSV handler starts immediately after the SysTick exception exits and carries out the context

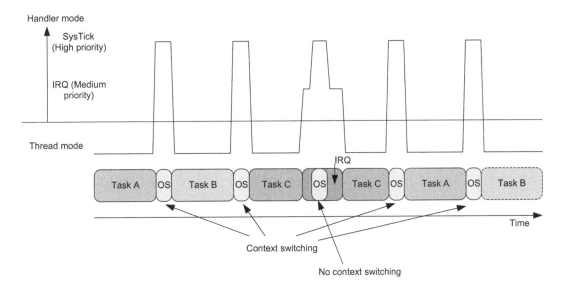

Figure 10.12:
Without PendSV. Context switching is not carried out if the OS detects that an ISR is running.

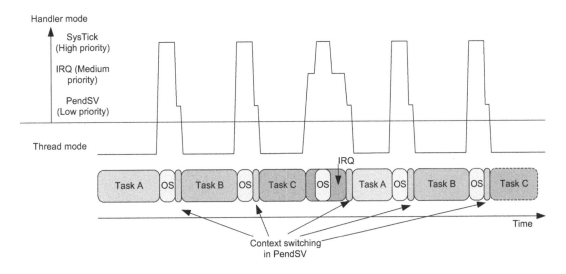

Figure 10.13:
With PendSV. Context switching can be carried out after the IRQ handler has completed its task.

switching. If an IRQ is running when the SysTick exception occurs, then the PendSV exception cannot start until the IRQ handler finishes, because the PendSV is programmed to the lowest priority level. When all the IRQ activities have been completed, the PendSV handler can then carry out the required context switching.

## CHAPTER 11

Low-Power Features

Low-Power Embedded System Overview

There are many types of low-power embedded systems, and different systems can have different low-power requirements. Typically, we can summarize these requirements into the categories outlined in Table 11.1.

In practice, most low-power embedded products will have to consider more than one of these low-power factors. For example, some of the embedded systems operate continuously and can be switched off completely when not in use. In this case, the low operating power would be the most important factor, whereas some systems might need to be in standby mode most of the time and only wake up to execute a program for a short period. In such a case, the low standby power would be the more critical requirement, especially if the battery life of the product is important.

As embedded products are getting more and more complex, the computation capability of the processor is becoming more important. Nowadays a lot of embedded systems are interrupt driven (Figure 11.1); when there is no interrupt request, no data processing is required and the processor can enter sleep mode. When a peripheral requires servicing, it generates an interrupt

Table 11.1: Typical Requirements of Low-Power Embedded Systems

Requirements	Typical Low-Power Considerations
Low operating power	During operation, the power consumption of a microcontroller is dominated by the transistor switching current. This includes the processor, memory, peripherals, clocking circuits, and other analog circuits on the chip.
Low standby power	The power consumption during standby is mostly caused by leakage circuits, clocking circuits, active peripherals, the analog systems and RAM retention power.
High energy efficiency	In many applications, the ratio between the processing capability and the power consumption is equally important. This figure can be measured based on a benchmark like Dhrystone (DMIPS), for example, in units of DMIPS/mW.
Wakeup latency	After entering sleep mode, there might be a short delay before the processor can resume operation when an interrupt request arrives. In some applications, the wakeup latency could be critical, and a system developer may need to decide if a lower power sleep mode is used, which might increase the wakeup latency, or if a basic sleep mode should be used, which provides a shorter wakeup latency but generates a higher standby current.

The Definitive Guide to the ARM Cortex-M0. DOI: 10.1016/B978-0-12-385477-3.10011-4

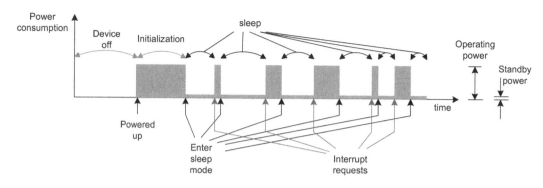

Figure 11.1:
Activities in an interrupt-driven system.

request and wakes up the processor. The processor then handles the required processing and returns to sleep. The better processing capability the processor has, the more time it can stay in sleep mode and hence improve battery life.

If you are using a slow processor for an interrupt-driven system, the interrupt service routine could take a lot longer to run and increase the duty cycles of the system. But if you are using a very powerful processor, despite being able to reduce the duty cycles, you might end up with much more operating power and standby power, which will increase power consumption. Depending on the data processing requirements, different applications require different processors for the optimum balance between performance and power (Figure 11.2).

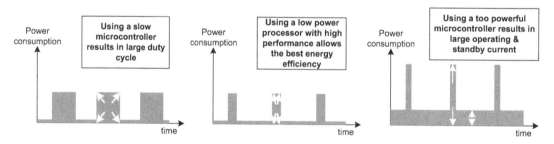

Figure 11.2:
Select the best processor for your low-power application.

For this reason, there are different types of ARM processors for different application requirements. The ARM processors are designed with low-power embedded systems in mind. Different ARM processors are optimized for different groups of applications based on processing requirements. The Cortex-M0 processor is developed to target small embedded systems with ultra-low-power requirements and mixed-signal applications, where the processor design complexity is limited by various constraints in mixed-signal semiconductor technologies.

Low-Power Advantages of the Cortex-M0 Processor

So how does the Cortex-M0 processor satisfy the ultra-low-power requirements in an embedded application?

First, the design of the Cortex-M0 processor is very small. For a minimum implementation, the design is only 12K gates. It is smaller than many 16-bit processors and much smaller than other 32-bit processors on the market. This reduces both operating power and static power (caused by the leakage current of transistors).

Second, the design of the Cortex-M0 processor utilizes many low-power design techniques to reduce the operating power consumption. ARM has more than 20 years of low-power processor design experience. The design of the Cortex-M0 has been extensively reviewed to ensure that the low-power design measures are utilized for lowest power consumption. This allows the operating power of the processor to be reduced.

Even with a tiny footprint, the Cortex-M0 is still able to deliver a performance that is much higher than 8-bit and 16-bit systems that have the same clock rate. The Cortex-M0 processor also has the best in class energy efficiency in the 32-bit processor market because of its small size and low power consumption.

The high code density of the Cortex-M processors also lower power by reducing the size of the flash memory required. In modern microcontroller designs, the majority of the silicon area is occupied by flash memory, SRAM, and peripherals. With traditional 8-bit microcontroller architectures, the lack of flexible addressing modes, the heavy use of accumulator registers, and limited data paths often result in large program code (Figure 11.3). By switching to Cortex-M0 or other Cortex-M processors, the program size can be greatly reduced and hence you can use a microcontroller with a smaller flash to save power and cost.

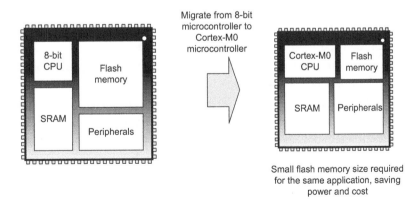

Figure 11.3:
Switch from 8-bit microcontrollers to the Cortex-M microcontroller can reduce flash size.

In addition, when comparing the Cortex-M0 microcontroller products to 8-bit and 16-bit microcontroller products, the Cortex-M0 offers much higher performance. As a result, an embedded developer can reduce power consumption by taking the following steps:

- Reducing the operating frequency of the device
- Reducing the duty cycle of the device (putting the core into sleep for a longer time)

Either way, or by combining both approaches, the energy efficiency characteristics of the Cortex-M0 processor allow longer battery life in portable products compared with 8-bit and 16-bit microcontrollers. Besides extending battery life, the low-power capability of the Cortex-M0 processor also:

- Reduces noise to allow better accuracy in analog applications like sensors
- Reduces interference in wireless and radio frequency applications
- Allows a simpler and cost-effective power supply design for the system

Overview of the Low-Power Features

A number of low-power features are available in the Cortex-M0 processor. In addition, microcontroller vendors usually also implement a number of low-power modes in their Cortex-M0 microcontroller products. This chapter focuses mostly on the low-power features provided by the Cortex-M0 processor. Details for microcontroller specific low-power features are usually available in user manuals or application notes available from the microcontroller vendor web sites or in example software packages.

In general the Cortex-M0 processor includes the following low-power features:

- *Two architectural sleep modes: normal sleep and deep sleep.* The sleep modes can be further extended with vendor-specific sleep control features. Within the processor, both sleep modes behave similarly. However, the rest of the microcontroller can typically reduce power by applying different methods to these two modes.
- *Two instructions for entering sleep modes. WFE (Wait for Event)* and *WFI (Wait for Interrupt).* Both can be used with normal sleep and deep sleep.
- *Sleep-on-exit (from exception) feature.* This feature allows interrupt-driven applications to stay in sleep mode as often as possible.
- *Optional Wakeup Interrupt Controller (WIC).* This optional feature allows the clocks of the processor to be completely removed during deep sleep. When this feature is used with state retention technology, found in certain modern silicon implementation processes, the processor can enter a power-down state with extremely low leakage power, and it is still able to wake up and resume operations almost immediately.
- *Low-power design implementation.* Various design techniques were used to reduce the power consumption as much as possible. Because the gate count is also very low, the static leakage power of the processor is tiny compared to most other 32-bit microcontrollers.

In addition, various characteristics of the Cortex-M0 also help reduce power consumption:

- *High performance.* The Cortex-M0 processor performance is several times higher than many popular 16-bit microcontrollers. This allows the same computational tasks to be carried out in shorter time, and the microcontroller can stay in sleep mode for longer period of time. Alternately, the microcontroller can run at a lower frequency to perform the same required processing to reduce power.
- *High code density.* By having a very efficient instruction set, the required program size can be reduced; as a result, you can use a Cortex-M0 microcontroller with smaller flash memory to reduce power consumption and cost.

Because the processor is only a small part of a microcontroller, to get the best energy efficiency and maximum battery life out of a microcontroller product, it is necessary to understand not only the processor but also the rest of the microcontroller. Most microcontroller vendors provide application note and software libraries to make this easier.

Sleep Modes

Most microcontrollers support at least one type of sleep mode to allow the power consumption to be reduced when no processing is required. In the Cortex-M0 processor, sleep mode support is included as part of the processor architecture.

The Cortex-M0 processor has two sleep modes:

- Normal sleep
- Deep sleep

The exact meaning and behaviors of these sleep modes depends on the implementation of the microcontroller. Microcontroller vendors can use various power-saving measures to reduce the power of the microcontroller during sleep. They can also further extend the sleep modes by adding extra power control capability. Typically, the following methods are used to reduce power during sleep:

- Stopping some or all of the clock signals
- Reducing the clock frequency to some parts of the microcontroller
- Reducing voltage to various parts of the microcontroller
- Turning off the power supply to some parts of the microcontroller

The sleep modes can be entered by three different methods:

- Execution of a Wait-for-Event (WFE) instruction
- Execution of a Wait-for-Interrupt (WFI) instruction
- Using the Sleep-on-Exit feature (this will be covered in detail later)

Whether the normal sleep mode or the deep sleep mode will be used is determined by a control bit called SLEEPDEEP. This bit is located in the System Control Register (SCR) of the System Control Block (SCB) region, which contains the control bits for the low-power features of the Cortex-M0 processor. Users of CMSIS-compliant device drivers can access to the System Control Register using the register name "SCB->SCR" (Table 11.2).

Different sleep modes and different sleep operation types can result in various combinations, as shown in Figure 11.4.

Table 11.2: System Control Register (0xE000ED10)

Bits	Field	Type	Reset Value	Descriptions
31:5	Reserved	—	—	Reserved.
4	SEVONPEND	R/W	0	When set to 1, an event is generated for each new pending of an interrupt. This can be used to wake up the processor if Wait-for-Event (WFE) sleep is used.
3	Reserved	—	—	Reserved.
2	SLEEPDEEP	R/W	0	When set to 1, deep sleep mode is selected when sleep mode is entered. When this bit is 0, normal sleep mode is selected when sleep mode is entered.
1	SLEEPONEXIT	R/W	0	When set to 1, enter sleep mode (Wait-for-Interrupt) automatically when exiting an exception handler and returning to thread level. When set to 0, this feature is disabled.
0	Reserved	—	—	Reserved.

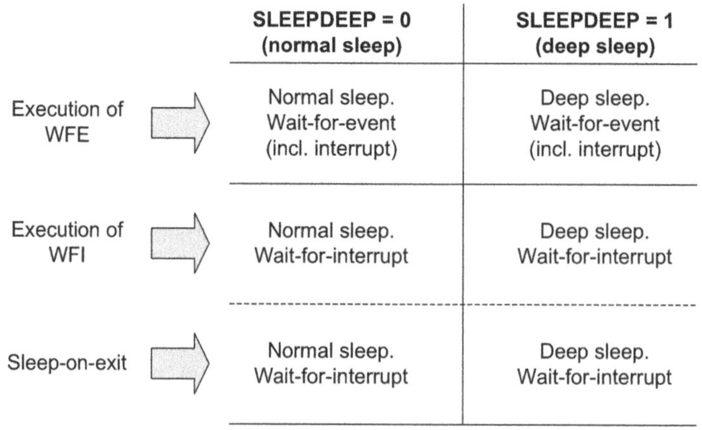

	SLEEPDEEP = 0 (normal sleep)	SLEEPDEEP = 1 (deep sleep)
Execution of WFE	Normal sleep. Wait-for-event (incl. interrupt)	Deep sleep. Wait-for-event (incl. interrupt)
Execution of WFI	Normal sleep. Wait-for-interrupt	Deep sleep. Wait-for-interrupt
Sleep-on-exit	Normal sleep. Wait-for-interrupt	Deep sleep. Wait-for-interrupt

Figure 11.4:
Combinations of sleep modes and sleep entering methods.

Wait-for-Event (WFE) and Wait-for-Interrupt (WFI)

There are two instructions that can cause the Cortex-M0 processor to enter sleep: WFE and WFI (Table 11.3). The WFE can be awakened by interrupt requests as well as events, whereas WFI can be awakened by interrupt requests or debug requests only.

Table 11.3: WFE and WFI Characteristics

Sleep Type	Wakeup Descriptions
WFE	• Wake up when an interrupt occurs and requires processing, or wake up when an event occurs (including debug requests), or the processor does not enter sleep because an event occurred before the WFE instruction executed, or termination of sleep mode by reset.
WFI	• Wake up when an interrupt occurs and requires processing, or wake up when there is a debug request, or termination of sleep mode by reset.

Wait for Event (WFE)

When WFE is used to enter sleep, it can be awakened by interrupts or a number of events, including the following:

- New pending interrupts (only when the SEVONPEND bit in System Control Register is set)
- External event requests
- Debug events

Inside the Cortex-M0 processor, there is a single bit event register. When the processor is running, this register can be set to 1 when an event occurs, and this information is stored until the processor executes a WFE instruction. The event register can be set by any of the following events:

- The arrival of an interrupt request that needs servicing
- An exception entrance and exception exit
- New pending interrupts (only when SEVONPEND bit in System Control Register is set), even if the interrupts are disabled
- An external event signal from on-chip hardware (device specific)
- Execution of a Send Event (SEV) instruction
- Debug event

When multiple events occur while the processor is awake, they will be treated as just one event.

This event register is cleared when the stored event is used to wake up the processor from a WFE instruction. If the event register was set when the WFE instruction is executed, the event register will be cleared and the WFE will be completed immediately without entering sleep.

If the event register was cleared when executing WFE, the processor will enter sleep, and the next event will wake up the processor, but the event register will remain cleared (Figure 11.5).

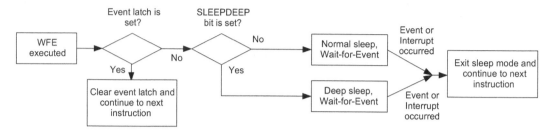

Figure 11.5:
WFE operation.

The WFE is useful for reducing power in polling loops. For example, a peripheral with event generation function can work with the WFE so that the processor wakes up upon completion of peripheral's task. As shown in Figure 11.6.

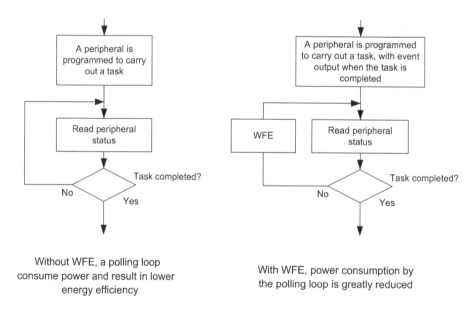

Figure 11.6:
WFE usage.

Because the processor can be awakened by different events, it must still check the peripheral status after being awakened to see if the task has completed.

If the SEVONPEND bit in the SCR is set, any new pending interrupts generate an event and wake up the processor. If an interrupt is already in pending state when the WFE is entered, a new interrupt request for the same interrupt does not cause the event to be generated and the processor will not be awakened.

Wait for Interrupt (WFI)

The WFI instruction can be awakened by interrupt requests that are a higher priority than the current priority level, or by debug requests (Figure 11.7).

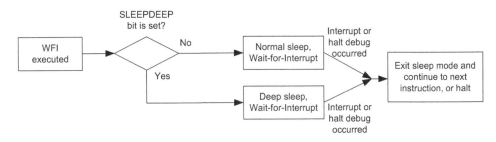

Figure 11.7:
WFI operation.

There is one special case of WFI operation. During WFI sleep, if an interrupt is blocked by PRIMASK but otherwise has a higher priority than the current interrupt, it can still wake up the processor, but the interrupt handler will not be executed until the PRIMASK is cleared.

This characteristic allows software to turn off some parts of the microcontroller (e.g., the peripheral bus clock), and the software can turn it back on after waking up before executing the interrupt service routine.

Wakeup Conditions

When a WFI instruction is executed or when the processor enters sleep mode using the Sleep-on-Exit feature, the processor stops instruction execution and wakes up when a (higher priority) interrupt request arrives and needs to be serviced. If the processor enters sleep in an exception handler, and if the newly arrived interrupt request has the same or lower priority as the current exception, the processor will not wake up and will remain in a pending state. The processor can also be awakened by a halt request from debugger or by a reset.

When the WFE instruction is executed, the action of the processor depends on the current state of an event latch inside the processor:

- If the event latch was set, it will be cleared and the WFE completes without entering sleep.
- If the event latch was cleared, the processor will enter sleep mode until an event takes place.

An event could be any of the following:

- An interrupt request arriving that needs servicing
- Entering or leaving an exception handler
- A halt debug request
- An external event signal from on-chip hardware (device specific)
- If the Send-Event-on-Pend (SEVONPEND) feature is enabled and a new pending interrupt occurs
- Execution of the Send Event (SEV) instruction

The event latch inside the processor can hold an event that happened in the past, so an old event can cause the processor to wake up from a WFE instruction. Therefore, usually the WFE is used in an idle loop or polling loop, as it might or might not cause entering of sleep mode.

WFE can also be awakened by interrupt requests if they have a higher priority than the current interrupt's priority level or when there is a new pending interrupt request and the SEVON-PEND bit is set. The SEVONPEND feature can wake up the processor from WFE sleep even if the priority level of the newly pended interrupt is at the same or lower level than the current interrupt. However, in this case, the processor will not execute the interrupt handler and will resume program execution from the instruction following the WFE.

The wakeup conditions of the WFE and WFI instructions are illustrated in Table 11.4.

Table 11.4: WFI and WFE Sleep Wakeup Behavior

WFI Behavior	Wakeup	ISR Execution
PRIMASK cleared		
IRQ priority > current level	Y	Y
IRQ priority ≤ current level	N	N
PRIMASK set (interrupt disabled)		
IRQ priority > current level	Y	N
IRQ priority ≤ current level	N	N
WFE Behavior	**Wakeup**	**ISR Execution**
PRIMASK cleared, SEVONPEND cleared		
IRQ priority > current level	Y	Y
IRQ priority ≤ current level	N	N
PRIMASK cleared, SEVONPEND set to 1		
IRQ priority > current level	Y	Y
IRQ priority ≤ current level, or IRQ disabled(SETENA = 0)	Y	N
PRIMASK set (interrupt disabled), SEVONPEND cleared		
IRQ priority > current level	N	N
IRQ priority ≤ current level	N	N
PRIMASK set (interrupt disabled), SEVONPEND set to 1		
IRQ priority > current level	Y	N
IRQ priority ≤ current level	Y	N

The wake up behavior of Sleep-on-Exit is same as WFI sleep.

Some of you might wonder why when PRIMASK is set, it allows the processor to wake up but without executing the interrupt service routine (Figure 11.8). This arrangement allows the processor to execute system management tasks (for example, restore the clock to peripherals) before executing the interrupt service routine.

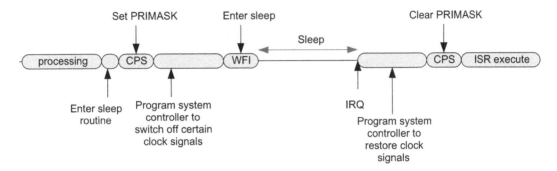

Figure 11.8:
Use of PRIMASK with sleep.

In summary, the differences between WFI and WFE included those described in Table 11.5.

Table 11.5: WFI and WFE Comparisons

	WFI and WFE
Similarities	Wake up on interrupt requests that are enabled and with higher priority than current level. Can be awakened by debug events. Can be used to produce normal sleep or deep sleep.
Differences	Execution of WFE does not enter sleep if the event register was set to 1, whereas execution of WFI always results in sleep. New pending of a disabled interrupt can wake up the processor from WFE sleep if SEVONPEND is set. WFE can be awakened by an external event signal. WFI can be awakened by an enabled interrupt request when PRIMASK is set.

Sleep-on-Exit Feature

One of the low-power features of the Cortex-M0 processor is the Sleep-on-Exit. When this feature is enabled, the processor automatically enters a Wait-for-Interrupt sleep mode when exiting an exception handler and if no other exception is waiting to be processed.

This feature is useful for applications where the processor activities are interrupt driven. For example, the software flow could be like the flowchart in Figure 11.9.

The resulting activities of the processor are illustrated in Figure 11.10.

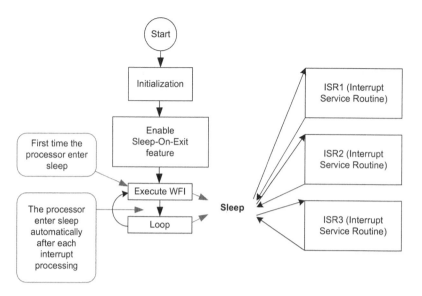

Figure 11.9:
Sleep-on-Exit program flow.

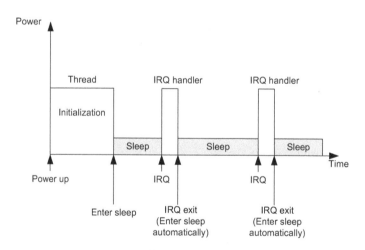

Figure 11.10:
Sleep-on-Exit operation.

The Sleep-on-Exit feature reduces the active cycles of the processor and reduces the energy consumed by the stacking and unstacking of processes between the interrupts. Each time the processor finishes an interrupt service routine and enters sleep, it does not have to carry out the unstacking process because it knows that these registers will have to be stacked again when another interrupt request arrives next time.

The Sleep-on-Exit feature is controlled by the SLEEPONEXIT bit in the System Control Register. Setting this bit in an interrupt-driven application is usually carried out as the last step of the initialization process. Otherwise, if an interrupt occurs during this stage, the processor might enter sleep during the initialization of the processor.

Wakeup Interrupt Controller

Designers of the Cortex-M0 microcontroller can optionally include a Wakeup Interrupt Controller (WIC) in their design. The WIC is a small interrupt detection logic that mirrors the interrupt masking function in the NVIC. The WIC allows the power consumption of the processor to be further reduced by stopping all the clock signals to the processor, or even putting the processor into a retention state. When an interrupt is detected, the WIC sends a request to a power management unit (PMU) in the microcontroller to restore power and clock signals to the processor, and then the processor can wake up and process the interrupt request.

The WIC itself does not contain any programmable registers; it has an interface that couples to the Cortex-M0 NVIC, and the interrupt mask information is transferred from the processor to the WIC automatically during sleep. The WIC is activated only in deep sleep mode (SLEEPDEEP bit is set), and you might also need to program additional control registers in the power management unit in the microcontroller to enable the WIC deep sleep mode.

The WIC enables the Cortex-M0 processor to reduce standby power consumption using a technology called State Retention Power Gating (SRPG). With SRPG, the leakage power of a sequential digital system during sleep can be minimized by powering off most parts of the logic, leaving a small memory element in each flip-flop to retain the current state. This is shown in Figure 11.11.

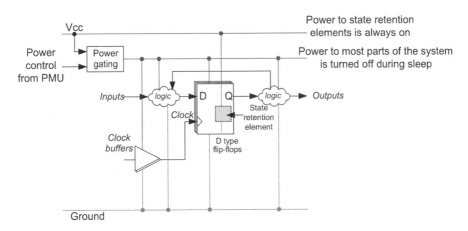

Figure 11.11:
SRPG technology allows most parts of a digital system to be powered down.

When working with the WIC, a Cortex-M0 processor implemented with SRPG technology can be powered down during deep sleep to minimize the leakage current of the microcontroller. During WIC mode deep sleep, the interrupt detection operation is handed over to the WIC (Figure 11.12). Because the state of the processor is retained in the flip-flops, the processor can wake up and resume operations almost immediately. In practice, the use of SRPG power down can increase the interrupt latency slightly, depending on how long it takes for the voltage on the processor to be stabilized after the power-up sequence.

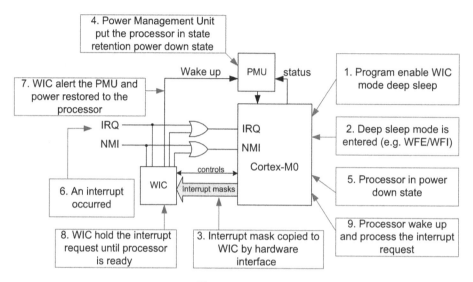

Figure 11.12:
Illustration of WIC mode deep sleep operations.

Not all Cortex-M0 microcontrollers support the WIC feature. The reduction of power using the WIC depends on the application and the semiconductor process being used. When the WIC mode deep sleep is used, the SysTick timer is stopped and you might need to set up a separate peripheral timer to wake up the processor periodically if your application requires an embedded OS and needs the OS to operate continuously. Also, when developing simple applications without any embedded OS and if WIC mode deep sleep is required, use a peripheral timer for periodic interrupt generation instead of the SysTick timer.

## CHAPTER 12

Fault Handling

Fault Exception Overview

In ARM processors, when a program goes wrong and if the processor detects the fault, a fault exception occurs. On the Cortex-M0 processor, there is only one exception type that handles faults: the hard fault handler.

The hard fault handler is almost the highest priority exception type, with a priority level of −1. Only the Nonmaskable interrupt (NMI) can preempt it. When it is executed, we know that the microcontroller is in trouble and corrective action is needed. The hard fault handler is also useful for debugging during the software development stage. When a breakpoint has been set in the hard fault handler, the program execution stops when a fault occurs. By examining the content of the stack, we can back trace the location of the fault and try to identify the reason for the failure.

This behavior is very different from that of most 8-bit and 16-bit microcontrollers. In these microcontrollers, often the only safety net is a watchdog timer. However, it takes time for a watchdog timer to be triggered, and often there is no way to tell how the program went wrong.

What Can Cause a Fault?

There are a number of possible reasons for a fault to occur. For the Cortex-M0 processor, we can group these potential causes into two areas, as described in Table 12.1.

For memory-related faults, the error response from the bus system can also have a number of causes:

- The address being accessed is invalid.
- The bus slave cannot accept the transfer because the transfer type is invalid (depending on bus slave).
- The bus slave cannot access the transfer because it is not enabled or initialized (for example, a microcontroller might generate an error response if a peripheral is accessed but the clock for the peripheral bus is turned off).

When the direct cause of the hard fault exception is located, it might still take some effort to locate the source of the problem. For example, a bus error fault can be caused by an incorrect pointer manipulation, a stack memory corruption, a memory overflow, an incorrect memory map setup, or other reasons.

The Definitive Guide to the ARM Cortex-M0. DOI: 10.1016/B978-0-12-385477-3.10012-6

Table 12.1: Fault That Triggers Hard Fault Exceptions

Fault Classification	Fault Condition

Memory related	• Bus error (can be program accesses or data accesses, also referred to as bus faults in Cortex-M3)
	    Bus error generated by bus infrastructure because of an invalid address during bus transaction
	    Bus error generated by bus slave
	• Attempt to execute the program from a memory region marked as nonexecutable (see the discussion of memory attributes in Chapter 7)

Program error (also referred to as usage faults in the Cortex-M3)	• Execution of undefined instruction • Trying to switch to ARM state (Cortex-M0 only supports Thumb instructions) • Attempt to generate an unaligned memory access (not allowed in ARMv6-M) • Attempt to execute an SVC when the SVC exception priority level is the same or lower than the current exception level • Invalid EXC_RETURN value during exception return • Attempt to execute a breakpoint instruction (BKPT) when debug is not enabled (no debugger attached)

Analyze a Fault

Depending on the type of fault, often it is straightforward to locate the instruction that caused the hard fault exception. To do that we need to know the register contents when the hard fault exception is entered and the register contents that were pushed to the stack just before the hard fault handler started. These values include the return program address, which usually tells us the instruction address that caused the fault.

If a debugger is available, we can start by creating a hard fault exception handler, with a breakpoint instruction that halts the processor. Alternatively, we can use the debugger to set a breakpoint to the beginning of the hard fault handler so that the processor halts automatically when a hard fault is entered. After the processor is halted because of a hard fault, we can then try to locate the fault by the flow shown in Figure 12.1.

To aid the analysis, we should also generate a disassembly listing of the compiled image and locate the fault using the stacked PC value found on the stack frame. If the faulting address is a memory access instruction, you should also check the register value (or stacked register value) to see if the memory access operated on the right address. In addition to checking the address range, we should also verify that the memory address is aligned correctly.

Apart from the stacked PC (return address) value, the stack frame also contains other stacked register values that can be useful for debugging. For example, the stacked IPSR (within the xPSR) indicates if the processor was running an exception, and the stacked EPSR shows the processor state (if the T bit of EPSR is 0, the fault is caused by accidentally switching to ARM state).

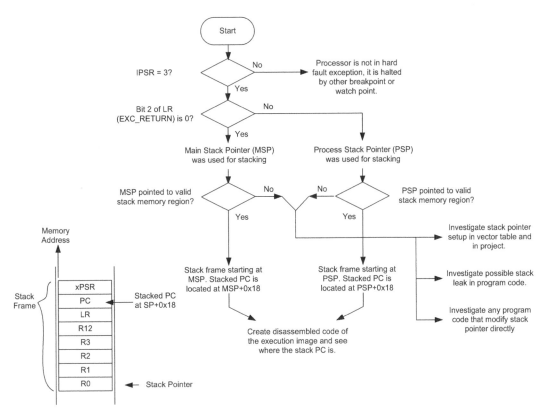

Figure 12.1:
Locating a fault.

The stacked LR might also provide information like the return address of the faulting function, or if the fault happened within an exception handler, or whether the value of the EXC_RETURN was accidentally corrupted.

Also, the current register values can provide various types of information that can help identify the cause of a fault. Apart from the current stack pointer values, the current Link Register (R14) value might also be useful. If the LR shows an invalid EXC_RETURN value, it could mean that the value of LR was modified incorrectly during a previous exception handler.

The CONTROL register can also be useful. In simple applications without an OS, the processor stack pointer (PSP) is not used and the CONTROL register should always be zero in such cases. If the CONTROL register value was set to 0x2 (PSP is used in Thread state), it could mean LR was modified incorrectly during a previous exception handler, or a stack corruption has taken place that resulted in an incorrect value for EXC_RETURN to be used.

Accidental Switching to ARM State

A number of common program errors that cause hard faults are related to the accidental switching to ARM state. Usually this can be detected by checking the values of the stacked xPSR. If the T (Thumb) bit is cleared, then the fault was caused by an accidental switching to ARM state.

Table 12.2 describes the common errors that cause this problem.

Table 12.2: Various Causes of Accidentally Switching to ARM State

Error	Descriptions
Use of incorrect libraries	The linking stage might have accidentally pulled in libraries compiled with ARM instructions (for ARM7TDMI). Check the linker script setting and disassembled the code of the compiled image to see if the C libraries are correct.
Functions not being declared correctly	If you are using GNU assembly tools and the project contains multiple files, you need to make sure functions being called from a different file are declared correctly. Otherwise any such calls might result in an accidental state change.
LSB of the vector in the vector table set to 0	The vector in the vector table should have the LSB set to 1 to indicate Thumb state. If the stacked PC is pointing to the beginning of an exception handler and the stacked xPSR has the T bit cleared to 0, the error is likely to be in the vector table.
Function pointer with LSB set to 0	If a function pointer is declared with the LSB set to 0, calling the function will also cause the processor to enter a hard fault.

Error Handling in Real Applications

In real applications, the embedded systems will be running without a debugger attached and stopping the processor is not acceptable for many applications. In most cases, the hard fault exception handler can be used to carry out safety actions and then reset the processor. For example, the following steps can be carried out:

- Perform application specific safety actions (e.g., performance shut-down sequence in a motor controller)
- Optionally the system can report the error within a user interface and then reset the system using the Application Interrupt and Reset Control Register (AIRCR; see Chapter 9, Table 9.8) or other system control methods specific to the microcontroller.

Because a hard fault could be caused by an error in the stack pointer value, a hard fault handler programmed in C language might not be able to perform correctly, as C-generated code might require stack memory to operate. Therefore, for safety-critical systems, ideally the hard fault handler should be programmed in assembly language, or use an assembly language wrapper to make sure that the stack pointer is in valid memory range before entering a C routine.

If a hard fault handler is written in C to report debug information like faulting a program address to a terminal display, we will also need an assembly wrapper (Figure 12.2). The wrapper code extracts the address of the exception stack frame and passes it on to the C hard fault handler for displaying. Otherwise, there is no easy way to locate the stack frame inside the C handler—although you can access the stack pointer value using inline assembly, embedded assembly, a named register variable, or an intrinsic function, the value of the stack pointer could have been changed by the C function itself.

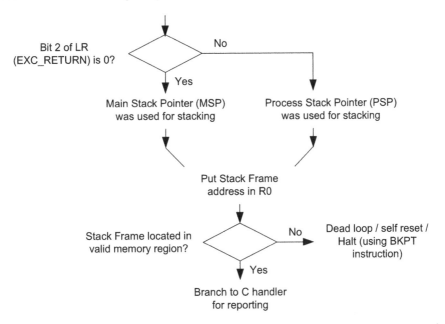

Figure 12.2:
Assembly wrapper for a hard fault handler.

The assembly code for such an assembly wrapper can be implemented using embedded assembly, for example:

Assembly wrapper using Embedded Assembler in Keil MDK

```
// Hard Fault handler wrapper in assembly
// It extracts the location of stack frame and passes it to handler
// in C as a pointer. We also extract the LR value as second
// parameter.
__asm void HardFault_Handler(void)
{
  MOVS    r0, #4
  MOV     r1, LR
  TST     r0, r1
  BEQ     stacking_used_MSP
  MRS     R0, PSP ; first parameter - stacking was using PSP
  B       get_LR_and_branch
stacking_used_MSP
  MRS     R0, MSP ; first parameter - stacking was using MSP
get_LR_and_branch
  MOV     R1, LR  ; second parameter is LR current value
  LDR     R2,=__cpp(hard_fault_handler_c)
  BX      R2
}
```

The handler in C accepts the parameters from the assembly wrapper and extracts the stack frame contents and LR values:

Hard Fault handler to report stacked register values

```
// Hard Fault handler in C, with stack frame location and LR value
// extracted from the assembly wrapper as input parameters
void hard_fault_handler_c(unsigned int * hardfault_args, unsigned lr_value)
{
  unsigned int stacked_r0;
  unsigned int stacked_r1;
  unsigned int stacked_r2;
  unsigned int stacked_r3;
  unsigned int stacked_r12;
  unsigned int stacked_lr;
  unsigned int stacked_pc;
  unsigned int stacked_psr;

  stacked_r0 = ((unsigned long) hardfault_args[0]);
  stacked_r1 = ((unsigned long) hardfault_args[1]);
  stacked_r2 = ((unsigned long) hardfault_args[2]);
  stacked_r3 = ((unsigned long) hardfault_args[3]);
  stacked_r12 = ((unsigned long) hardfault_args[4]);
  stacked_lr  = ((unsigned long) hardfault_args[5]);
  stacked_pc  = ((unsigned long) hardfault_args[6]);
  stacked_psr = ((unsigned long) hardfault_args[7]);

  printf ("[Hard fault handler]\n");
  printf ("R0  = %x\n", stacked_r0);
  printf ("R1  = %x\n", stacked_r1);
  printf ("R2  = %x\n", stacked_r2);
  printf ("R3  = %x\n", stacked_r3);
  printf ("R12 = %x\n", stacked_r12);
  printf ("Stacked LR  = %x\n", stacked_lr);
  printf ("Stacked PC  = %x\n", stacked_pc);
  printf ("Stacked PSR = %x\n", stacked_psr);
  printf ("Current LR  = %x\n", lr_value);

  while(1); // endless loop
}
```

The C handler can only work if the stack is still in a valid memory region because it tries to extract debug information from the stack, and the program codes generated from C compilers often require stack memory. Alternatively, you can carry out the debug information reporting entirely in assembly code. Doing this in assembly language is relatively easy when you have an assembly routine for text output ready. Examples of assembly text outputting routines can be found in Chapter 16. Details about embedded assembly programming (used in the assembly wrapper) can also be found in that chapter.

Lockup

The Cortex-M0 processor can enter a lockup state if another fault occurs during the execution of a hard fault exception handler or when a fault occurs during the execution of an NMI handler. This is because when these two exception handlers are executing, the priority level does not allow the hard fault handler to preempt.

During the lockup state, the processor stops executing instructions and asserts a LOCKUP status signal. Depending on the implementation of the microcontroller, the LOCKUP status signal can be programmed to reset the system automatically, rather than waiting for a watchdog timer to time out and reset the system.

The lockup state prevents the failed program from corrupting more data in the memory or data in the peripherals. During software development, this behavior can help us debug the problem, as the memory contents might contain vital clues about how the software failed.

Causes of Lockup

A number of conditions can cause lockup in the Cortex-M0 processor (or ARMv6-M architecture):

- A fault occurred during the execution of the NMI handler
- A fault occurred during the execution of the hard fault handler (double fault)
- There was an SVC execution inside the NMI handler or the hard fault handler (insufficient priority)
- A bus error response during reset sequence (e.g. when reading initial SP value)
- There was a bus fault during the unstacking of the xPSR during the exception return using the main stack pointer (MSP) for the unstacking

Besides fault conditions, the use of an SVC in an NMI or hard fault handler can also cause a lockup because the SVC priority level is always lower than these handlers and therefore blocked. Because this program error cannot be handled by the hard fault exception (the priority level is already -1 or -2), the system enters lockup state.

The lockup state can also be caused by a bus system error during the reset sequence. When the first two words of the memory are fetched and if a bus fault happens in one of these accesses, it means the processor cannot determine the initial stack pointer value (the hard fault handler might need the stack as well), or the reset vector is unknown. In these cases, the processor cannot continue normal operation and must enter a lockup state.

If a bus error response occurs at exception, entrance (stacking) does not cause a lockup, even it is entering hard fault or entering NMI exception (Figure 12.3). However, once the hard fault exception or NMI exception handlers are entered, a bus error response can cause lockup. As a result, in safety-critical systems, a hard fault handler written in C might not be the best

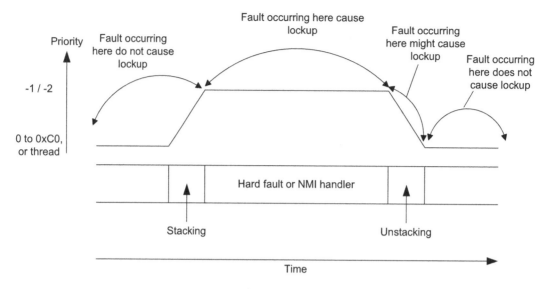

Figure 12.3:
Lockup condition during exception sequences.

arrangement because the C compiler might insert stack operations right at the beginning of the handler code:

```
HardFault_Handler
    PUSH{R4, R5} ; This can cause lock up if the MSP is corrupted
    ...
```

For an exception exit (unstacking), it is possible to cause a lockup if a bus error response is received during the unstacking process of the xPSR using MSP. In such cases, the xPSR cannot be determined and therefore the correct priority level of the system is unknown. As a result, the system is locked up and cannot be recovered apart from resetting it or halting it for debug.

What Happens during a Lockup?

If the lockup is caused by a double fault, the priority level of the system stays at -1. If an NMI exception has occurred, it is possible for the NMI to preempt and execute. After the NMI is completed, the exception handler is terminated and the system returns to lockup state.

Otherwise, in other lockup scenarios the system cannot be recovered and must be reset or restored using a debugger attached to it. Microcontroller designers or system-on-chip designers can use the LOCKUP signal to reset the system via a configurable setting in the reset controller.

Preventing Lockup

Lockup and hard fault exceptions might look scary to some embedded developers, but embedded systems can go wrong for various reasons and the lockup and hard fault mechanisms can be used to keep the problem from getting worse. Various sources of errors or problems can cause an embedded system to crash in any microcontrollers, for example:

- Unstable power supply or electromagnetic interferences
- Flash memory corruption
- An error in external interface signals
- Component damage that results from operating conditions or the natural aging process
- An incorrect clock generation arrangement or poor clock signal quality
- Software errors

The hard fault and lockup behaviors allow error conditions to be detected and help debugging. Although we cannot fully prevent all the potential issues listed, we can take various measures in software to improve the reliability of an embedded system.

First, we should keep the NMI exception handler and hard fault exception handler as simple as possible. Some tasks associated with the NMI exception or hard fault exception can be separated into a different exception like PendSV and executed after the urgent parts of the exception handling have completed. By making the NMI and hard fault handler shorter and easier to understand, we can also reduce the risk of accidentally using SVC instructions in these handlers (this can be caused by calling a function which contains an SVC instruction).

Second, for safety-critical applications, you might want to use an assembly wrapper to check the SP value before entering the hard fault handler in C (Figure 12.4).

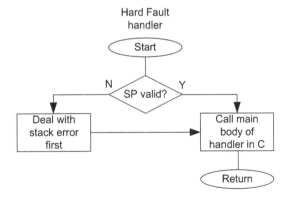

Figure 12.4:
Adding of SP checking in assembly.

If necessary, we can program the entire hard fault handler in assembly. In such cases, we can avoid some stack memory accesses to prevent lockup if the stack pointer is corrupted and pointing to an invalid memory location.

Similarly, if the NMI handler is simple, we can program the NMI handler in assembly language and use just R0 to R3 and R12 if we want to avoid stack memory accesses because these registers are already stacked. But in most cases, a stack pointer error would be likely to trigger the hard fault exception fairly quickly, so there is no need to worry about programming the NMI in C language.

## CHAPTER 13

Debug Features

Software Development and Debug Features

During software development, we often need to examine the operation of program execution in detail to understand why a program does not work as expected or to ensure correct operation. In some cases, we can output a small amount of program operation detail using various interfaces such as a UART. Often this does not provide sufficient data to debug the program. We often need to add breakpoints, add data watchpoints, view memory and registers, and so on. These debug architectural features are now part of modern processor design.

In this chapter we will cover a number of debug terms. Note that these terms are not standardized across all microcontroller architectures, so the terms used by some microcontroller vendors can be different from those listed here (Table 13.1).

Table 13.1: Common Debug Features on ARM Microcontrollers

Terms	Descriptions
Halt	Stopping of program execution due to a debug event (e.g., breakpoint or watchpoint), or due to user debug request.
Breakpoint	Program execution reach an address marked as a breakpoint, causing a debug event to be generated which halts the processor.
Hardware breakpoint	A hardware comparator is used to compare the current program address to a reference address setup by the debugger. When the processor fetches and executes an instruction from this address, the comparator generates a debug event signal to stop the processor.
Software breakpoint	A breakpoint instruction (BKPT) is inserted to the program memory so that program execution halts when it get to this address.
Watchpoint	A data or peripheral address can be marked as a watched variable, and an access to this address causes a debug event to be generated, which halts program execution.
Debugger	A piece of software running on a debug host (e.g., a personal computer) that communicates with the debug system in a microcontroller, usually via a USB adaptor (or an in-circuit debugger), so that debug features of the microcontroller can be accessed.
In-circuit debugger	A piece of hardware that connects between the debug host (e.g., a personal computer) and the microcontroller. Usually the connection to the debug host is a USB or an Ethernet, and the connection to the microcontroller is a JTAG or a serial wire protocol. Various terminologies are used for in-circuit debuggers: USB-JTAG adaptor, in-circuit emulator (ICE), JTAG/SW emulator, Run Time Control Unit, and so on.
Profiling	A feature in the debugger that collects statistics of program execution. This is useful for performance analysis and software optimization.

The Definitive Guide to the ARM Cortex-M0. DOI: 10.1016/B978-0-12-385477-3.10013-8

In most microcontrollers, the debug features include the following:

- Halting program execution through user requests via a debugger or by debug events
- Resuming program execution
- Examining and modifying system status

These features are all supported on the Cortex-M0 processor and can be performed on a target platform via a low-pin-count serial link. This is different from some older-generation microcontrollers, which require an emulator to emulate the microcontroller, or other micro-controller products that require the microcontroller to be programmed before insertion in the targeted platform (in-system programmable).

Another difference between ARM-based microcontrollers and some other microcontrollers is that there is no need for a debug agent (a small piece of debug support software) running on the processor to carry out the debug operations. When a debug feature is accessed, the feature is carried out entirely by the debug support hardware inside the processor. As a result, it does not require any program size overhead and does not affect any data in memory including the stack.

Debug Features Overview

The Cortex-M0 processor supports a number of useful debug features:

- Halting, resuming, and single stepping of program execution
- Access to processor core registers and special registers
- Hardware breakpoints (up to four comparators)
- Software breakpoints (BKPT instruction)
- Data watchpoints (up to two comparators)
- On-the-fly memory access (system memory can be accessed without stopping the processor)
- PC sampling for basic profiling
- Support of JTAG or serial wire debug protocol

These debug features are vital for software development and can be used for other tasks like flash programming and product testing.

The debug features of the Cortex-M0 processor are based on the ARM CoreSight debug architecture. They are consistent among all Cortex-M processors, making it easy for a debug tool to support all Cortex-M processors with little modification. The debug architecture is also very scalable, making it possible to build complex multiprocessor products using the CoreSight debug architecture.

The design of the Cortex-M0 processor allows the debug features to be configurable. For example, system-on-chip designers can remove some or all of the debug features to reduce the circuit size for ultra-low-power applications like wireless sensors. If a debug interface is implemented, debugger software can also read various registers to detect which debug features were implemented.

Debug Interface

To access the debug features on the microcontroller, a debug interface is needed (Figure 13.1). For ARM Cortex-M0 microcontrollers, this interface can either be in Joint Test Action Group (JTAG) protocol or serial wire debug protocol. Both protocols transfer control information and data in serial bit sequences.

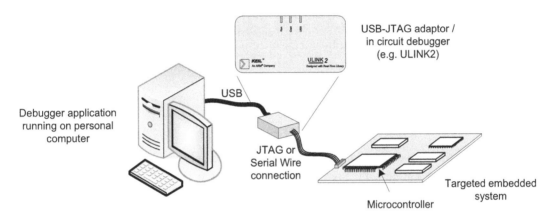

Figure 13.1:
Using the debug interface on the Cortex-M0 microcontroller.

The debug interface allows the following:

- The flash memory to be reprogrammed easily without the need to remove it from the circuit board
- Applications to be tested
- Production testing (e.g., self-test application can be downloaded to the microcontroller memory and executed, or a boundary scan could be carried out via a JTAG connection if it is implemented in the microcontroller)

Unlike most other processors, in ARM Cortex processors the debug interface and the debug features are separated. The processor design contains a generic parallel bus interface that allows all the debug features to be accessed. A separated debug interface block (called the Debug Access Port in ARM documentation) is used to convert a debug interface protocol to the parallel bus interface (Figure 13.2). This arrangement is part of the CoreSight debug architecture, and it makes the ARM Cortex processors' debug solution flexible.

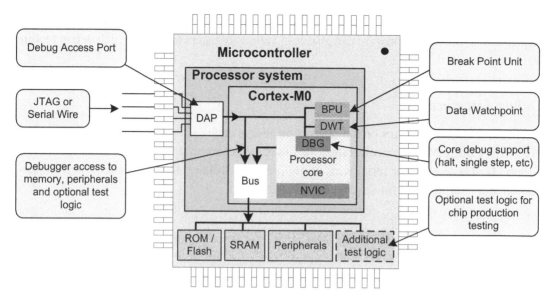

Figure 13.2:
Debug interface (DAP) in Cortex-M0.

JTAG is a four-pin or five-pin serial protocol that is commonly used for digital component testing. The interface contains the signals described in Table 13.2.

Table 13.2: Signals connection for the JTAG debug.

JTAG Signal	Descriptions
TCK	Clock signal
TMS	Test Mode Select signal—controls the protocol state transition
TDI	Test Data In—serial data input
TDO	Test Data Out—serial data output
nTRST	Test reset—active low asynchronous reset for a JTAG state control unit called the TAP controller (The nTRST signal is optional. Without nTRST, the TAP controller can be reset with five cycles of TMS pulled high.)

Although the JTAG interface is commonly used and well supported, using four or five pins for debug operations is too many for some microcontrollers with low pin counts. As a result, ARM developed the serial wire debug protocol, which uses only two pins (Table 13.3).

Although only two signals are required, the serial wire debug protocol can offer better performance than JTAG and can provide the same processor debug functionality. Most

Table 13.3: Signal Connection for the Serial Wire Debug

Serial Wire Signal	Descriptions
SWCLK	Clock signal
SWDIO	Data input/output—bidirectional data and control communication

in-circuit debuggers and debugger software tools that support the ARM Cortex-M processor family already support the serial wire debug protocol.

The use of the CoreSight debug architecture brings a number of advantages to the Cortex-M0 processor and other processors in the Cortex-M processor family:

• By separating the debug interface from the main processor logic, the choice of debug interface protocol becomes much more flexible, without affecting the underlying debug features on the main processor logic.
• Multiple processors can share the same debug interface block, allowing a more scalable debug system. Other test logic can also be added to the system easily, as the internal connection is a simple parallel bus interface.
• The design is consistent among all Cortex-M processors, making it easy for tool vendors to support the whole Cortex-M processor family with one tool chain.

Many microcontroller products have the JTAG or serial wire interface pin shared with the peripheral interface or other I/O pins. When the debug interface pins are used for I/O, usually by programming certain peripheral control registers to switch the usage to I/O, the debugger cannot connect to the processor. Therefore, when designing an embedded system, you should avoid using the debug interface pins as I/O if you want to allow the system to be debugged easily.

In some cases, if the pins are switched from debug mode to I/O quickly after the program starts; this could end up locking out the debugger completely because the debugger will not have enough time to connect and halt the processor before the pin usage is switched. As a result, you cannot debug the application and cannot reprogram the flash memory. From another point of view, you might be able to use it as a feature to block other people from accessing the program code in the chip. However, this arrangement is not guaranteed to be secure and can be worked around if the microcontroller's design has a special boot mode that can disable the application. For secure firmware protection, please refer to the documentation from your microcontroller vendor.

Details of the CoreSight debug architecture can be found on the ARM web site. A document called "CoreSight Technology System Design Guide" (ARM DGI 0012B, reference 6) provides a good overview of the CoreSight debug architecture. In addition, the ARM Debug Interface v5 (ARM IHI 0031A, reference 7) provides detailed information on the serial wire debug protocol.

Halt Mode and Debug Events

The Cortex-M0 processor has a halt mode, which stops program execution and allows the debugger to access processor registers and memory space. During halt mode, the following activities occur:

- Instruction execution stops.
- The SysTick timer stops counting.
- If the processor was in sleep mode, it wakes up from the sleep mode before halt.
- Registers in the processor's register bank, as well as special registers, can be accessed (both read and write).
- Memory and peripheral contents can be accessed (this can be done without halting the processor).
- Interrupts can still enter pending state.
- You can resume program execution, carry out single-step operation, or reset the microcontroller.

When a debugger is connected to the Cortex-M0, it first programs a debug control register in the processor to enable the debug system. This cannot be done by the application running on the microcontroller. After the debug system is enabled, the debugger can then stop the processor, download the application to the microcontroller flash memory if required, and reset the microcontroller; we can then test the application.

The Cortex-M0 processor enters halt mode when:

- Debug is enabled by a debugger
- A debug event occurs

There are various sources of debug events. They can be generated by either hardware or software (Figure 13.3).

A debugger can stop program execution by writing to debug control registers. On an embedded system with multiple processors, it is also possible to stop multiple processors at the same time from using a hardware debug request signal and an on-chip debug event communication system.

The program execution can be stopped by hardware breakpoints, software breakpoints, watchpoints, or vector catch event. The vector catch is a mechanism that allows the core to be halted when certain exceptions take place. On the Cortex-M0 processor, two vector catch conditions are provided:

- Reset
- Hard fault

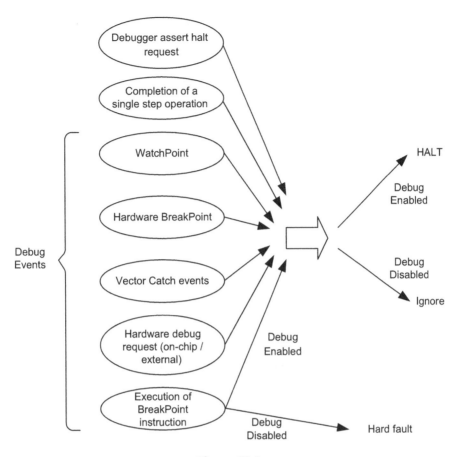

Figure 13.3:
Debug events on the Cortex-M0 processor.

The vector catch feature is controlled by debug registers in the Cortex-M0 processor, allowing the processor to be stopped automatically upon a reset or when a hard fault execution takes place (e.g., because of a software error). When the vector catch operation takes place, the processor stops before execution of the first instruction in the reset or hard fault exception handler.

Once the debugger application detects that the processor is halted, it then checks a Debug Fault Status Register inside the System Control Block (SCB) of the Cortex-M0 processor to determine the reason for halting. Then it can inform the user that the processor is halted. After the processor has been halted, you can then access to the registers inside the processor's register bank and special registers, access the data in memories or peripherals, or carry out a single-step operation.

A halted Cortex-M0 processor can resume program execution by writing to the debug register through the debugger connection, by a hardware debug restart interface (e.g., use in multiprocessor systems so that multiple processors can resume program execution at the same time), or by reset.

Debug System

The debug features on the Cortex-M0 are controlled by a number of debug components. These debug components are connected via an internal bus system. However, application code running on the Cortex-M0 processor cannot access these components (this is different from the Cortex-M3/M4 processor, where software can access the debug components). The debug components can only be accessed by the debugger connected to the microcontroller (Figure 13.4).

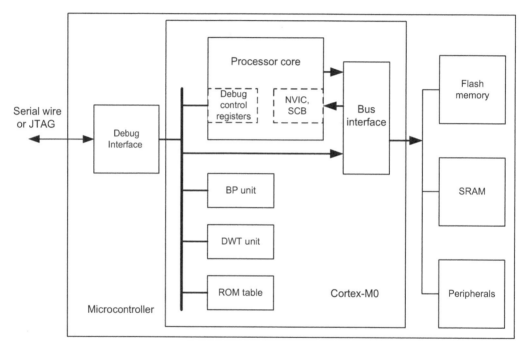

Figure 13.4:
Debug components in the Cortex-M0.

There are a number of debug components in the Cortex-M0, as describe in Table 13.4.

The debug system also allows access to the system's memory map including flash, SRAM, and peripherals. The accesses to the system memory can be carried out even if the processor is running. By accessing the Application Interrupt and Reset Control Register (AIRCR) in the System Control Block (SCB), the debugger can also request a system reset to reset the microcontroller.

Additional information about the debug components is covered in Appendix E.

Table 13.4: Debug Components in the Cortex-M0

Debug Components	Descriptions
Processor core debug registers	Debug features inside the processor core are accessible by a few debug control registers. They provide the following: —Halting, single step, resume execution —Access to the core's registers when the processor is halted —Control of vector catch
BP unit	The breakpoint unit provides up to four breakpoint address comparators.
DWT unit	The data watchpoint unit provides up to two data address comparators. It also allows the debugger to sample the program counter regularly for profiling.
ROM table	This small lookup table allows the debugger to locate available debug components in the system. It lists the addresses of debug components, and the debugger can then identify the available debug features by checking the identification registers of these components.

## CHAPTER 14

Getting Started with Keil MDK

Introduction to Keil MDK

The ARM Keil Microcontroller Development Kit (MDK) is one of the most popular development suites for ARM microcontrollers. The Keil MDK is a Windows-based development suite and provides the following components:

- The µVision integrated development environment (IDE)
- C compiler, assembler, linker, and utilities
- Debugger
- Simulator
- RTX Real-Time Kernel, an embedded OS for microcontrollers
- Startup code for various microcontrollers
- Flash programming algorithms for various microcontrollers
- Program examples and development board support files

The debugger in µVision IDE works with Keil USB-JTAG adaptors like ULINK2 and ULINK Pro, and a number of third-party adaptors like the Signum JtagJet/JtagJet-Trace, SEGGER J-Link, ST-Link, LuminaryMicro's Evaluation Board, and Altera Blaster Cortex Debugger. In addition, you can also use other in-circuit debuggers if a third-party plug-in for Keil MDK is available. Even if you don't have an in-circuit debugger, you can generate the program image and program the microcontroller using third-party programming tools. But, of course, having a supported in-circuit debugger allows you to debug the system though the µVision IDE, which is much easier and more effective.

The C compiler used in the Keil MDK is based on the same compiler engine in the ARM RealView Development Suit, which provides excellent performance and code density. Another advantage of using the Keil MDK is that it supports a huge number of ARM microcontrollers on the market. In addition to standard compiler and debug support, it also provides configuration files such as startup code and RTX OS configuration files, making software development easier and quicker.

The Keil MDK includes a simulator that can emulate the operations of the processor and some of the peripherals. This enables users to build and run ARM applications without requiring the actual hardware.

An evaluation version of the Keil MDK can be downloaded from the Keil web site (www.keil. com). The evaluation version is limited to 32 KB of program memory. This memory size is

The Definitive Guide to the ARM Cortex-M0. DOI: 10.1016/B978-0-12-385477-3.10014-X sufficient for most simple applications. You might also receive the evaluation version of Keil MDK from various development kits from the microcontroller vendors. If you decide to use Keil MDK for commercial projects, you can purchase a license on the Keil web site and obtain a software license number. This license number can then be used to convert the evaluation version to a full version.

First Step of Using Keil MDK

Create the Blinky Project

Here is what you need to follow the examples presented in this chapter:

1. Keil MDK (either the full version or an evaluation version can be used, but it needs to be version 4.10 or later) installed on your personal computer.
2. Access to a Cortex-M0 development board would be ideal. If it is not available, most the examples can still be tested using simulation support in MDK.
3. An in-circuit debug adaptor supported by Keil MDK.

The setups used in the following the examples in this chapter are based on the NXP LPC1114 microcontroller, ULINK-2 USB-JTAG adaptor (Figure 4.12, presented in chapter 4), and Keil MDK version 4.10.

When the μVision IDE starts, you will see a screen similar to the one shown in Figure 14.1.

We start by creating a new project. This can be done by using the pull-down menu: select Project → New μVision Project, as shown in Figure 14.2.

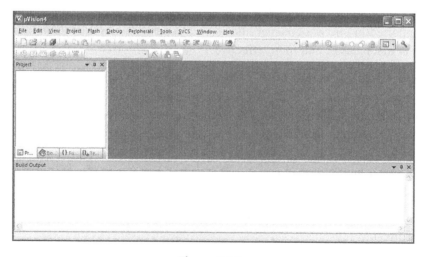

Figure 14.1:
μVision IDE starting screen.

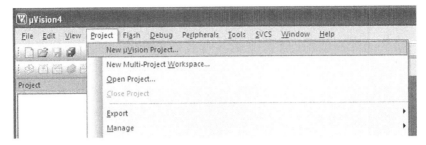

Figure 14.2:
Creating a new project

For the first project, we are going to create a simple program that toggles an LED. We will call this project "blinky." The location of the project depends on your preference; in this demonstration we'll put the project in "C:\CortexM0\blinky," as shown in Figure 14.3.

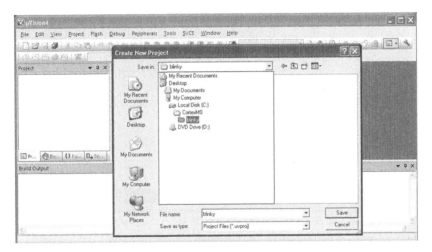

Figure 14.3:
Create the blinky project.

The next step of the project creation wizard defines the microcontroller to be used for the project. In this example we'll select the NXP LPC1114-301, as shown in Figure 14.4.

The last step of the project creation wizard will ask if you want to copy the default startup code for NXP LPC11xx to your project, as shown in Figure 14.5. Select yes, as this will save us a lot of time in preparing the startup code.

Once this is done, we will have a blinky project set up with just the startup code, as shown in Figure 14.6.

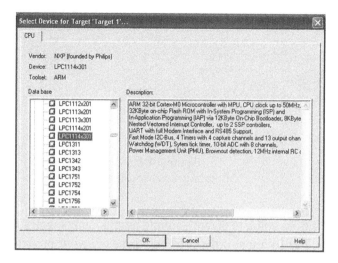

Figure 14.4:
Device selection.

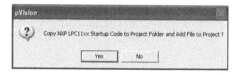

Figure 14.5:
Option to copy startup code.

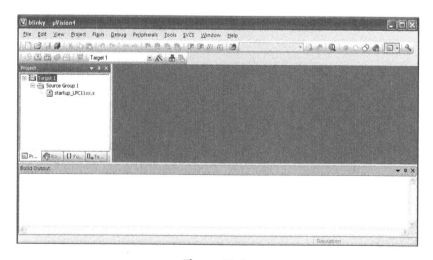

Figure 14.6:
Blinky project created.

Create the Project Code

The next stage of the project is to create the program code. This can be done by using the pull-down menu: File → New. Then put the following program code in the new file, and save it as blinky.c. The operation of the program is illustrated in the flowchart in Figure 14.7.

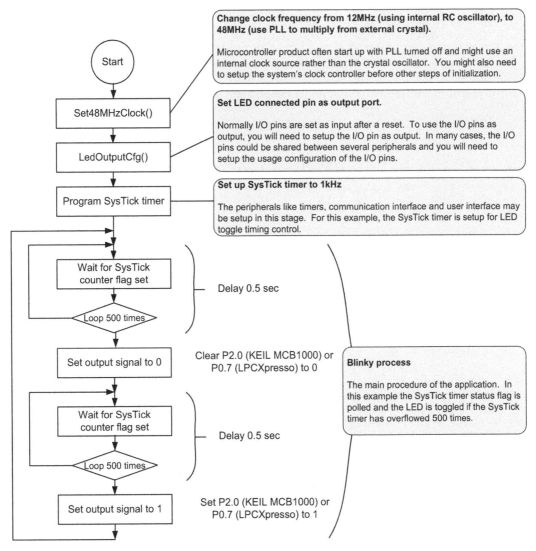

Figure 14.7:
Flowchart of the blinky application.

blinky.c

```c
#include "LPC11XX.h"
#define KEIL_MCB1000_BOARD

// Function declarations
void Set48MHzClock(void); // Program PLL to generate 48MHz clock
void LedOutputCfg(void);  // Set I/O pin connected to LED as output

int main(void)
{
  int i;
  // Switch Clock to 48MHz
  Set48MHzClock();
#ifdef KEIL_MCB1000_BOARD
  // Initialize LED (GPIO #2, bit[0]) output
#else
  // Initialize LED (GPIO #0, bit[7]) output
#endif
  LedOutputCfg();

  // Program SysTick timer at 1KHz.
  // At 48MHz, SysTick trigger every 48000 CPU cycles
  SysTick->LOAD = (48000-1); // Count from 47999 to 0
  SysTick->VAL  = 0;   // Clear SysTick value
  SysTick->CTRL = 0x5; // Enable, using core clock

  while(1){     // Blink at 1Hz
    for (i=0;i<500;i++) { // Wait for 0.5 seconds
        while ((SysTick->CTRL & 0x10000)==0); // Wait for counter underflow
        }
#ifdef KEIL_MCB1000_BOARD
    // For Keil MCB1000, use P2.0 for LED output
    LPC_GPIO2->MASKED_ACCESS[1] = 0;
#else
    // For LPCXpresso, use P0.7 for LED output
        LPC_GPIO0->MASKED_ACCESS[1<<7] = (0<<7); // Clear bit 7
          // Alternatively, we can use "LPC_GPIO0->DATA = 0x00;"
          // as the other bits are not used
#endif
    for (i=0;i<500;i++) { // Wait for 0.5 seconds
        while ((SysTick->CTRL & 0x10000)==0); // Wait for counter underflow
        }
#ifdef KEIL_MCB1000_BOARD
    // For Keil MCB1000, use P2.0 for LED output
    LPC_GPIO2->MASKED_ACCESS[1] = 1;
#else
    // For LPCXpresso, use P0.7 for LED output
    LPC_GPIO0->MASKED_ACCESS[1<<7] = (1<<7); // Set bit 7
      // Alternatively, we can use "LPC_GPIO0->DATA = 0x80;"
      // as the other bits are not used
```

```
#endif
  } // end while
} // end main

// Switch LED signal to output port with no pull up or pulldown
void LedOutputCfg(void)
{
  // Enable clock to IO configuration block (bit[16] of AHBCLOCK Control register)
  // and enable clock to GPIO (bit[6] of AHBCLOCK Control register
  LPC_SYSCON->SYSAHBCLKCTRL = LPC_SYSCON->SYSAHBCLKCTRL | (1<<16) | (1<<6);
#ifdef KEIL_MCB1000_BOARD
  // For Keil MCB1000, use P2.0 for LED output
  // PIO2_0 IO output config
  //   bit[5]   - Hysteresis (0=disable, 1 =enable)
  //   bit[4:3] - MODE(0=inactive, 1 =pulldown, 2=pullup, 3=repeater)
  //   bit[2:0] - Function (0 = IO, 1=DTR, 2=SSEL1)
  LPC_IOCON->PIO2_0 = (0<<5) + (0<<3) + (0x0);

  // Initial bit 0 output is 0
  LPC_GPIO2->MASKED_ACCESS[1] = 0;
  // Set pin 7 to 0 as output
  LPC_GPIO2->DIR = LPC_GPIO2->DIR | 0x1;
#else
  // For LPCXpresso, use P0.7 for LED output
  // PIO0_7 IO output config
  //   bit[5]   - Hysteresis (0=disable, 1 =enable)
  //   bit[4:3] - MODE(0=inactive, 1 =pulldown, 2=pullup, 3=repeater)
  //   bit[2:0] - Function (0 = IO, 1=CTS)
  LPC_IOCON->PIO0_7 = (0x0) + (0<<3) + (0<<5);
  // Initial bit[7] output is 0
  LPC_GPIO0->MASKED_ACCESS[1<<7] = 0;
  // Set pin 7 as output
  LPC_GPIO0->DIR = LPC_GPIO0->DIR | (1<<7);
#endif
  return;
} // end LedOutputCfg

// Switch the CPU clock frequency to 48MHz
void Set48MHzClock(void)
{
  // Power up the PLL and System oscillator
  // (clear the powerdown bits for PLL and System oscillator)
  LPC_SYSCON->PDRUNCFG = LPC_SYSCON->PDRUNCFG & 0xFFFFFF5F;

  // Select PLL source as crystal oscillator
  //   0 - IRC oscillator
  //   1 - System oscillator
  //   2 - WDT oscillator
  LPC_SYSCON->SYSPLLCLKSEL = 1;
```

(Continued)

```
blinky.c—Cont'd
    // Update SYSPLL setting (0->1 sequence)
    LPC_SYSCON->SYSPLLCLKUEN = 0;
    LPC_SYSCON->SYSPLLCLKUEN = 1;
    // Set PLL to 48MHz generate from 12MHz
    //   M = 48/12 = 4 (MSEL = 3)
    //   FCCO (must be between 156 to 320MHz, and is 2x, 4x, 8x or 16x of Clock)
    //   Clock freq out selected as 192MHz
    //   P = 192MHz/48MHz/2 = 2 (PSEL = 1)
    //   bit[8]   - BYPASS
    //   bit[7]   - DIRECT
    //   bit[6:5] - PSEL  (1,2,4,8)
    //   bit[4:0] - MSEL  (1-32)
    LPC_SYSCON->SYSPLLCTRL = (3 + (1<<5)); // M = 4, P = 2
    // wait until PLL is locked
    while(LPC_SYSCON->SYSPLLSTAT == 0);
    // Switch main clock to PLL clock
    //   0 - IRC
    //   1 - Input clock to system PLL
    //   2 - WDT clock
    //   3 - System PLL output
    LPC_SYSCON->MAINCLKSEL = 3;
    // Update Main Clock Select setting (0->1 sequence)
    LPC_SYSCON->MAINCLKUEN = 0;
    LPC_SYSCON->MAINCLKUEN = 1;

    return;
} // end Set48MHzClock
```

The program code performs a number of tasks before starting the LED toggling process. Although the exact details are dependent on the microcontroller used, the blinky example can be reproduced on other Cortex-M0 microcontrollers using the flowchart shown in Figure 14.7.

In this example program, either the bit 7 of port 0 or bit 0 of port 2 is used to drive the LED output. The program code contains a preprocessing option called KEIL_MCB1000_BOARD. By setting this option, one of the LEDs (bit 0 of port 2) on the Keil MCB1000 board will be used. Otherwise the bit 7 of port 2 would be used for LED output, which is the arrangement for the NXP LPCXpresso board. The system tick timer is used for timing control, and a polling loop is used to detect every 500th SysTick overflow and toggle the LED output.

The file LPC11xx.h is provided as part of the Keil MDK installation. This header file is used for the CMSIS device driver and contains the peripheral register definitions used in the application.

After the blinky project is created, we can then add the file to the project. This can be done by right-clicking on the "Source Group 1" of the project window and selecting "Add files to Group 'Source Group 1'..." as shown in Figure 14.8.

Figure 14.8:
Add the program file to the project.

Project Settings

After the program file is created, it might be necessary to adjust a few project settings before the application can be downloaded to the microcontroller's flash memory and be tested. In most cases (including LPC1114), the Keil µVision IDE will set up all the required project settings automatically once the device is selected. However, it is useful to understand what settings are available and what settings are needed to get a project to work.

Many project settings are available. First we will introduce the settings that are essential for getting the program code downloaded to the flash and executing it. The project settings menu can be accessed by doing the following:

- Select the Target option button on the toolbar.
- On the pull-down menu, select Project → Option for Target.
- Right-click on the project target name (e.g., "Target 1") in the project window, and select options for target.
- Select hot key Alt-F7.

The project option menu contains a number of tabs. Figure 14.9 shows the list of option tabs.

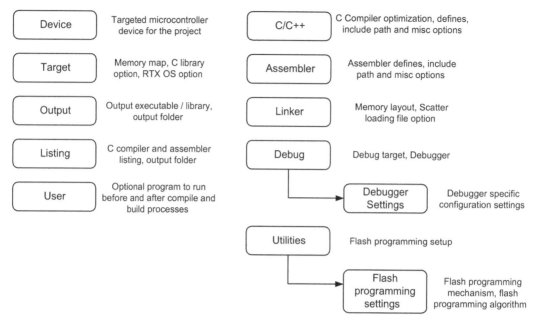

Figure 14.9:
Project options tabs in a Keil MDK project.

By default, the Keil μVision IDE automatically sets up the memory map for us when we select the microcontroller device. In most cases we do not need to change the memory settings. However, if the program operation fails or if flash programming is not functioning correctly, we need to go through the settings to make sure that they were not accidentally changed to incorrect values.

Some settings have to be set up manually. An example would be the debugger configuration, because μVision IDE does not know which in-circuit debugger you will be using. First we look at the debug options shown in Figure 14.10. Here we selected "ULINK Cortex Debugger." You can change the settings to use another supported debugger.

Next, click on the Settings button next to the debugger selected. This will bring us to a configuration menu, which is dependent on the debugger selected. For the ULINK 2 debugger, you will see a screen like the one shown in Figure 14.11.

Because the NXP LPC1114 device does not support JTAG, the debug setting must be set to use the serial wire debug protocol. The serial wire clock frequency being used depends on the microcontroller device, the circuit board (PCB) design, and the debug cable length.

In the ULINK Cortex-M configuration window, there is a tab for flash download (programming). Here we can define the flash programming algorithm as shown in Figure 14.12. This should have been set up automatically by the Keil μVision IDE.

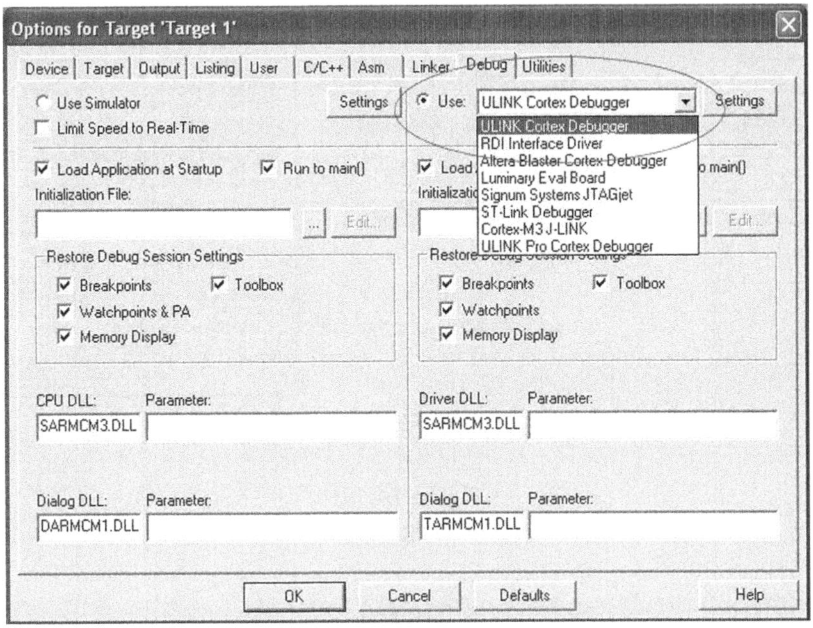

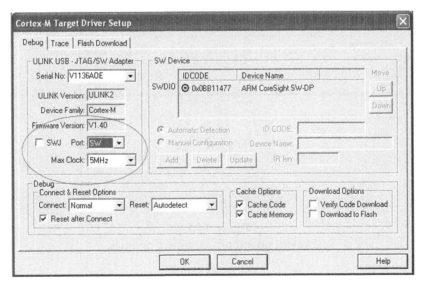

Figure 14.10:
Debug options.

Figure 14.11:
ULINK debug options.

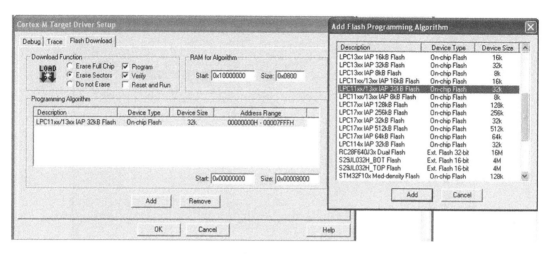

Figure 14.12:
Flash programming algorithm option.

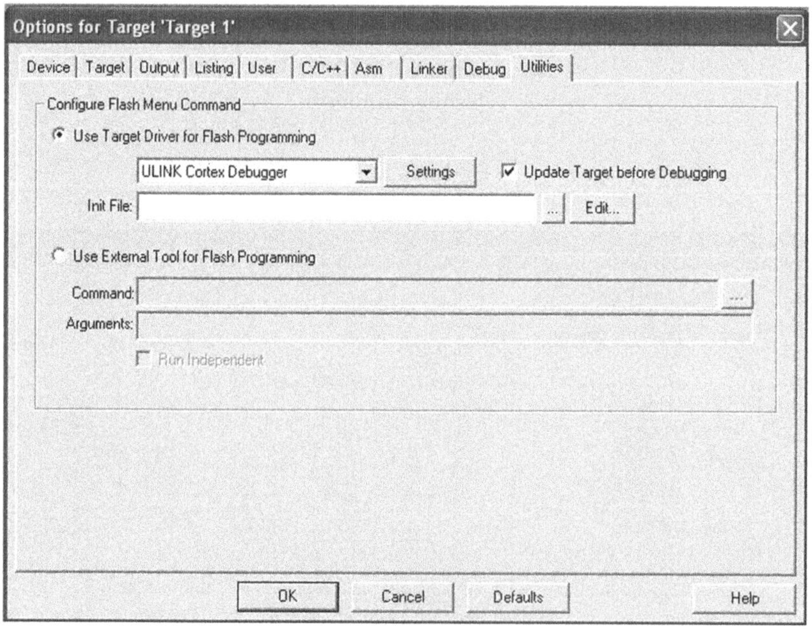

Figure 14.13:
Utilities option tab in project options.

Finally, it might also be necessary to set up the Utilities options in the project to select ULINK 2 (or another in-circuit debugger of your choice) as the programming method. This is shown in Figure 14.13. The "Update Target before Debugging" option automatically updates the program to the flash memory and then starts the debug session. This saves the users from having to remember to reprogram the flash each time the program image is rebuilt.

In the utilities tab, we can also access the flash algorithm settings from the setting button.

If you are using NXP LPC11xx (Cortex-M0), LPC13xx (Cortex-M3), or LPC17xx (Cortex-M3) and if you are not using Keil's ULINK in-circuit debugger for flash memory programming, an extra step might be needed. In these NXP products, address 0x1C-0x1F (32-bit) is used as a checksum and is generated automatically during flash memory programming. Because the program image generated by the linker does not have this checksum, the flash programmer would report an error as it tries to verify the programmed image and compares the read-back values to the original image.

To solve this problem, Keil MDK has included a utility call ElfDwt, a signature creator that inserts the required checksum to the program image generated by the linker. This feature is introduced in MDK 4.10. To use this feature, you can run ELFDWT directly in the folder that contains the program image:

```
C:\KEIL\ARM\BIN\ELFDWT elf_file.AXF
```

Alternatively, this can be set up as an automatic process by adding the command line "$K\ARM\BIN\ELFDWT #L" in the User option in the project setting, in the "Run User Program After Build/Rebuild" field (Figure 14.14).

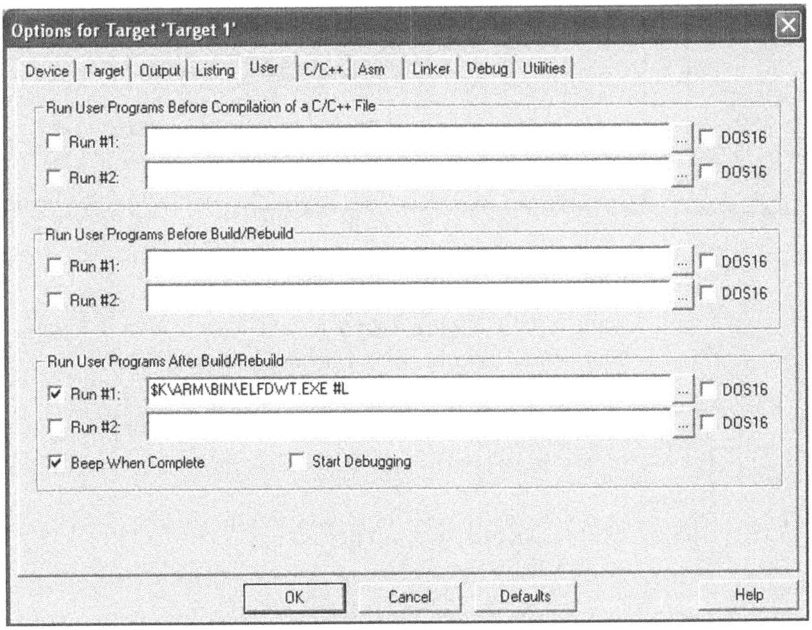

Figure 14.14:
Using ELFDWT for NXP LPC11xx/LPC13xx/LPC17xx product.

When this option is used, there is an additional output message during the build process:

```
Build target 'Demonstration'
assembling startup_LPC11xx.s...
compiling blinky.c...
linking...
Program Size: Code=540 RO-data=228 RW-data=0 ZI-data=608
User command #1: C:\Keil\\ARM\BIN\ELFDWT.EXE C:\CortexM0\2_blinky\blinky.axf
ELFDWT - Signature Creator V1.00
COPYRIGHT Keil - An ARM Company, Copyright (C) 2010
*** Updated Signature over Range[ 32] (0x00000000 - 0x00000018): @0x0000001C = 0xEFFFF9AF
*** Processing completed, no Errors.
"blinky.axf" - 0 Error(s), 0 Warning(s).
```

Please note that this setting is not necessary if a ULINK product is used for flash programming or if the project does not use a NXP LPC11xx/LPC13xx/LPC17xx product.

Once completed, the application you have created is ready to be compiled and tested.

Compile and Build the Program

The compile process can be carried out by a number of buttons on the toolbar as shown in Figure 14.15. Simply click on the "Build Target" button to start the compile process, use the pull-down menu (in the Project menu → Build Target), or use hot key F7. After the program is compiled and linked, we will see the compile status message as shown in Figure 14.16.

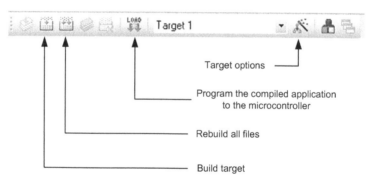

Figure 14.15:
Frequently used buttons on the toolbar.

```
Build Output
Build target 'Target 1'
assembling startup_LPC11xx.s...
compiling blinky.c...
linking...
Program Size: Code=528 RO-data=240 RW-data=0 ZI-data=608
"blinky.axf" - 0 Error(s), 0 Warning(s).
```

Figure 14.16:
Compile result for the blinky project on the Build Output window.

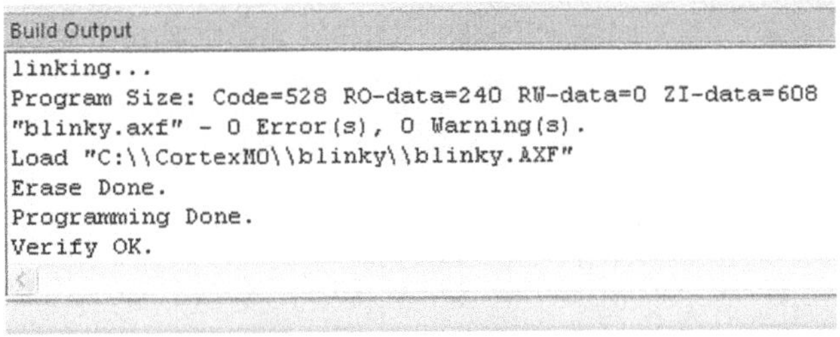

```
Build Output
linking...
Program Size: Code=528 RO-data=240 RW-data=0 ZI-data=608
"blinky.axf" - 0 Error(s), 0 Warning(s).
Load "C:\\CortexM0\\blinky\\blinky.AXF"
Erase Done.
Programming Done.
Verify OK.
```

Figure 14.17:
Flash programming status output.

It is now possible to program the application to the microcontroller's flash memory using the "Load" button on the toolbar (Figure 14.17).

The program can then be tested by starting a debug session by using the pull-down menu (Debug → Start/Stop Debug session), by clicking on the debug session button ⓠ on the toolbar, or using the hot key Ctrl-F5 (Figure 14.18). When the debug session starts, by default it will start at the boot loader inside the NXP LPC111x. This behavior is specific to the LPC111x design and can be completely different in other Cortex-M0 products.

Now click the run button; the LED (connected to port 2 bit 0 or port 0 bit 7) starts to blink. Congratulations! You have successfully created and tested your first Cortex-M0 project.

Using the Debugger

The debugger in µVision IDE provides a lot of useful features. For example, in the C source code window, a breakpoint can be inserted by simply right-clicking on a line of C code and then selecting insert breakpoint (Figure 14.19).

The following descriptions cover some of the commonly used debug features:

- *Processor registers access.* When the processor core is halted, the current register values are displayed in the register window. You can modify the value by double-clicking on the value you want to change.
- *Examine memory contents.* The memory contents can be examined by entering the address value in the memory window. The display can be configured to display the values as bytes, half words (16-bit), or 32 bit words. You can also modify the values in the memory. The memory window can be used even when the processor is running.
- *Single stepping.* You can carry out single stepping in the C source code window, as well as in the assembly window. When single stepping in the C source code window, each line of C source code will perform the single stepping. When single stepping in the disassembly window, you can single step the assembly instructions one by one.

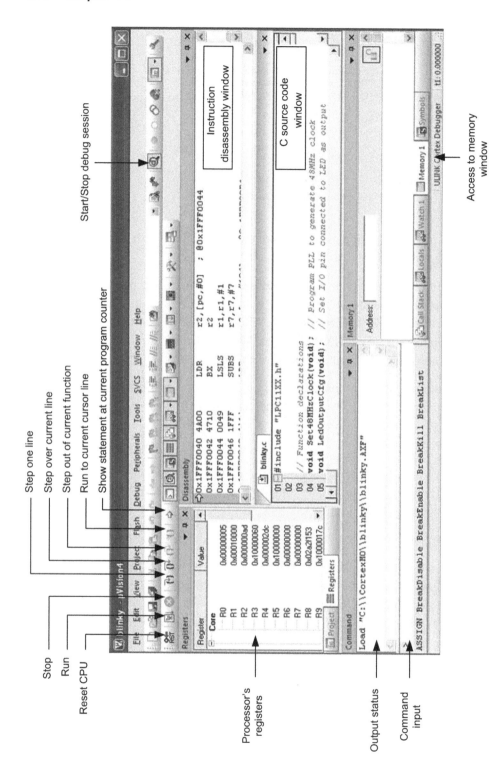

Figure 14.18:
Screen shot of debug sessions.

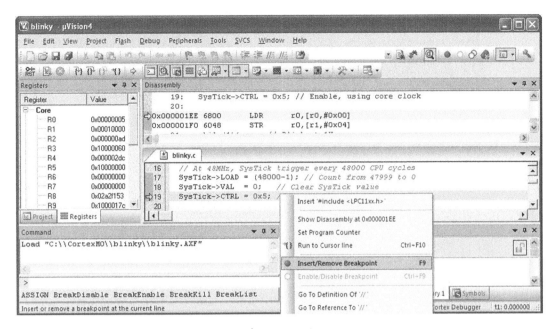

Figure 14.19:
Breakpoint can be set by right-click on a line of code and select insert breakpoint.

- *Run to main.* This feature allows the processor to start program execution after reset and halt at "main()," instead of halting at the beginning of the internal boot loader or startup code. This is done by checking the "Run to main" option in the project's debug option as shown in Figure 14.20. However, if the program does not start properly, for example, because of an incorrect memory map configuration or incorrect stack size setup, you should disable this option so that you can debug the reset handler.
- *Peripheral pull-down menu.* This pull-down menu provides access to the status of NVIC, System Control Block register, and the registers in the SysTick timer.

Other Project Configurations

In the μVision IDE, there are a number of other useful project options.

Target, Source Groups

In the project window, by default the project created consists of only one target, "Target 1," which contains a source group called "Source Group 1." You can rename the target and source groups to make it more clear how the files in the project are organized. These items can be renamed by clicking on the "Target 1" or "Source Group 1" to highlight the text and clicking on it again to edit the text directly. You can also add other groups to the project.

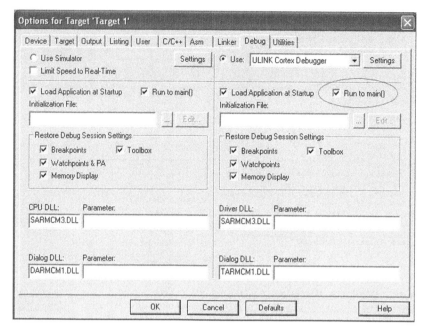

Figure 14.20:
Run to the main option.

For example, you can divide the source code files into groups like "startup" and "application," and you can even add documentation into the project by adding text files. This is useful when a project contains a large number of files. By organizing files into various groups, it is easier to locate the files you want to edit. An example of project file arrangement is shown in Figure 14.21.

A µVision project can contain multiple targets. Different targets can have different compiler and debug settings. By creating multiple targets, switching between multiple project configurations is easy. An additional target can be added by right-clicking on the target name and selecting manage components.

A typical usage of having multiple targets is to allow the same applications to be compiled with debug symbols and without debug symbols (for release). If there are multiple targets in a project, you can switch between targets using the target selection box on the toolbar (at the right-hand side of the flash download button).

Compiler and Code-Generation Options

A number of compiler and code-generation options are available to allow different optimizations. The first group of the options is composed of the C compiler options, as shown in Figure 14.22.

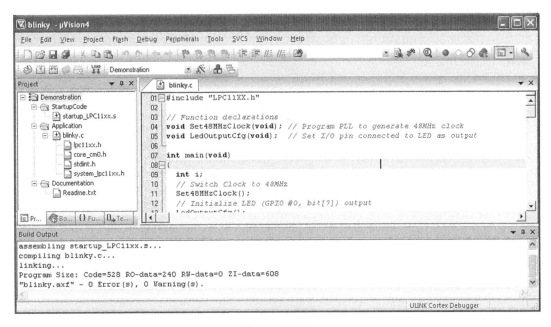

Figure 14.21:
Target name and group renamed.

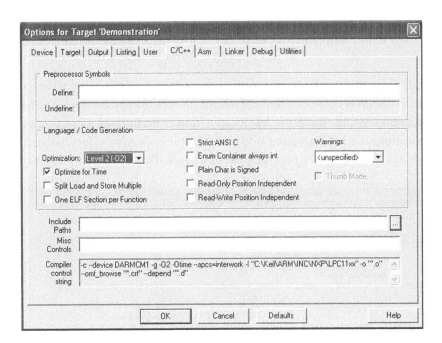

Figure 14.22:
C compiler options.

Table 14.1: Optimization Levels

Optimization Level	Descriptions
-O0	*Applies minimum optimization.* Most optimizations are switched off, and the code generated has the best debug view.
-O1	*Applies restricted optimization.* Unused inline functions, unused static functions, and redundant codes are removed. Instructions can be reordered to avoid interlock situations. The code generated is reasonably optimized with a good debug view.
-O2	*Applies high optimization.* Optimizes the program code according to the processor specific behavior. The code generated is highly optimized, with limited debug view.
-O3	*Applies the most aggressive optimization.* Optimizes in accordance with the time/space option. By default, multifile compilation is enabled at this level. This gives the highest level of optimization but takes longer compilation time and has lower software debug visibility.

The C compiler options allow you to select optimization levels (0 to 3) through a drop-down menu (Table 14.1). When using level 3, optimization is set for code size by default unless the tick box "Optimize for Time" is set. There are various other compiler configurations, which can be set by the tick boxes. All of these settings will appear in the "Compiler control string" text box. You can also add more compiler switches directly in the "Misc Controls" text box. For example, if you are using a Cortex-M0 product with a 32-cycle multiplier (e.g., Cortex-M0 in minimum size configuration), you can add the *--multiply_latency=32* option so that the C compiler can optimize the generated code accordingly.

A second group of useful options can be found in the target options window as shown in Figure 14.23.

The MicroLIB C library is optimized for microcontrollers and other embedded applications. If the MicroLIB option is not selected, the standard ISO C libraries are used. The MicroLIB has a smaller program memory footprint, but it has a slower performance and a few limitations. In most applications that are migrating from 8-bit/16-bit microcontrollers to the ARM Cortex-M0 processor, the slightly lower performance of MicroLIB is unlikely to be an issue because

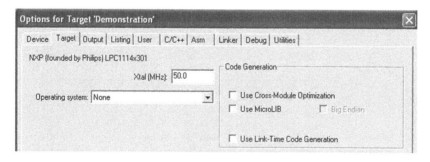

Figure 14.23:
Code-generation options.

the Cortex-M0 processor provides a much higher performance than most 8-bit or 16-bit processors.

In some cases, the cross module optimization and link-time code-generation options can also improve the program size and performance:

- *Cross module optimization.* This technique takes information from a prior build and uses it to place UNUSED functions into their own ELF section in the compiled object file. In this way, the linker can remove unused functions to reduce code size.
- *Link-time code generation.* The objects in the compiled output will be in an intermediate format so that the linker can perform further code optimizations during the linkage stage. This can reduce code size and allow the applications to run faster.

More details of optimization techniques can be found in Keil Application Note 202—MDK-ARM Compiler Optimizations (reference 8).

Simulator

The μVision IDE includes a simulator. The simulator provides instruction set level simulation, and for some microcontroller devices, a device-level simulation feature (including peripheral simulation) is also available. To enable this feature, change the debug option to use simulation as shown in Figure 14.24.

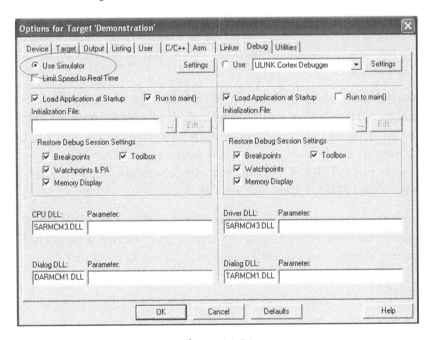

Figure 14.24:
Simulation option in debug.

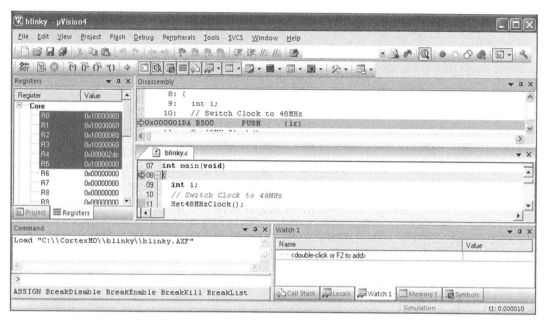

Figure 14.25:
Debug simulation window.

After this is set, you can start your debug session using the pull-down menu (Debug → Start/ Stop Debug session), or by clicking on the debug session button ⓠ on the toolbar, or by using the hot key Ctrl-F5. A debug window will then be displayed as shown in Figure 14.25, from here it is possible to execute the program, single step through the program, and also examine the system status.

In some cases, depending on the microcontroller product you are using, the debug simulator might not be able to fully simulate all the peripherals available on the microcontroller. Also, it may be necessary to adjust the memory map of the simulated device. This can be done by accessing the memory configuration via the pull-down menu: Debug → Memory. This is not required if you are using NXP LPC111x microcontrollers because the Keil MDK provides full device simulation for LPC111x from version 4.10.

A very useful feature in the simulator is the execution profiling. By enabling the profiling function from the pull-down menu (Debug → Execution Profiling) as shown in Figure 14.26, you can measure the execution time of a function as shown in Figure 14.27.

Execution in RAM

In addition to downloading the program to flash memory, you can also download a program to RAM and test it without changing the content inside the flash memory. To do this, we need to change a number of options in the project. First, we need to specify the new memory map

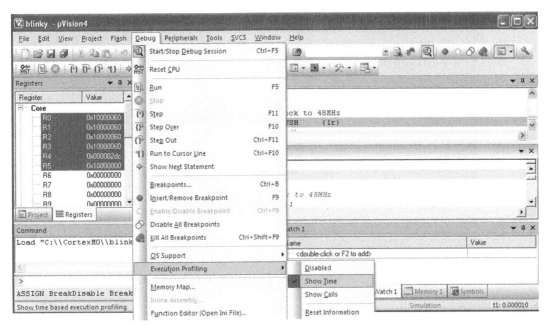

Figure 14.26:
Enabling the profiling feature.

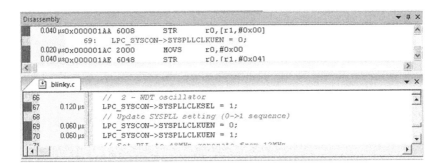

Figure 14.27:
Timing information of program execution.

for the compiled image, as shown in the example in Figure 14.28. The memory layout depends on the microcontroller used for the project. For this example, the 8 KB of RAM on LPC1114 is divided into two halves, one for the program code and the other for data and stack.

Then, the flash programming option is modified to remove flash programming, as shown in Figure 14.29.

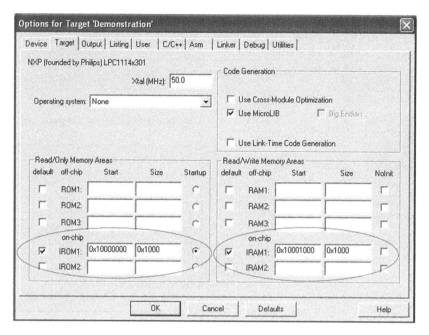

Figure 14.28:
Example memory configuration for program execution from RAM.

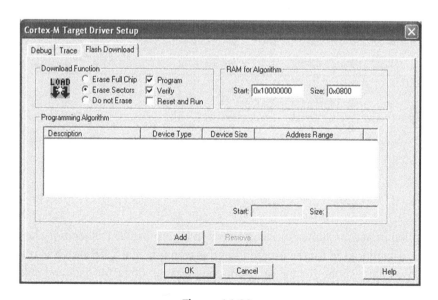

Figure 14.29:
Flash programming algorithm is not required if the program is to be tested in RAM.

The next step is to create a simple debug startup script to load the initial stack pointer and program counter to the right location. For this example, a file called ram_debug.ini is created with the following text:

```
ram_debug.ini

reset
// System memory remap register in LPC1114
// User RAM mode  : Remap interrupt vectors to SRAM
_WDWORD(0x40008000, 0x00000001);

LOAD blinky.axf INCREMENTAL  // Download image to board

SP = _RDWORD(0x10000000);    // Setup Stack Pointer
PC = _RDWORD(0x10000004);    // Setup Program Counter

g, main                      // Goto Main
```

We then need to set up the debug option so that this debug startup script is used when the debug session starts. The debug option changes for this example are shown in Figure 14.30.

Now we can start the debug session by using the start debug button on the toolbar, using the pull-down menu (Debug → Start/Stop Debug session), or using hot key Ctrl-F5. Note that

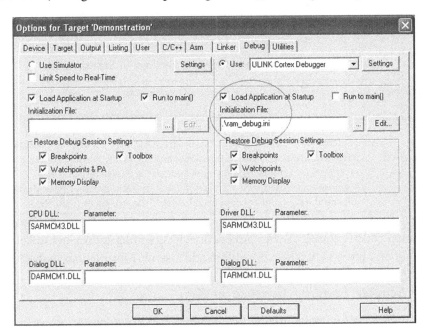

Figure 14.30:
Adding a debug startup script file in debug options.

we do not use the flash programming button for this example because the program is running from SRAM.

When the debug session starts, it will download the program to SRAM and set the program counter to the correct starting point in our program image automatically. The application can then be started.

Testing a program image from RAM can have a number of limitations. First, it is necessary to use a debugger script to change the program counter and initial stack pointer to the right locations. Otherwise, the reset vector and initial stack pointer value in the flash memory will be used after the processor is reset. The second issue is that additional hardware is required to use the exception vector table in RAM. The vector table normally resides in the flash memory from address 0x0. Because of the inclusion of some additional memory remapping hardware (as in the case of NXP LPC1114), the vector table in the program image in SRAM is remapped to this address. Therefore, in the debug script, ram_debug.ini, we programmed the system memory remap register to enable this remap function. If such remapping hardware is not available, the vector table in the flash memory would be used, making testing of interrupts more difficult.

Customizing the Startup Code in Keil

The startup code in the Keil μVision IDE provides the vector table as well as a great deal of configuration information. For example, the stack size and heap size are defined here. For applications that require more stack space or heap space, you can edit the startup code to get the required stack or heap memory size (Table 14.2).

Table 14.2: Stack Size and Heap Size Settings in Startup Code

Defines in Startup Code	Descriptions
Stack_Size	Memory space allocated for stack memory. The stack can be used for saving register contents when calling a function, as space for local variable in a function, for parameter passing between functions, and for saving register contents during exceptions/interrupts. The usual default size is 512 bytes (0x200).
Heap_Size	Memory space used by a number of C library functions that need dynamic memory allocation (e.g., malloc). The usual default size is 0 bytes (0x000).

After the startup code is copied to your project directory, editing it does not affect the default startup code for new projects. It is also possible to add assembly instructions to this file if needed.

Using the Scatter Loading Feature in Keil

Besides using the memory layout dialog (see Figure 14.28) to define a project's memory layout, you can also use a scatter file to define the memory layout in the microcontroller. This allows a more complex memory layout to be created.

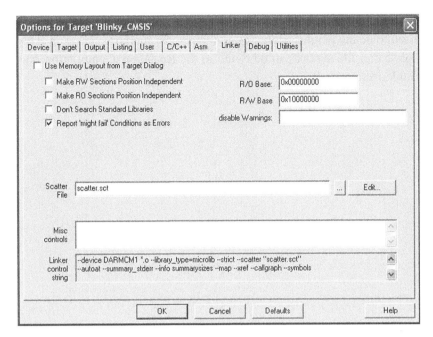

Figure 14.31:
Scatter file option in Keil MDK.

In the linker settings of project options, after disabling the "Use Memory Layout from Target Dialog" option, you can then define a scatter file to use the scatter loading feature (Figure 14.31).

A simple scatter file is as follows:

```
LOAD_REGION 0x00000000 0x00200000
{ ; flash memory start at 0x00000000
 ;; Maximum of 48 exceptions (48* 4 bytes == 0xC0)
 VECTORS 0x0 0xC0
 {
    ; Provided by the user in startup_LPC11xx.s
    * (RESET,+FIRST)
 }

 CODE 0xC0 FIXED
 {  ; The rest of the program code start from 0xC0
    * (+RO)
 }

 DATA 0x10000000 0x2000
 {  ; In LPC1114, the SRAM start at 0x10000000
    * (+RW, +ZI)
 }
}
```

In addition to defining the memory layout available in the microcontroller, the scatter file can also be used to define pointer addresses and to reserve memory spaces in RAM for special use. Details of scatter file syntax can be found in the RealView Compilation Tools Developer Guide (reference 9).

## CHAPTER 15

Simple Application Programming

Using CMSIS

In the blinky example presented in the previous chapter, we used a CMSIS header file (LPC11xx.h) so that we did not have to re-create the register definitions ourselves. The benefit of CMSIS does not stop there. In the next example, we will see how CMSIS can greatly simplify our blinky example program.

In the CMSIS software package from NXP website, or from Keil installation (C:\Keil\ ARM\Startup\NXP\LPC11xx), we can find a file called system_LPC11xx.c, which is specific to the NXP LPC11xx microcontrollers. Create a new blinky project in a new directory and copy the file system_LPC11xx.c to the new directory. Use the µVision IDE to add this file to the new blinky project (Figure 15.1).

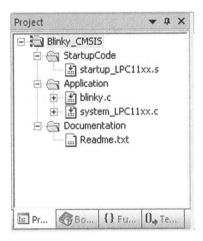

Figure 15.1:
Project window for blinky example with CMSIS.

The system_LPC11xx.c contains a system initialization function for LPC11xx. Instead of using a custom clock initialization function ("Set48MHzClock()"), the CMSIS system initialization function in this file (called "SystemInit()") is used instead. The system_LPC11xx.c has a number of parameters that can be easily customized for different clock frequency requirements. The settings in Table 15.1 are used to generate a 48MHz clock from a 12MHz crystal.

The Definitive Guide to the ARM Cortex-M0. DOI: 10.1016/B978-0-12-385477-3.10015-1

Table 15.1: Parameter Examples in System_LPC11xx.c

Parameter	Value	Descriptions
__XTAL	12000000	Crystal clock frequency
__IRC_OSC_CLK	12000000	Internal RC oscillator frequency
CLOCK_SETUP	1	Enable clock setup processing in SystemInit()
SYSCLK_SETUP	1	Enable system clock setup in SystemInit()
SYSOSC_SETUP	1	Enable external crystal oscillator setup in SystemInit()
SYSPLL_SETUP	1	Enable system PLL setup process in SystemInit()
SYSPLLCTRL_Val	0x23	System PLL MSEL (M = 4, P = 2)
MAINCLKSEL_Val	3	PLL use crystal oscillator as source
SYSAHBCLKDIV_Val	1	AHB clock frequency ratio is 1

In the example that follows, we modified the blinky example from the previous chapter to demonstrate the usage of CMSIS functions. Three CMSIS functions are used here: SystemInit() for clock initialization, SysTick_Handler() to toggle the LED, and SysTick_Config() to set up the SysTick timer for regular SysTick interrupts.

blinky.c (with CMSIS)

```
#include "LPC11XX.h"
#define KEIL_MCB1000_BOARD


// Function declarations
void LedOutputCfg(void);  // Set I/O pin connected to LED as output

int main(void)
{
  // Switch Clock to 48MHz
  SystemInit();
  // Initialize LED output
  LedOutputCfg();
  // Program SysTick timer interrupt at 1KHz.
  // At 48MHz, SysTick trigger every 48000 CPU cycles
  SysTick_Config(48000);

  while(1);
} // end main

// SysTick handler to toggle LED every 500 ticks
void SysTick_Handler(void)
{
static short int TickCount = 0;
if ((TickCount++) == 500) { // for every 500 counts, toggle LED
  TickCount = 0; // reset counter to 0
#ifdef KEIL_MCB1000_BOARD
  // For Keil MCB1000, use P2.0 for LED output
  LPC_GPIO2->MASKED_ACCESS[1] = ~LPC_GPIO2->MASKED_ACCESS[1]; // Toggle bit 0
```

```
#else
  LPC_GPIO0->MASKED_ACCESS[1<<7] = ~LPC_GPIO0->MASKED_ACCESS[1<<7];
  // Toggle bit 7
#endif
  }
return;
}
// Switch LED signal (P0_7) to output port with no pull up or pulldown
void LedOutputCfg(void)
{
  // Enable clock to IO configuration block (bit[16] of AHBCLOCK Control register)
  // and enable clock to GPIO (bit[6] of AHBCLOCK Control register
  LPC_SYSCON->SYSAHBCLKCTRL = LPC_SYSCON->SYSAHBCLKCTRL | (1<<16) | (1<<6);

#ifdef KEIL_MCB1000_BOARD
  // For Keil MCB1000, use P2.0 for LED output
  // PIO2_0 IO output config
  //  bit[5]   - Hysteresis (0=disable, 1 =enable)
  //  bit[4:3] - MODE(0=inactive, 1 =pulldown, 2=pullup, 3=repeater)
  //  bit[2:0] - Function (0 = IO, 1=DTR, 2=SSEL1)
  LPC_IOCON->PIO2_0 = (0<<5) + (0<<3) + (0x0);

  // Initial bit 0 output is 0
  LPC_GPIO2->MASKED_ACCESS[1] = 0;
  // Set pin 7 to 0 as output
  LPC_GPIO2->DIR = LPC_GPIO2->DIR | 0x1;
#else
  // For LPCXpresso, use P0.7 for LED output
  // PIO0_7 IO output config
  //  bit[5]   - Hysteresis (0=disable, 1 =enable)
  //  bit[4:3] - MODE(0=inactive, 1 =pulldown, 2=pullup, 3=repeater)
  //  bit[2:0] - Function (0 = IO, 1=CTS)
  LPC_IOCON->PIO0_7 = (0x0) + (0<<3) + (0<<5);
  // Initial bit[7] output is 0
  LPC_GPIO0->MASKED_ACCESS[1<<7] = 0;
  // Set pin 7 as output
  LPC_GPIO0->DIR = LPC_GPIO0->DIR | (1<<7);
#endif
  return;
} // end LedOutputCfg
```

Instead of polling the SysTick status for timing control, this example uses the SysTick exception. The startup code "startup_LPC11xx.s" already has the SysTick_Handler (the CMSIS standardized name for SysTick handler) defined in the vector table, so we only need to create the handler code for SysTick exception in C. The SysTick exception handler increments the "TickCount" variable each time it is executed, and it toggles the LED for every 500 times of execution. Unlike most peripheral timers, the SysTick exception does not need to be cleared by software within the handler.

You might wonder why we do not need to define the other interrupt handlers listed in the vector table in the startup code. The startup code provided in the Keil µVision IDE already contains dummy versions of the other handlers. Because these default handlers are given the "[WEAK]" property, they can be overridden if another implementation of the handler is presented.

The initialization of the SysTick timer has been replaced by a function called SysTick_Config(). The "SysTick_Config" is a CMSIS function, and it configures the SysTick timer to generate SysTick exceptions regularly. Because the system clock frequency is 48MHz and we would like a SysTick exception for every 1 ms, program the SysTick to be triggered every 48 millions/1000 = 48,000 clock cycles.

After the modifications are made, compile and download the CMSIS blinky example to the microcontroller and test it.

Using the SysTick Timer as a Single Shot Timer

In the previous example, we used SysTick to generate the SysTick exception regularly. In this example, we use SysTick for a single-shot operation. In this way, once the SysTick exception is triggered and the handler is entered, the SysTick timer is then disabled so that the SysTick handler is only executed once.

For this example, we also edit the system_LPC11xx.c so that the SystemInit() function does not switch the clock to 48 MHz so that we can observe the delay of LED activity caused by the SysTick exception. This is done by setting CLOCK_SETUP to 0. As a result, the processor clock in the LPC1114 is running at 12 MHz using the internal RC oscillator.

---

blinky.c (with CMSIS)

```
#include "LPC11XX.h"
#define KEIL_MCB1000_BOARD

// Function declarations
void LedOutputCfg(void);   // Set I/O pin connected to LED as output

int main(void)
{
  // Clock remain at 12MHz
  SystemInit();
  // Initialize LED output
  LedOutputCfg();
  // Program SysTick timer to generate an interrupt after 0xFFFFFF cycles.
  // At 12MHz, SysTick trigger after 1.4 second
  SysTick->LOAD = 0xFFFFFF;
  SysTick->VAL  = 0x0;
  SysTick->CTRL = 0x7; // Enable SysTick with exception generation
                       // and use core clock as source
  while(1);
```

```
} // end main

// SysTick handler to toggle LED and disable SysTick
void SysTick_Handler(void)
{
#ifdef KEIL_MCB1000_BOARD
  // For Keil MCB1000, use P2.0 for LED output
  LPC_GPIO2->MASKED_ACCESS[1] = ~LPC_GPIO2->MASKED_ACCESS[1]; // Toggle bit 0
#else
  LPC_GPIO0->MASKED_ACCESS[1<<7] = ~LPC_GPIO0->MASKED_ACCESS[1<<7];
  // Toggle bit 7
#endif

  // Disable SysTick
  SysTick->CTRL = 0;
  // Clear SysTick pending status in case it has already been triggered
  SCB->ICSR = SCB->ICSR | (1<<25); // Set PENDSTCLR
  return;
}
// Switch LED signal (P0_7) to output port with no pull up or pulldown
void LedOutputCfg(void)
{
  // Enable clock to IO configuration block (bit[16] of AHBCLOCK Control
  register)
  // and enable clock to GPIO (bit[6] of AHBCLOCK Control register
  LPC_SYSCON->SYSAHBCLKCTRL = LPC_SYSCON->SYSAHBCLKCTRL | (1<<16) | (1<<6);

#ifdef KEIL_MCB1000_BOARD
  // For Keil MCB1000, use P2.0 for LED output
  // PIO2_0 IO output config
  //  bit[5]  - Hysteresis (0=disable, 1 =enable)
  //  bit[4:3] - MODE(0=inactive, 1 =pulldown, 2=pullup, 3=repeater)
  //  bit[2:0] - Function (0 = IO, 1=DTR, 2=SSEL1)
  LPC_IOCON->PIO2_0 = (0<<5) + (0<<3) + (0x0);

  // Initial bit 0 output is 0
  LPC_GPIO2->MASKED_ACCESS[1] = 0;
  // Set pin 7 to 0 as output
  LPC_GPIO2->DIR = LPC_GPIO2->DIR | 0x1;
#else
  // For LPCXpresso, use P0.7 for LED output
  // PIO0_7 IO output config
  //  bit[5]  - Hysteresis (0=disable, 1 =enable)
  //  bit[4:3] - MODE(0=inactive, 1 =pulldown, 2=pullup, 3=repeater)
  //  bit[2:0] - Function (0 = IO, 1=CTS)
  LPC_IOCON->PIO0_7 = (0x0) + (0<<3) + (0<<5);
  // Initial bit[7] output is 0
  LPC_GPIO0->MASKED_ACCESS[1<<7] = 0;
  // Set pin 7 as output
  LPC_GPIO0->DIR = LPC_GPIO0->DIR | (1<<7);
#endif
  return;
} // end LedOutputCfg
```

When the program is executed, approximately 1.4 seconds after the program starts, the LED is turned on. During this delay period, the SysTick timer decreases from 0xFFFFFF and reaches 0. To generate short timing delays, you might want to factor in the interrupt latency during the SysTick timer exception entry. For example, to have the delay of 100 cycles, program the SysTick reload value to $99 - 16$ (interrupt latency) $= 83$.

UART Examples

A single blinking LED provides very little information to the outside world. In most applications, microcontrollers need to communicate with their environment using a more efficient method. This example demonstrates how to use a Universal Asynchronous Receiver/Transmitter (UART) interface on a Cortex-M0 microcontroller to communicate with a terminal application running on a personal computer.

A UART is a simple serial communication protocol that can be used to transfer text as well as binary data. It is widely supported by microcontrollers and is often available in older generations of personal computers (usually referred as COM ports or serial ports). Although UARTs have been around for long time, they are still commonly used in embedded systems.

To test the program, connect the UART interface of the microcontroller to the serial port of your PC through an RS232 level shifter (Figure 15.2). If the PC does not have a serial port, a USB-to-serial adaptor can be purchased from most electronic stores.

On the personal computer, open a terminal program, such as the Window's Hyper Terminal utility, to send and receive data between the computer and the microcontroller. Data received from the microcontroller are displayed, and user's keystrokes are sent back to the microcontroller. It is also possible to use other terminal software to carry out these tests. The example code here contains additional features to allow better handling of new line and carriage return characters for the Hyper Terminal.

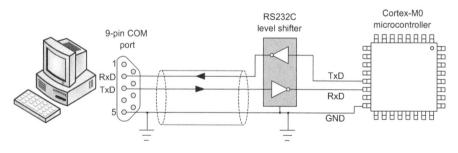

Figure 15.2:
Connecting the UART of the Cortex-M0 microcontroller to a personal computer.

Simple Input/Output

Similar to the examples in the previous chapter, we created a new project targeted at NXP
LPC1114 and added the CMSIS function to the project. The main program is called
simple_uart.c. This program carries out the following steps:

- Sets up the system clock (SystemInit)
- Initializes the UART interface to a baud rate of 38400, 8-bit data, no parity, and no flow
  control (UartConfig)
- Prints a "Hello" message (UartPuts)
- Echos any key that you send to the UART

The program code is written as follows:

```c
simple_uart.c

#include "LPC11XX.h"
// Function declarations
void          UartConfig(void);          // UART configuration
unsigned char UartPutc(unsigned char my_ch);    // UART character output
void          UartPuts(unsigned char * mytext); // UART string output
int           UartGetRxDataAvail(void); // Detect if new data is received
unsigned char UartGetRxData(void);       // Get received data from UART

int main(void)
{
  SystemInit();
  UartConfig();

  UartPuts("Hello\n");
  while(1){
    if (UartGetRxDataAvail()) {
        UartPutc(UartGetRxData());  // echo received data
        }    // end if
    } // end while
}

// Uart string output
void UartPuts(unsigned char * mytext)
{
  unsigned char CurrChar;
  CurrChar = *mytext;
  while (CurrChar != (char) 0x0){
    UartPutc(CurrChar);  // Normal data
    mytext++;
    CurrChar = *mytext;
    }
  return;
}
```

(Continued)

simple_uart.c—Cont'd

```c
void UartConfig(void)
{
  // UART interface are : PIO1_7 (TXD) and PIC1_6 (RXD)
  // Other UART signals (DTR, DSR, CTS, RTS, RI) are not used

  // Enable clock to IO configuration block
  // (bit[16] of AHBCLOCK Control register)
  LPC_SYSCON->SYSAHBCLKCTRL = LPC_SYSCON->SYSAHBCLKCTRL | (1<<16);

  // PIO1_7 IO output config
  //  bit[5]  - Hysteresis (0=disable, 1 =enable)
  //  bit[4:3] - MODE(0=inactive, 1 =pulldown, 2=pullup, 3=repeater)
  //  bit[2:0] - Function (0 = IO, 1=TXD, 2=CT32B0_MAT1)
  LPC_IOCON->PIO1_7 = (0x1) + (0<<3) + (0<<5);
  // PIO1_6 IO input config
  //  bit[5]  - Hysteresis (0=disable, 1 =enable)
  //  bit[4:3] - MODE(0=inactive, 1 =pulldown, 2=pullup, 3=repeater)
  //  bit[2:0] - Function (0 = IO, 1=RXD, 2=CT32B0_MAT0)
  LPC_IOCON->PIO1_6 = (0x1) + (2<<3) + (1<<5);

  // Enable clock to UART (bit[12] of AHBCLOCK Control register
  LPC_SYSCON->SYSAHBCLKCTRL = LPC_SYSCON->SYSAHBCLKCTRL | (1<<12);
  // UART_PCLK divide ratio = 1
  LPC_SYSCON->UARTCLKDIV = 1;

  // UART_PCLK = 48MHz, Baudrate = 38400, divide ratio = 1250
  // Line Control Register
  LPC_UART->LCR = (1<<7) |   // Enable access to Divisor Latches
      (0<<6) |   // Disable Break Control
      (0<<4) |   // Bit[5:4] parity select (odd, even, sticky-1, sticky-0)
      (0<<3) |   // parity disabled
      (0<<2) |   // 1 stop bit
      (3<<0);    // 8-bit data
  LPC_UART->DLL = 78;  // Divisor Latch Least Significant Byte
                       // 48MHz/38400/16 = 78.125
  LPC_UART->DLM = 0;   // Divisor Latch Most Significant Byte  : 0
  LPC_UART->LCR = (0<<7) |   // Disable access to Divisor Latches
      (0<<6) |   // Disable Break Control
      (0<<4) |   // Bit[5:4] parity select (odd, even, sticky-1, sticky-0)
      (0<<3) |   // parity disabled
      (0<<2) |   // 1 stop bit
      (3<<0);    // 8-bit data

  LPC_UART->FCR = 1; // Enable FIFO
  return;
}
// Get received data
__inline unsigned char UartGetRxData(void)
{
  return ((char)LPC_UART->RBR);
}
```

```
// Detect if new received data is available
__inline int  UartGetRxDataAvail(void){
  return (LPC_UART->LSR & 0x1);
}
// Output a character, with additional formatting for HyperTerminal
unsigned char UartPutc(unsigned char my_ch)
{
  if (my_ch == '\n') {
    while ((LPC_UART->LSR & (1<<5))==0);
        // Wait if Transmit Holding register is not empty
    LPC_UART->THR = 13;
        // Output carriage return (for Windows Hyperterminal)
    }
  while ((LPC_UART->LSR & (1<<5))==0);
      // Wait if Transmit Holding register is not empty
  LPC_UART->THR = my_ch; // write to transmit holding register

  if (my_ch == 13) {
    while ((LPC_UART->LSR & (1<<5))==0);
        // Wait if Transmit Holding register is not empty
    LPC_UART->THR = 10;
        // Output new line (for Windows Hyperterminal)
    }
  return (my_ch);
}
```

After creating the program file, set up the rest of the files in the project, as shown in Figure 15.3.

Do not forget that it may also be necessary to set the debug, flash programming, and utilities options in the project to enable debugging and flash programming.

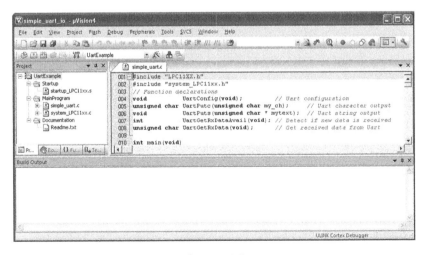

Figure 15.3:
Project for simple UART interface test.

When the program is executed, a "Hello" message will be displayed on the terminal application on the personal computer. The microcontroller then echoes the keystrokes that are typed in to the terminal program.

Retargeting

In user interface designs, it is common to use library functions, such as the "printf" function, to handle the formatting of the output string in message displays. This practice is commonly known as "retargeting" in embedded software programming. To enable "retargeting" of the input and output to the UART interface, a few more functions need to be implemented. For Keil MDK or ARM RealView C compiler, the C library uses the function "fputc" for output redirection and "fgetc" for input redirection. For convenience, these functions are grouped together in a file called retarget.c. Example implementations of retarget.c can be found in various examples in the Keil MDK installation. The following example has been modified to call the UART functions implemented in the UART application program file.

retarget.c

```c
/***************************************************************************/
/* RETARGET.C: 'Retarget' layer for target-dependent low level functions  */
/***************************************************************************/
/* This file is part of the uVision/ARM development tools.                */
/* Copyright (c) 2005-2009 Keil Software. All rights reserved.            */
/* This software may only be used under the terms of a valid, current,    */
/* end user licence from KEIL for a compatible version of KEIL software   */
/* development tools. Nothing else gives you the right to use this software. */
/***************************************************************************/
#include <stdio.h>
#include <time.h>
#include <rt_misc.h>
#pragma import(__use_no_semihosting_swi)

extern unsigned char UartGetc(void);
extern unsigned char UartPutc(unsigned char my_ch);
struct __FILE { int handle; /* Add whatever you need here */ };
FILE __stdout;
FILE __stdin;
int fputc(int ch, FILE *f) {
  return (UartPutc(ch));
}

int fgetc(FILE *f) {
  return (UartPutc(UartGetc()));
}

int ferror(FILE *f) {
  /* Your implementation of ferror */
  return EOF;
}
```

```
void _ttywrch(int ch) {
  UartPutc (ch);
}


void _sys_exit(int return_code) {
label:  goto label;  /* endless loop */
}
```

The main application code is similar to that in the previous example, but this time C library functions are used for user inputs and message display. The flow chart of this example project is shown in Figure 15.4. Compared to the last example, the project for this example has the addition of "retarget. c" (Figure 15.5).

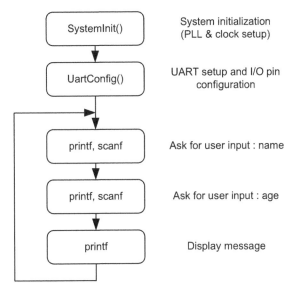

Figure 15.4:
Flowchart of a simple retarget example program.

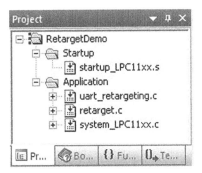

Figure 15.5:
Project for the UART retargeting demonstration.

The program code for the uart_retargeting.c is as follows:

```
uart_retargeting.c
#include "LPC11XX.h"
#include <stdio.h>

// Function declarations
void          UartConfig(void);           // UART configuration
unsigned char UartPutc(unsigned char my_ch);      // UART character output
void          UartPuts(unsigned char * mytext);  // UART string output
int           UartGetRxDataAvail(void); // Detect if new data is received
unsigned char UartGetRxData(void);        // Get received data from UART
unsigned char UartGetc(void);             // UART character input

int main(void)
{
  char UserName[40];
  int  UserAge;

  SystemInit();  // System Initialization
  UartConfig();  // Initialize UART

  while(1){
    printf ("Please enter your name: ");
    scanf ("%s", &UserName[0]);
    printf ("Please enter your age: ");
    scanf ("%d", &UserAge);
    printf ("\n  Hi %s, you are %d years old\n\n", UserName, UserAge);
    } // end while
}

void UartConfig(void)
{
  // UART interface are : PIO1_7 (TXD) and PIC1_6 (RXD)
  // Other UART signals (DTR, DSR, CTS, RTS, RI) are not used

  // Enable clock to IO configuration block
  // (bit[16] of AHBCLOCK Control register)
  LPC_SYSCON->SYSAHBCLKCTRL = LPC_SYSCON->SYSAHBCLKCTRL | (1<<16);

  // PIO1_7 IO output config
  //  bit[5]   - Hysteresis (0=disable, 1 =enable)
  //  bit[4:3] - MODE(0=inactive, 1 =pulldown, 2=pullup, 3=repeater)
  //  bit[2:0] - Function (0 = IO, 1=TXD, 2=CT32B0_MAT1)
  LPC_IOCON->PIO1_7 = (0x1) + (0<<3) + (0<<5);
  // PIO1_6 IO input config
  //  bit[5]   - Hysteresis (0=disable, 1 =enable)
  //  bit[4:3] - MODE(0=inactive, 1 =pulldown, 2=pullup, 3=repeater)
  //  bit[2:0] - Function (0 = IO, 1=RXD, 2=CT32B0_MAT0)
  LPC_IOCON->PIO1_6 = (0x1) + (2<<3) + (1<<5);

  // Enable clock to UART (bit[12] of AHBCLOCK Control register
```

```
LPC_SYSCON->SYSAHBCLKCTRL = LPC_SYSCON->SYSAHBCLKCTRL | (1<<12);
// UART_PCLK divide ratio = 1
LPC_SYSCON->UARTCLKDIV = 1;


// UART_PCLK = 48MHz, Baudrate = 38400, divide ratio = 1250
// Line Control Register
LPC_UART->LCR = (1<<7) |   // Enable access to Divisor Latches
    (0<<6) |   // Disable Break Control
    (0<<4) |   // Bit[5:4] parity select (odd, even, sticky-1, sticky-0)
    (0<<3) |   // parity disabled
    (0<<2) |   // 1 stop bit
    (3<<0);    // 8-bit data
LPC_UART->DLL = 78;  // Divisor Latch Least Significant Byte
                     // 48MHz/38400/16 = 78.125
LPC_UART->DLM = 0;   // Divisor Latch Most Significant Byte  : 0
LPC_UART->LCR = (0<<7) |   // Disable access to Divisor Latches
    (0<<6) |   // Disable Break Control
    (0<<4) |   // Bit[5:4] parity select (odd, even, sticky-1, sticky-0)
    (0<<3) |   // parity disabled
    (0<<2) |   // 1 stop bit
    (3<<0);    // 8-bit data


LPC_UART->FCR = 1; // Enable FIFO


return;
}
// Get received data
__inline unsigned char UartGetRxData(void)
{
return ((char)LPC_UART->RBR);
}
// Detect if new received data is available
__inline int  UartGetRxDataAvail(void)
{
return (LPC_UART->LSR & 0x1);
}
// Output a character, with additional formatting for HyperTerminal
unsigned char UartPutc(unsigned char my_ch)
{
if (my_ch == '\n') {
  while ((LPC_UART->LSR & (1<<5))==0);
      // Wait if Transmit Holding register is not empty
  LPC_UART->THR = 13;
      // Output carriage return (for Windows Hyperterminal)
  }
while ((LPC_UART->LSR & (1<<5))==0);
    // Wait if Transmit Holding register is not empty
LPC_UART->THR = my_ch; // write to transmit holding register

if (my_ch == 13) {
  while ((LPC_UART->LSR & (1<<5))==0);
```

(*Continued*)

```
uart_retargeting.c—Cont'd
        // Wait if Transmit Holding register is not empty
    LPC_UART->THR = 10;
        // Output new line (for Windows Hyperterminal)
    }
  return (my_ch);
}
// Get a character from UART, if no data available then wait
unsigned char UartGetc(void)
{
  while (UartGetRxDataAvail()==0); // wait if receive buffer empty
  return ((char)LPC_UART->RBR);
}
```

When the program is executed, it asks for your name and your age, and then it displays the results (Figure 15.6).

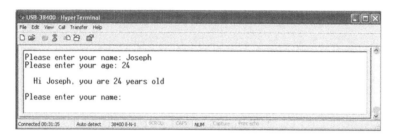

Figure 15.6:
Retargeting example.

(Alright, I admit that I lied in the example!:)

The retarget functions "fputc" and "fgetc" can be customized to use other interfaces rather than the UART. For example, you can output text to a character LCD display and use a keypad for data input. Note that the "fputc" and "fgetc" retarget functions are specific to ARM/Keil development tools. Other development tools can have difference retarget function definitions.

Developing Your Own Input and Output Functions

The C libraries provided a number of functions for text output formatting and text input; however, in some cases it is necessary to use custom input and output functions, for a couple of reasons:

- It might help to reduce program size.
- It gives complete control on the program's behavior.

One might wonder why it is important to have total control over program behavior. For one reason, if a "scanf" function is used in your program and a hacker deliberately entered a string that is much bigger than the size of the text buffer, the memory would be corrupted and the application could crash, and this might give the hacker the chance to take control of the system. Another reason is that the input device could require additional processing for detecting a user's input (e.g., a simple keypad that needs key matrix scanning). In addition, you might want to add features to allow extra capabilities in the input and output functions.

In the first UART example, a simple function called "UartPuts" was used to output a text string:

UartPuts function – display of text string by UART

```
// Uart string output
void UartPuts(unsigned char * mytext)
{
  unsigned char CurrChar;
  do {
    CurrChar = *mytext;
    if (CurrChar != (char) 0x0) {
      UartPutc(CurrChar);   // Normal data
      }
    *mytext++;
  } while (CurrChar != 0);
  return;
}
```

A simple function for outputting numeric values in hexadecimal can also be created:

UartPutHex function – display of unsigned hexadecimal value by UART

```
void UartPutHex(unsigned int din)
{
unsigned int nmask = 0xF0000000U;
unsigned int nshift = 28;
unsigned short int data4bit;
  do {
    data4bit = (din & nmask) >> nshift;
    data4bit = data4bit+48; // convert data to ASCII
    if (data4bit>57) data4bit = data4bit+7;
    UartPutc((char) data4bit);
    nshift = nshift - 4;
    nmask = nmask >> 4;
  } while (nmask!=0);
  return;
}
```

And a simple function for outputting numeric values in decimal number format can be written as follows:

UartPutDec function – display of unsigned decimal value by UART, up to 10 digits

```c
void UartPutDec(unsigned int din)
{
const unsigned int DecTable[10] = {
  1000000000,100000000,10000000,1000000,
  100000, 10000, 1000, 100, 10, 1};

int count=0;// digital count
int n;       // calculation for each digital
// Remove preceding zeros
  while ((din < DecTable[count]) && (din>10)) {count++;}

  while (count<10) {
    n=0;
    while (din >= DecTable[count]) {
      din = din - DecTable[count];
      n++;
      }
    n = n + 48; // convert to ascii 0 to 9
    UartPutc((char) n);
    count++;
  };
  return;
}
```

Similarly, it is also possible to create input functions for strings and numbers. Unlike the "scanf" function in the C library, we pass two input parameters to the function: the first parameter is a pointer of the text buffer, and the second parameter is the maximum length of text that can be input.

UartGets function – Get a user input string from UART

```c
int UartGets(char dest[], int length)
{
unsigned int textlen=0; // Current text length
char ch; // current character
do {
  ch = UartGetc(); // Get a character from UART
  switch (ch) {
    case 8: // Back space
      if (textlen>0) {
        textlen--;
        UartPutc(ch); // Back space
        UartPutc(' '); // Replace last character with space on console
```

```
        UartPutc(ch); // Back space again to adjust cursor position
      }
    break;
  case 13: // Enter is pressed
    dest[textlen] = 0; // null terminate
    UartPutc(ch); // echo typed character
    break;
  case 27: // ESC is pressed
    dest[textlen] = 0; // null terminate
    UartPutc('\n');
    break;
  default: // if input length is within limit and input is valid
    if ((textlen<length) &
      ((ch >= 0x20) & (ch < 0x7F))) // valid characters
      {
      dest[textlen] = ch; // append character to buffer
      textlen++;
      UartPutc(ch); // echo typed character
      }
    break;
  } // end switch
} while ((ch!=13) && (ch!=27));
if (ch==27) {
  return 1; // ESC key pressed
} else {
  return 0; // Return key pressed
  }
}
```

Unlike "scanf", the "UartGets" function we created allows us to determine if the user completed the input process by pressing the ENTER or ESC key. To use this function, declare a text buffer as an array of characters, and pass its address to this function.

Example of using the UartGets function

```
int main(void)
{
  char textbuf[20];
  int  return_state;

  // System Initialization
  SystemInit();
  // Initialize UART
  UartConfig();

  while (1) {
    UartPutc('\n');
    UartPuts ("String input test : ");
    return_state = UartGets(&textbuf[0], 19);
```

(*Continued*)

Example of using the UartGets function—Cont'd

```
   if (return_state!=0) {
     UartPuts ("\nESC pressed :");
     } else {
     UartPuts ("\nInput was    :");
     }
   UartPuts (textbuf);
   UartPutc('\n');
   };
};
```

By modifying the case statement in the "UartGets" function, you can create input functions that only accept numeric value inputs, or other types of text input functions required for your application. You can even change the implementation so that it gets input from different interfaces than the UART.

Simple Interrupt Programming

General Overview of Interrupt Programming

Interrupts are essential for majority of embedded systems. For example, user inputs can be handled by an interrupt service routine so that the processor does not have to spend time checking the input interface status. Whereas in the previous UART, example polling loops were used to check if a character was received. In addition to handling user inputs, interrupts can also be used for other hardware interface blocks, peripherals, or by software.

In the Cortex-M0, the interrupt feature is very easy to use. In general, we can summarize the configuration of an interrupt service as follows:

*   Set up the vector table (this is done by the startup code from a CMSIS-compliant device driver library).
*   Set up the priority level of the interrupt. This step is optional; by default the priority levels of interrupts are set to level 0 (the highest programmable level).
*   Define an interrupt service routine (ISR) in your application. This can be a C function.
*   Enable the interrupt (e.g., using the NVIC_EnableIRQ() function).

By default the global interrupt mask PRIMASK is cleared after reset, so there is no need to enable interrupts globally.

The CMSIS has made these steps much easier, as the priority level and enabling of the interrupt can be carried out by functions provided in the CMSIS. The interrupt service routine is application dependent and will have to be created by a software developer. In most cases, you can find example code from the microcontroller vendors that will make software development easier. Depending on the peripheral design on the microcontroller, you might have to clear the interrupt requests inside the interrupt service routines. Please note that global variables used by the interrupt service routines need to be defined as volatile.

Dial Control Interface Example

In addition to switches, UARTs, and keypads, there are many other types of input devices. In recent years, dial controls have been used in many electronic gadgets. In this example we use a simple dial control to demonstrate simple interrupt programming in the Cortex-M0.

There are a number of ways to detect movement on a dial control interface. One simple implementation is to use two optical sensors with a rotary disc connected to a dial, as shown in Figure 15.7.

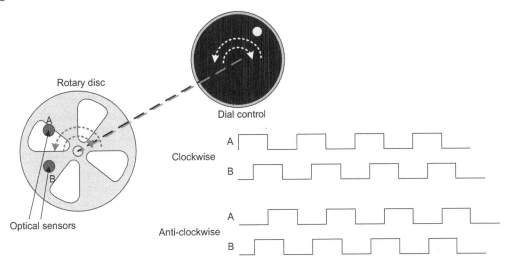

Figure 15.7:
Simple dial control implementation.

By detecting the edge transitioning of the two sensors, we can detect the rotation of the dial. This detection can be implemented on a Cortex-M0 microcontroller like the NXP LPC1114 quite easily. In this application example, we connect the two optical sensors to port 3 (P3_2 connected to sensor A, P3_3 connected to sensor B). We can then use the interrupt feature to detect when the dial is rotated. The NXP LPC1114 allows interrupt generation from port 3 on both rising and falling edges of signal transition. The interrupt handler needs to detect the direction of the dial movement from the previous state and the new state of the sensor's outputs, and then it will update software variables to inform the main application that the dial has been activated.

The program flow for the dial control example is shown in the flowchart in Figure 15.8. In this example, we use the UART interface to output the position of the dial range from 0 to 255. In real applications, the position value of the dial control can be used in other ways and the UART setup code might not be required and could be removed if that is the case.

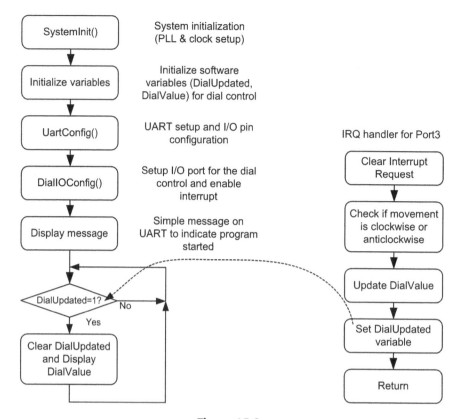

Figure 15.8:
Flowchart for simple dial control application.

The application contains the following files:

- startup_LPC11xx.s (the assembly startup code, as in previous example)
- system_LPC11xx.c (the system initialization function, as in previous examples)
- dial_ctrl.c (the dial control application)
- retarget.c (for text display functions, reuse the previous retarget example code)

Most of the files are identical to previous project examples. The only new file is the dial control code. To get the interrupt to work, the "DialIOCfg()" function uses a number of CMSIS functions. The rest of the code in "DialIOcfg" is used for configuring the pins for the input function and to generate interrupt on both rising and falling edges (Figure 15.9).

The interrupt handler is also simple. It takes the previous state of the input pins and merges with the new state of the pins to produce a 4-bit index value. This index value is then used to obtain an increment (1)/decrement (−1)/unchanged (0) value from a lookup table to adjust the "DialValue" variable. It then sets the software flag, "DialUpdated", to indicate to the main program that movement has been detected (Figure 15.10).

```
void DialIOcfg(void)
{ // The inputs are P3.2 and P3.3
  // Enable clock to GPIO block (bit[6] of AHBCLOCK Control register
  LPC_SYSCON->SYSAHBCLKCTRL = LPC_SYSCON->SYSAHBCLKCTRL | (1<<6);
  // PIO1_7 IO output config
  // bit[5]   - Hysteresis (0=disable, 1 =enable)
  // bit[4:3] - MODE(0=inactive, 1 =pulldown, 2=pullup, 3=repeater)
  // bit[2:0] - Function (0 = IO, 1=nDCD)
  LPC_IOCON->PIO3_2 = (0x0) + (0<<3) + (1<<5);
  // PIO1_6 IO output config
  // bit[5]   - Hysteresis (0=disable, 1 =enable)
  // bit[4:3] - MODE(0=inactive, 1 =pulldown, 2=pullup, 3=repeater)
  // bit[2:0] - Function (0 = IO, 1=RI)
  LPC_IOCON->PIO3_3 = (0x0) + (0<<3) + (1<<5);
  // Set direction of P3.2 and P3.3 as input
  LPC_GPIO3->DIR   = LPC_GPIO3->DIR & ~(0x0C); // Clear bit [3:2] to 0
  // Set interrupt of P3.2 and P3.3 as edge sensitive
  LPC_GPIO3->IS    = LPC_GPIO3->IS  & ~(0x0C); // Clear bit [3:2] to 0
  // Set interrupt of P3.2 and P3.3 for both rising and falling edge
  LPC_GPIO3->IBE   = LPC_GPIO3->IBE | (0x0C); // Set bit [3:2] to 1
  // Set interrupt mask of P3.2 and P3.3
  LPC_GPIO3->IE    = LPC_GPIO3->IE | (0x0C); // Set bit [3:2] to 1
  // Clear any previous interrupt of P3.2 and P3.3
  LPC_GPIO3->IC    = LPC_GPIO3->IC | (0x0C); // write bit [3:2] to 1
  // Clear any previous occurred interrupt for port 3
  NVIC_ClearPendingIRQ(EINT3_IRQn);    ◄────────── CMSIS Function to clear
  // Set priority of port 3 interrupt                         pended interrupt
  NVIC_SetPriority(EINT3_IRQn, 0);  ◄─────
  // Enable interrupt at NVIC
  NVIC_EnableIRQ(EINT3_IRQn);  ◄─────
  return;
}
```

Device specific interrupt number assignment. (Defined in LPC11XX.h)

CMSIS Function to enable interrupt

CMSIS Function to set interrupt priority

Input	Level	
0	0x00	Highest priority
1	0x40	
2	0x80	
3	0xC0	Lowest priority

Figure 15.9:
Use of CMSIS for simple interrupt configuration.

When the program is executed, the software variable "DialValue" is displayed when the dial is moved. For example, if the dial rotates clockwise, the "DialValue" increases as shown in Figure 15.11.

The complete listing of the dial_ctrl.c is as follows:

dial_ctrl.c

```
#include "LPC11XX.h"
#include <stdio.h>


// Function declarations
void          DialIOcfg(void);  // Configure Port 3 for dial interface
void          UartConfig(void); // Uart configuration
unsigned char UartPutc(unsigned char my_ch);    // Uart character output
void          UartPuts(unsigned char * mytext);  // Uart string output
```

(Continued)

dial_ctrl.c—Cont'd

```c
// Global variable for communicating between main program and ISR
volatile char DialUpdated; // Set to 1 if the dial value is updated
volatile int  DialValue;   // Dial value (0 to 0xFF)
short    int  last_state;  // Last state of I/O port signals

// Start of main program
int main(void)
{
  SystemInit();  // System Initialization

  DialUpdated=1; // Software variable initialization
  DialValue=0;
  last_state=(LPC_GPIO3->DATA & 0xC)>>2;
      // capture and save signal levels for next compare

  UartConfig();  // Initialize UART
  DialIOcfg();   // IO port and interrupt setup
  printf ("\nDial test\n"); // Test message

  while(1){
    if (DialUpdated) {
        DialUpdated = 0;
        printf("%d\n",DialValue);
      } // end if
    } // end while
} // end main


void DialIOcfg(void)
{ // The inputs are P3.2 and P3.3
  // Enable clock to GPIO block (bit[6] of AHBCLOCK Control register
  LPC_SYSCON->SYSAHBCLKCTRL = LPC_SYSCON->SYSAHBCLKCTRL | (1<<6);
  // PIO1_7 IO output config
  //  bit[5]   - Hysteresis (0=disable, 1 =enable)
  //  bit[4:3] - MODE(0=inactive, 1 =pulldown, 2=pullup, 3=repeater)
  //  bit[2:0] - Function (0 = IO, 1=nDCD)
  LPC_IOCON->PIO3_2 = (0x0) + (0<<3) + (1<<5);
  // PIO1_6 IO output config
  //  bit[5]   - Hysteresis (0=disable, 1 =enable)
  //  bit[4:3] - MODE(0=inactive, 1 =pulldown, 2=pullup, 3=repeater)
  //  bit[2:0] - Function (0 = IO, 1=RI)
  LPC_IOCON->PIO3_3 = (0x0) + (0<<3) + (1<<5);
  // Set direction of P3.2 and P3.3 as input
  LPC_GPIO3->DIR    = LPC_GPIO3->DIR & ~(0x0C); // Clear bit [3:2] to 0
  // Set interrupt of P3.2 and P3.3 as edge sensitive
  LPC_GPIO3->IS     = LPC_GPIO3->IS  & ~(0x0C); // Clear bit [3:2] to 0
  // Set interrupt of P3.2 and P3.3 for both rising and falling edge
  LPC_GPIO3->IBE    = LPC_GPIO3->IBE |  (0x0C); // Set bit [3:2] to 1
  // Set interrupt mask of P3.2 and P3.3
  LPC_GPIO3->IE     = LPC_GPIO3->IE  |  (0x0C); // Set bit [3:2] to 1
  // Clear any previous interrupt of P3.2 and P3.3
  LPC_GPIO3->IC     = LPC_GPIO3->IC  |  (0x0C); // write bit [3:2] to 1
  // Clear any previous occurred interrupt for port 3
```

```
   NVIC_ClearPendingIRQ(EINT3_IRQn);
   // Set priority of port 3 interrupt
   NVIC_SetPriority(EINT3_IRQn, 0);
   // Enable interrupt at NVIC
   NVIC_EnableIRQ(EINT3_IRQn);
   return;
}
// Interrupt handler for port 3
void PIOINT3_IRQHandler(void)
{
   short int new_state;
   // Pattern for determine the direction of changes
   // Clock wise       pattern is 00 -> 01 -> 11 -> 10 -> 00 -> ...
   // Anti Clock wise pattern is 00 -> 10 -> 11 -> 01 -> 00 -> ...
   // After merging the new_state and last_state, clock wise can be
   // pattern b0100(4), b1101(13), b1011(11) and b0010(2)
   // anti-clockwise can be pattern b1000(8), b1110(14),b0111(7) and b0001(1)
   const signed char Pattern[] = { 0,-1,1,0,  1,0,0,-1,  -1,0,0,1,  0,1,-1,0};
   // Clear asserted interrupt P3.2 or P3.3
   LPC_GPIO3->IC = LPC_GPIO3->MIS & (0x0C); // write bit [3:2] to 1
   // Extract bit 3 and 2 and combine with last state
   new_state = (LPC_GPIO3->DATA & 0xC) | last_state;
   // Obtain increment/decrement info from new_state and calculate new DialValue
   DialValue = (DialValue + Pattern[new_state]) & 0xFF;
   // Save the current state for next time
   last_state  = (new_state & 0xC) >> 2;
   DialUpdated = 1;// Set software flag for display
   return;
}
void UartConfig(void)
{
   // UART interface are : PIO1_7 (TXD) and PIC1_6 (RXD)
   // Other UART signals (DTR, DSR, CTS, RTS, RI) are not used

   // Enable clock to IO configuration block (bit[16] of AHBCLOCK Control
   // register
   LPC_SYSCON->SYSAHBCLKCTRL = LPC_SYSCON->SYSAHBCLKCTRL | (1<<16);

   // PIO1_7 IO output config
   //  bit[5]   - Hysteresis (0=disable, 1 =enable)
   //  bit[4:3] - MODE(0=inactive, 1 =pulldown, 2=pullup, 3=repeater)
   //  bit[2:0] - Function (0 = IO, 1=TXD, 2=CT32B0_MAT1)
   LPC_IOCON->PIO1_7 = (0x1) + (0<<3) + (0<<5);
   // PIO1_6 IO input config
   //  bit[5]   - Hysteresis (0=disable, 1 =enable)
   //  bit[4:3] - MODE(0=inactive, 1 =pulldown, 2=pullup, 3=repeater)
   //  bit[2:0] - Function (0 = IO, 1=RXD, 2=CT32B0_MAT0)
   LPC_IOCON->PIO1_6 = (0x1) + (2<<3) + (1<<5);

   // Enable clock to UART (bit[12] of AHBCLOCK Control register
   LPC_SYSCON->SYSAHBCLKCTRL = LPC_SYSCON->SYSAHBCLKCTRL | (1<<12);
   // UART_PCLK divide ratio = 1
```

(Continued)

dial_ctrl.c—Cont'd

```c
  LPC_SYSCON->UARTCLKDIV = 1;

  // UART_PCLK = 48MHz, Baudrate = 38400, divide ratio = 1250
  // Line Control Register
  LPC_UART->LCR = (1<<7) |   // Enable access to Divisor Latches
    (0<<6) |    // Disable Break COntrol
    (0<<4) |    // Bit[5:4] parity select (odd, even, sticky-1, sticky-0)
    (0<<3) |    // parity disabled
    (0<<2) |    // 1 stop bit
    (3<<0);     // 8-bit data
  LPC_UART->DLL = 78;// Divisor Latch Least Significant Byte :
                    // 48MHz/38400/16=78.125
  LPC_UART->DLM = 0; // Divisor Latch Most Significant Byte  : 0
  LPC_UART->LCR = (0<<7) |   // Disable access to Divisor Latches
    (0<<6) |    // Disable Break Control
    (0<<4) |    // Bit[5:4] parity select (odd, even, sticky-1, sticky-0)
    (0<<3) |    // parity disabled
    (0<<2) |    // 1 stop bit
    (3<<0);     // 8-bit data

  LPC_UART->FCR = 1; // Enable FIFO

  return;
}
// Output a character
unsigned char UartPutc(unsigned char my_ch)
{
  if (my_ch == '\n') {
    while ((LPC_UART->LSR & (1<<5))==0);
    // Wait if Transmit Holding register is not empty
    LPC_UART->THR = 13;
    // Output carriage return (for Windows Hyperterminal)
  }
  while ((LPC_UART->LSR & (1<<5))==0);
    // Wait if Transmit Holding register is not empty
  LPC_UART->THR = my_ch;
    // write to transmit holding register

  if (my_ch == 13) {
    while ((LPC_UART->LSR & (1<<5))==0);
    // Wait if Transmit Holding register is not empty
    LPC_UART->THR = 10;
    // Output new line (for Windows Hyperterminal)
  }

  return (my_ch);
}
// Detect if new received data is available
// (only required if retarget.c is used because UartGetc is reference in that file)
__inline int  UartGetRxDataAvail(void)
{
  return (LPC_UART->LSR & 0x1);
}
```

```
// Get a character from UART, if no data available then wait
// (only required if retarget.c is used because it is referenced in that file)
unsigned char UartGetc(void)
{
  while (UartGetRxDataAvail()==0); // wait if receive buffer empty
  return ((char)LPC_UART->RBR);
}
```

```
// Interrupt handler for port 3                      Name of interrupt handler need to
void PIOINT3_IRQHandler(void)  ◄───────────── match the name specified in the
{                                                  vector table
  short int new_state;
  // Pattern for determine the direction of changes
  // Clock wise     pattern is 00 -> 01 -> 11 -> 10 -> 00 -> ...
  // Anti Clock wise pattern is 00 -> 10 -> 11 -> 01 -> 00 -> ...
  // After merging the new_state and last_state, clock wise can be
  // pattern b0100(4), b1101(13), b1011(11) and b0010(2)
  // anti-clockwise can be pattern b1000(8), b1110(14),b0111(7) and b0001(1)
  const signed char Pattern[] = { 0,-1,1,0,  1,0,0,-1,  -1,0,0,1,  0,1,-1,0};   The Interrupt handler is implement
  // Clear asserted interrupt P3.2 or P3.3                                       as a normal C subroutine.
  LPC_GPIO3->IC = LPC_GPIO3->MIS & (0x0C); // write bit [3:2] to 1
  // Extract bit 3 and 2 and combine with last state
  new_state = (LPC_GPIO3->DATA & 0xC) | last_state;
  // Obtain increment/decrement info from  new_state and calculate new DialValue
  DialValue = (DialValue + Pattern[new_state]) & 0xFF;
  // Save the current state for next time
  last_state  = (new_state & 0xC) >> 2;
  DialUpdated = 1;// Set software flag for display
  return;
}
```

Figure 15.10:
The interrupt handler for the dial control example.

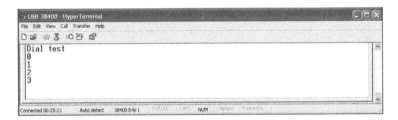

Figure 15.11:
Result screen of the dial control example.

Interrupt Control Functions

There are a number of interrupt control functions in CMSIS. Most of them have been described in Chapter 9. Table 15.2 summarizes the CMSIS functions for general interrupt controls.

Table 15.2: CMSIS Interrupt Control Functions

Function	Descriptions
void NVIC_EnableIRQ (IRQn_Type IRQn);	Enable an interrupt. This function does not apply to system exceptions.
void NVIC_DisableIRQ (IRQn_Type IRQn);	Disable an interrupt. This function does not apply to system exceptions.
void NVIC_SetPendingIRQ (IRQn_Type IRQn);	Set the pending status of an interrupt. This function does not apply to system exceptions.
void NVIC_ClearPendingIRQ (IRQn_Type IRQn);	Clear the pending status of an interrupt. This function does not apply to system exceptions.
uint32_t NVIC_GetPendingIRQ (IRQn_Type IRQn);	Obtain the interrupt pending status of an interrupt. This function does not apply to system exceptions.
void NVIC_SetPriority (IRQn_Type IRQn, uint32_t priority);	Set up the priority level of an interrupt or system exception. The priority level value is automatically shifted to the implemented bits in the priority level register.
uint32_t NVIC_GetPriority (IRQn_Type IRQn);	Obtain the priority level of an interrupt or system exception. The priority level is automatically shifted to remove unimplemented bits in the priority level values.
void __enable_irq(void);	Clear PRIMASK. Enable interrupts and system exceptions.
void __disable_irq(void);	Set PRIMASK. Disable all interrupts including system exceptions (apart from hard fault and NMI).

The input parameter "IRQn_Type IRQn" is defined in the header file for the device. For example, for the NXP LPC1114, the IRQn types are defined in an enumeration list in "LPC11XX.h":

IRQn_Type defined in LPC11xx.h

```
typedef enum IRQn
{
/******  Cortex-M0 Processor Exceptions Numbers
***************************************************/
  NonMaskableInt_IRQn     = -14,    /*!< 2 Non Maskable Interrupt          */
  HardFault_IRQn          = -13,    /*!< 3 Cortex-M0 Hard Fault Interrupt */
  SVCall_IRQn             = -5,     /*!< 11 Cortex-M0 SV Call Interrupt    */
  PendSV_IRQn             = -2,     /*!< 14 Cortex-M0 Pend SV Interrupt    */
  SysTick_IRQn            = -1,     /*!< 15 Cortex-M0 System Tick Interrupt */

/******  LPC11xx Specific Interrupt Numbers ********************************/
  WAKEUP0_IRQn     = 0,   /*!< All I/O pins can be used as wakeup source.  */
  WAKEUP1_IRQn     = 1,   /*!< There are 13 pins in total for LPC11xx       */
  WAKEUP2_IRQn     = 2,
  WAKEUP3_IRQn     = 3,
  WAKEUP4_IRQn     = 4,
  WAKEUP5_IRQn     = 5,
  WAKEUP6_IRQn     = 6,
  WAKEUP7_IRQn     = 7,
```

```
    WAKEUP8_IRQn     = 8,
    WAKEUP9_IRQn     = 9,
    WAKEUP10_IRQn    = 10,
    WAKEUP11_IRQn    = 11,
    WAKEUP12_IRQn    = 12,
    SSP1_IRQn        = 14,     /*!< SSP1 Interrupt                      */
    I2C_IRQn         = 15,     /*!< I2C Interrupt                       */
    TIMER_16_0_IRQn  = 16,     /*!< 16-bit Timer0 Interrupt            */
    TIMER_16_1_IRQn  = 17,     /*!< 16-bit Timer1 Interrupt            */
    TIMER_32_0_IRQn  = 18,     /*!< 32-bit Timer0 Interrupt            */
    TIMER_32_1_IRQn  = 19,     /*!< 32-bit Timer1 Interrupt            */
    SSP0_IRQn        = 20,     /*!< SSP0 Interrupt                      */
    UART_IRQn        = 21,     /*!< UART Interrupt                      */
    ADC_IRQn         = 24,     /*!< A/D Converter Interrupt            */
    WDT_IRQn         = 25,     /*!< Watchdog timer Interrupt           */
    BOD_IRQn         = 26,     /*!< Brown Out Detect(BOD) Interrupt    */
    EINT3_IRQn       = 28,     /*!< External Interrupt 3 Interrupt     */
    EINT2_IRQn       = 29,     /*!< External Interrupt 2 Interrupt     */
    EINT1_IRQn       = 30,     /*!< External Interrupt 1 Interrupt     */
    EINT0_IRQn       = 31,     /*!< External Interrupt 0 Interrupt     */
} IRQn_Type;
```

Note that the comments in this file and in various CMSIS files contain Doxygen tags (e.g., /*!< *comments* */). Doxygen is a tool for automatic documentation generation.

The first group of the IRQn is made up of system exceptions; they are available in all versions of the Cortex-M0 CMSIS device driver library. The exception numbers 0 to 31 are device-specific interrupt types. They are defined according to the interrupt request connection from the peripherals to the NVIC in the Cortex-M0. In our previous dial control examples, we used the following code to enable interrupts from port 3:

```
NVIC_EnableIRQ(EINT3_IRQn); // Enable External Interrupt 3 Interrupt
```

The constant `EINT3_IRQn` is defined as 28 in the enumeration. We use `EINT3_IRQn` rather than 28 to make the program code more readable and improve software reusability.

If necessary, we can disable all peripheral interrupts and system exceptions using the PRIMASK feature in the Cortex-M0 processor. Typically this is carried out when we need to perform a time-critical task and we do not want the control timing to be affected by any interrupt. The CMSIS provides two functions to access the PRIMASK feature. For example,

```
__disable_irq(); // Set PRIMASK — disable interrupts
... ;            // time critical tasks
__enable_irq();  // clear PRIMASK — enable interrupts
```

Please note that the PRIMASK does not block the nonmaskable interrupt (NMI) and the HardFault exception. Also, if PRIMASK is set inside an interrupt handler, you should clear it in the handler. Otherwise the interrupts will be remains disabled. This is different from ARM7TDMI where an interrupt return can reenable the interrupt.

Different Versions of CMSIS

The CMSIS project is in continuous development. A Cortex-M0 device driver library from a microcontroller vendor could be in version 1.2, 1.3, 2.0 or later. At the moment, many CMSIS device drivers are already based on version 1.3. The examples used in this book should work with versions 1.2, 1.3, and later versions.

There are several differences between CMSIS version 1.2 and version 1.3 that apply to the uses of CMSIS on the Cortex-M0:

- The SystemInit() function is different. In CMSIS version 1.2, the SystemInit() function is called at the start of the main code. In CMSIS version 1.3, the System-Init() function could be called from the reset handler.
- The "SystemCoreClock" variable has been added. The "SystemCoreClock" variable is used instead of "SystemFrequency". The "SystemCoreClock" definition is clearer—processor clock speed—whereas "SystemFrequency" could be unclear because many microcontrollers have multiple clocks for different parts of the system.
- A core register bit definition has been added.

In December 2010, CMSIS version 2 was released. CMSIS version 2 includes support for the Cortex-M4 processor and some changes in the file organizations. For example, the contents in "core_cm0.h" are divided into multiple files in CMSIS version 2 for easier management (shown in Figure 4.18).

In most cases, software device driver packages from microcontroller vendors should already contain the files needed. If necessary, you can download a preferred version of CMSIS from the OnARM web site (www.onarm.com).

## CHAPTER 16

Assembly Projects and Mixed-Assembly and C Projects

Project Development in Assembly

In addition to C language projects, you can also program the Cortex-M0 microcontrollers in assembly language. The same development tools for C programming are used; for example, the Keil MDK, the ARM RealView Development Suite (RVDS), or the GNU tool chain can all be used to develop assembly projects.

There are a number of reasons for using assembly language for programming, or using assembly in a part of a project:

- To allow the direct manipulation of stack memory (e.g., embedded OS development) for program operation that requires it
- To optimize the maximum speed/performance for specific hardware
- To reuse the assembly code from other projects
- To learn about processor architecture

However, the use of assembly language for an entire project is less common for embedded product developments. This is due to the following shortcomings of assembly language programming:

- It is more difficult to program in assembly language, especially when the application involves a lot of complicated data processing.
- It takes time to learn assembly, and mistakes are not easy to spot. As a result, it can take longer to complete a project.
- Assembly program files are less portable. For instance, different development tools can have different assembly directives and syntax.
- Modern C compilers can generate very efficient code—in many cases, better than assembly code written by inexperience engineers.
- Most microcontroller vendors provide libraries and header files for C development. If assembly is used for accessing peripherals, you will need to create your own device driver code and header files.

Nevertheless, some developers do build embedded applications in assembly. In this chapter, we will see how this can be done, as well as how to develop mixed-language projects. The examples in this chapter are targeted at the Keil MDK or ARM RVDS development environments. For other development suites, the assembler directives and the syntax can be different.

The Definitive Guide to the ARM Cortex-M0. DOI: 10.1016/B978-0-12-385477-3.10016-3

Recommended Practice in Assembly Programming

Before we actually start doing assembly language programming, we need to cover a few recommended practices. ARM has a document called the ARM Architecture Procedure Call Standard (AAPCS, reference 4), which describes how programming code should work on an ARM processor. By following the programming convention set out in this document, various software components can work together, allowing better software reusability and avoiding problems with integrating your assembly code with program code generated by compilers or program codes from third parties.

The AAPCS covers the following main areas:

*   *Register usage in function calls.* A function or a subroutine should retain the values in R4 through R11. If these registers are changed during the function or the subroutine, the values should be saved onto the stack and be restored before return to the calling code.
*   *Parameters and return result passing.* For simple cases, input parameters can be passed on to a function using R0 (first parameter), R1 (second parameter), R2 (third parameter), and R3 (fourth parameter). Usually the return value of a function is stored in R0, R1 might also be used if the result is 64-bit. If more than four parameters have to be passed on to a function, the stack would be used (details can be found in the AAPCS).
*   *Stack alignment.* If an assembly function needs to call a C function, it should ensure that the current selected stack pointer points to a double-word-aligned address location (e.g., 0x20002000, 0x20002008, 0x20002010, etc). This is a requirement for the EABI standard. Program code generated from an EABI-compliant C compiler can assume that the stack pointer is pointing to a double-word-aligned location. If the assembly code does not call any C function (either directly or indirectly), this is not strictly required.

For example, when developing assembly functions to be used by C, or calling C functions from assembly code, we also need to ensure that data contents in a register bank will not be accidentally erased (Table 16.1).

If the function call requires input parameters, or if the function returns a parameter, this can be handled with registers R0 to R3 (Table 16.2).

In ARM/Keil development tools, the assembler provides the REQUIRE8 directive to indicate if the function requires double-word-stack alignment and the PRESERVE8 directive to indicate that a function preserves the double-word alignment. This directive can help the assembler to analyze your code and generate warnings if a function that requires a double-word-aligned stack frame is called by another function that does not guarantee double-word-stack alignment. Depending on your application, these directives might not be required, especially for projects built entirely with assembly code.

Table 16.1: Register Usages and Requirements in Function Calls

Register	Function Call Behavior
R0–R3, R12	Caller saved register. Contents in these registers can be changed by a function. Assembly code calling a function might need to save the values in these registers if they are required for operations in later stages.
R4–R11	Caller saved register. Contents in these registers must be retained by a function. If a function needs to use these registers for processing, they need to be saved on to the stack memory and restored before the function returns.
R14 (LR)	Content in the Link Register needs to be saved to stack if the function contains a "BL"/"BLX" instruction (calling another function) because the value in LR will be overwritten when "BL"/"BLX" is executed.
R13 (SP), R15 (PC)	Should not be used for normal processing.

Table 16.2: Simple Parameter Passing and Returning Value in a Function Call

Register	Input Parameter	Return Value
R0	First input parameter	Function return value
R1	Second input parameter	—, or return value (64-bit result)
R2	Third input parameter	—
R3	Fourth input parameter	—

In this chapter we will only cover simple cases. For more details, please refer to the AAPCS document on the ARM web site.

Structure of an Assembly Function

An assembly function can be very simple. For example, a function to add two input parameters can be as simple as

```
My_Add   ADDS   R0, R0, R1 ; Add R0 and R1, result store in R0
         BX     LR         ; Return
```

To help improve clarity, we can add further directives to indicate the start and end of a function. The FUNCTION directive indicates the start of a function, and the ENDFUNC directive indicatesthe end of the function:

```
My_Add   FUNCTION
         ADDS   R0, R0, R1 ; Add R0 and R1, result store in R0
         BX     LR         ; Return
         ENDFUNC
```

A similar pair of directives is PROC and ENDP, which are synonyms for FUNCTION and ENDFUNC. Each FUNCTION directive must have a matching ENDFUNC directive, and they must not be nested.

In a simple assembly file, in addition to the assembly code, you need additional directives to indicate the start of the program code and type of the memory where it is stored—for example, a short assembly file with the `My_Add` function:

```
            PRESERVE8 ; Indicate the code here preserve
                      ; 8 byte stack alignment
            THUMB     ; Indicate THUMB code is used
            AREA    |.text|, CODE, READONLY   ; Start of CODE area
My_Add      FUNCTION
            ADDS    R0, R0, R1  ; Add R0 and R1, result store in R0
            BX      LR          ; Return
            ENDFUNC


            END                 ; End of file
```

In more complex assembly functions, more steps might be required. In general, the structure of an assembly function can be divided into the following stages:

- prolog (saving register contents to the stack memory if necessary)
- allocate stack space memory for local variables (decrement SP)
- copy some of R0 to R3 (input parameters) to high registers (R8-R12) for later use (optional)
- carry out processing/calculation
- store result in R0 if a result is to be returned, R1 might also be used if return value is 64-bit
- stack adjustment to free space for local variables (increment SP)
- epilog (restore register values from stack)
- return

Most of these steps are optional—for example, prolog and epilog are not required if the function does not corrupt the contents in R4 to R11. The stack adjustments are also not required if there are sufficient registers for the processing. The following assembly function template illustrates some of these steps:

```
My_Func   FUNCTION
          PUSH    {R4-R6, LR} ; 4 registers are pushed to stack
                              ; double word stack alignment is
                              ; preserved
          SUB     SP, SP, #8  ; Reserve 8 bytes for local variables
          ; Now local variables can be accessed with SP related
          ; addressing mode
          ...                 ; Carry out processing
          MOVS    R0, R5      ; Store result in R0 for return value
          ADD     SP, SP, #8  ; Restore SP to free stack space
          POP     {R4-R6, PC} ; epilog and return
          ENDFUNC
```

In some cases, it can be useful to copy some of the contents in R0 to R3 (input parameters) to high registers at the beginning of the function because most 16-bit THUMB instructions can only use low registers. Moving the input parameters to high registers for later use allow more registers to be available for data processing, and making it easier to develop function code.

If the function is calling another assembly or C function, the values in registers R0 to R3, and R12 could be changed after the function call. So unless you are certain that the function being called will not change these registers, you need to save the contents of these registers if they will be used later. Alternatively, you might need to avoid using these registers for the data processing in your function.

Simple Assembly Project Example

In this example, we will reproduce the blinky project (the LED toggling example in Chapter 14) with assembly code. Just as we do when creating a C project, we create a new project using the project wizard and click YES when the Keil project wizard asks us if we want to copy the default startup code.

We then need to modify the startup code slightly to remove the stack and heap initialization functions, which are used by the C startup code and at the end, and modify the call to "__main" to our "Blinky" program code.

```
startup_LPC11xx.s for simple assembly project

; <h> Stack Configuration
;   <o> Stack Size (in Bytes) <0x0-0xFFFFFFFF:8>
; </h>

Stack_Size      EQU     0x00000200

                AREA    STACK, NOINIT, READWRITE, ALIGN=3
Stack_Mem       SPACE   Stack_Size
__initial_sp

                PRESERVE8
                THUMB

; Vector Table Mapped to Address 0 at Reset

                AREA    RESET, DATA, READONLY
                EXPORT  __Vectors
```

(Continued)

startup_LPC11xx.s for simple assembly project—Cont'd

```
__Vectors       DCD     __initial_sp            ; Top of Stack
                DCD     Reset_Handler           ; Reset Handler
                DCD     NMI_Handler             ; NMI Handler
                DCD     HardFault_Handler       ; Hard Fault Handler
                DCD     0                       ; Reserved
                DCD     0                       ; Reserved
                DCD     0                       ; Reserved
                DCD     0                       ; Reserved
                DCD     0                       ; Reserved
                DCD     0                       ; Reserved
                DCD     0                       ; Reserved
                DCD     SVC_Handler             ; SVCall Handler
                DCD     0                       ; Reserved
                DCD     0                       ; Reserved
                DCD     PendSV_Handler          ; PendSV Handler
                DCD     SysTick_Handler         ; SysTick Handler

                ; External Interrupts
                DCD     WAKEUP_IRQHandler       ; 16+ 0: Wakeup PIO0.0
                DCD     WAKEUP_IRQHandler       ; 16+ 1: Wakeup PIO0.1
                DCD     WAKEUP_IRQHandler       ; 16+ 2: Wakeup PIO0.2
                DCD     WAKEUP_IRQHandler       ; 16+ 3: Wakeup PIO0.3
                DCD     WAKEUP_IRQHandler       ; 16+ 4: Wakeup PIO0.4
                DCD     WAKEUP_IRQHandler       ; 16+ 5: Wakeup PIO0.5
                DCD     WAKEUP_IRQHandler       ; 16+ 6: Wakeup PIO0.6
                DCD     WAKEUP_IRQHandler       ; 16+ 7: Wakeup PIO0.7
                DCD     WAKEUP_IRQHandler       ; 16+ 8: Wakeup PIO0.8
                DCD     WAKEUP_IRQHandler       ; 16+ 9: Wakeup PIO0.9
                DCD     WAKEUP_IRQHandler       ; 16+10: Wakeup PIO0.10
                DCD     WAKEUP_IRQHandler       ; 16+11: Wakeup PIO0.11
                DCD     WAKEUP_IRQHandler       ; 16+12: Wakeup PIO1.0
                DCD     0                       ; 16+13: Reserved
                DCD     SSP1_IRQHandler         ; 16+14: SSP1
                DCD     I2C_IRQHandler          ; 16+15: I2C
                DCD     TIMER16_0_IRQHandler    ; 16+16: 16-bit Counter-Timer 0
                DCD     TIMER16_1_IRQHandler    ; 16+17: 16-bit Counter-Timer 1
                DCD     TIMER32_0_IRQHandler    ; 16+18: 32-bit Counter-Timer 0
                DCD     TIMER32_1_IRQHandler    ; 16+19: 32-bit Counter-Timer 1
                DCD     SSP0_IRQHandler         ; 16+20: SSP0
                DCD     UART_IRQHandler         ; 16+21: UART
                DCD     0                       ; 16+22: Reserved
                DCD     0                       ; 16+24: Reserved
                DCD     ADC_IRQHandler          ; 16+24: A/D Converter
                DCD     WDT_IRQHandler          ; 16+25: Watchdog Timer
                DCD     BOD_IRQHandler          ; 16+26: Brown Out Detect
                DCD     0                       ; 16+27: Reserved
                DCD     PIOINT3_IRQHandler      ; 16+28: PIO INT3
                DCD     PIOINT2_IRQHandler      ; 16+29: PIO INT2
                DCD     PIOINT1_IRQHandler      ; 16+30: PIO INT1
                DCD     PIOINT0_IRQHandler      ; 16+31: PIO INT0
```

```
                   IF      :LNOT::DEF:NO_CRP
                   AREA    |.ARM.__at_0x02FC|, CODE, READONLY tt
CRP_Key            DCD     0xFFFFFFFF
                   ENDIF

                   AREA    |.text|, CODE, READONLY
; Reset Handler
Reset_Handler      PROC
                   EXPORT  Reset_Handler              [WEAK]
                   IMPORT  Blinky
                   ENTRY
                   LDR     R0, =Blinky
                   BX      R0
                   ENDP

; Dummy Exception Handlers (infinite loops which can be modified)
NMI_Handler        PROC
                   EXPORT  NMI_Handler                [WEAK]
                   B       .
                   ENDP
HardFault_Handler\
                   PROC
                   EXPORT  HardFault_Handler          [WEAK]
                   B       .
                   ENDP
SVC_Handler        PROC
                   EXPORT  SVC_Handler                [WEAK]
                   B       .
                   ENDP
PendSV_Handler     PROC
                   EXPORT  PendSV_Handler             [WEAK]
                   B       .
                   ENDP
SysTick_Handler    PROC
                   EXPORT  SysTick_Handler            [WEAK]
                   B       .
                   ENDP


Default_Handler    PROC
                   EXPORT  WAKEUP_IRQHandler          [WEAK]
                   EXPORT  SSP1_IRQHandler            [WEAK]
                   EXPORT  I2C_IRQHandler             [WEAK]
                   EXPORT  TIMER16_0_IRQHandler       [WEAK]
                   EXPORT  TIMER16_1_IRQHandler       [WEAK]
                   EXPORT  TIMER32_0_IRQHandler       [WEAK]
                   EXPORT  TIMER32_1_IRQHandler       [WEAK]
                   EXPORT  SSP0_IRQHandler            [WEAK]
                   EXPORT  UART_IRQHandler            [WEAK]
                   EXPORT  ADC_IRQHandler             [WEAK]
```

(Continued)

---

startup_LPC11xx.s for simple assembly project—Cont'd

```
                EXPORT  WDT_IRQHandler          [WEAK]
                EXPORT  PIOINT2_IRQHandler      [WEAK]
                EXPORT  PIOINT1_IRQHandler      [WEAK]
                EXPORT  PIOINT0_IRQHandler      [WEAK]


WAKEUP_IRQHandler


SSP1_IRQHandler
I2C_IRQHandler
TIMER16_0_IRQHandler
TIMER16_1_IRQHandler
TIMER32_0_IRQHandler
TIMER32_1_IRQHandler
SSP0_IRQHandler
UART_IRQHandler
ADC_IRQHandler
WDT_IRQHandler
BOD_IRQHandler
PIOINT3_IRQHandler
PIOINT2_IRQHandler
PIOINT1_IRQHandler
PIOINT0_IRQHandler
                B       .
                ENDP
                ALIGN
                END
```

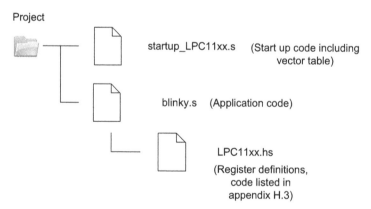

Figure 16.1:

Blinky project in assembly.

We then create our application code "blinky.s" (Figure 16.1) and an assembly header file "LPC11xx.hs." The file "LPC11xx.hs" contains register addresses definitions and is shown in Appendix H.

The file "blinky.s" is shown here:

blinky.s

```
        PRESERVE8 ; Indicate the code here preserve
                  ; 8 byte stack alignment
        THUMB     ; Indicate THUMB code is used
        AREA    |.text|, CODE, READONLY   ; Start of CODE area

        INCLUDE LPC11xx.hs
        EXPORT Blinky
Blinky  FUNCTION
        ; Switch Clock to 48MHz
        BL Set48MHzClock


        ; Initialize LED (GPIO #0, bit[7]) output
        BL LedOutputCfg


        ; Program SysTick timer at 1KHz.
        ; At 48MHz, SysTick trigger every 48000 CPU cycles
        ;  SysTick->LOAD = (48000-1); // Count from 47999 to 0
        ;  SysTick->VAL  = 0;   // Clear SysTick value
        ;  SysTick->CTRL = 0x5; // Enable, using core clock
        LDR    R0,=SysTick_BASE
        LDR    R1,=47999
        STR    R1,[R0,#SysTick_LOAD]
        MOVS   R1, #0
        STR    R1,[R0,#SysTick_VAL]
        MOVS   R1, #5
        STR    R1,[R0,#SysTick_CTRL]


        ; while(1){         // Blink at 1Hz
Blinky_loop
        ; for (i=0;i<500;i++) { // Wait for 0.5 seconds
        ;   while ((SysTick->CTRL & 0x10000)==0); // Wait for counter underflow
        ; }
        LDR    R1, =500
Blinky_inner_loop1
Blinky_inner_loop2
        LDR    R2, =0x10000
        LDR    R3, [R0,#SysTick_CTRL]
        TST    R3, R2
        BEQ    Blinky_inner_loop2
        SUBS   R1, R1, #1
        BNE    Blinky_inner_loop1

        ; Toggle bit 7
        ; LPC_GPIO0->MASKED_ACCESS[1<<7] = ~LPC_GPIO0->MASKED_ACCESS[1<<7];

        LDR    R1,=LPC_GPIO0_BASE
        LDR    R2,=0x200 ; (0x80 * 4)
        LDR    R3, [R1, R2]
        MVNS   R3, R3
        STR    R3, [R1, R2]

        ; } // end while
```

(Continued)

blinky.s—Cont'd

```
        B        Blinky_loop
        ENDFUNC
; --------------------------------------------------------
Set48MHzClock FUNCTION
        LDR      R0,=LPC_SYSCON_BASE

        ; Power up the PLL and System oscillator
        ; (clear the powerdown bits for PLL and System oscillator)
        ; LPC_SYSCON->PDRUNCFG = LPC_SYSCON->PDRUNCFG & 0xFFFFFF5F;

        LDR      R3, =PDRUNCFG
        LDR      R1,[R0, R3]
        MOVS     R2, #0xA0
        BICS     R1, R1, R2
        STR      R1,[R0, R3]


        ; Select PLL source as crystal oscillator
        ;  0 - IRC oscillator
        ;  1 - System oscillator
        ;  2 - WDT oscillator
        ; LPC_SYSCON->SYSPLLCLKSEL = 1;
        MOVS     R1, #0x1
        STR      R1,[R0, #SYSPLLCLKSEL]


        ; Update SYSPLL setting (0->1 sequence)
        ; LPC_SYSCON->SYSPLLCLKUEN = 0;
        MOVS     R1, #0x0
        STR      R1,[R0, #SYSPLLCLKUEN]


        ; LPC_SYSCON->SYSPLLCLKUEN = 1;
        MOVS     R1, #0x1
        STR      R1,[R0, #SYSPLLCLKUEN]


        ; Set PLL to 48MHz generate from 12MHz
        ;  M = 48/12 = 4 (MSEL = 3)
        ;  FCCO (must be between 156 to 320MHz, and is 2x, 4x, 8x or 16x of Clock)
        ;  Clock freq out selected as 192MHz
        ;  P = 192MHz/48MHz/2 = 2 (PSEL = 1)
        ;  bit[8]   - BYPASS
        ;  bit[7]   - DIRECT
        ;  bit[6:5] - PSEL  (1,2,4,8)
        ;  bit[4:0] - MSEL  (1-32)
        ; LPC_SYSCON->SYSPLLCTRL = (3 + (1<<5)); // M = 4, P = 2
        MOVS     R1, #0x23
        STR      R1,[R0, #SYSPLLCTRL]


        ; wait until PLL is locked
        ; while(LPC_SYSCON->SYSPLLSTAT == 0);
Set48MHzClock_waitloop1
        LDR      R1,[R0,#SYSPLLSTAT]
```

```
        CMP     R1, #0
        BEQ     Set48MHzClock_waitloop1

        ; Switch main clock to PLL clock
        ;    0 - IRC
        ;    1 - Input clock to system PLL
        ;    2 - WDT clock
        ;    3 - System PLL output
        ; LPC_SYSCON->MAINCLKSEL = 3;
        MOVS    R1, #0x3
        STR     R1,[R0, #MAINCLKSEL]

        ; Update Main Clock Select setting (0->1 sequence)
        ; LPC_SYSCON->MAINCLKUEN = 0;
        MOVS    R1, #0x0
        STR     R1,[R0, #MAINCLKUEN]

        ; LPC_SYSCON->MAINCLKUEN = 1;
        MOVS    R1, #0x1
        STR     R1,[R0, #MAINCLKUEN]
        BX      LR
        ENDFUNC
; -------------------------------------------------
LedOutputCfg FUNCTION
        ; Enable clock to IO configuration block (bit[16] of AHBCLOCK Control
register)
        ; and enable clock to GPIO (bit[6] of AHBCLOCK Control register)
        ; LPC_SYSCON->SYSAHBCLKCTRL=LPC_SYSCON->SYSAHBCLKCTRL|(1<<16)|(1<<6);
        LDR     R0,=(LPC_SYSCON_BASE+SYSAHBCLKCTRL)
        LDR     R2,=0x10040  ; (1<<16) | (1<<6)
        LDR     R1, [R0]
        ORRS    R1, R1, R2
        STR     R1, [R0]

        ; PIO0_7 IO output config
        ; bit[5]  - Hysteresis (0=disable, 1 =enable)
        ; bit[4:3] - MODE(0=inactive, 1 =pulldown, 2=pullup, 3=repeater)
        ; bit[2:0] - Function (0 = IO, 1=CTS)
        ; LPC_IOCON->PIO0_7 = (0x0) + (0<<3) + (0<<5);
        LDR     R0,=LPC_IOCON_BASE
        MOVS    R1, #0x0
        STR     R1, [R0,#PIO0_7]

        ; Initial bit[7] output is 0
        ; LPC_GPIO0->MASKED_ACCESS[1<<7] = 0;
        LDR     R0,=LPC_GPIO0_BASE
        MOVS    R1, #0
        MOVS    R2, #0x80  ; (1<<7)
        LSLS    R2, R2, #2 ; (R2 = (1<<7) x 4 (bytes)
        STR     R1, [R0, R2]
```

(Continued)

```
blinky.s—Cont'd
            ; Set pin 7 as output
            ; LPC_GPIO0->DIR = LPC_GPIO0->DIR | (1<<7);
            LDR     R0,=LPC_GPIO0_REGBASE
            MOVS    R1, #0x80
            STR     R1, [R0,#GPIO_DIR]

            BX      LR  ; Return
            ENDFUNC
;   ---------------------------------------------------
            END
;   ---------------------------------------------------
```

After the files are created, we can then add the blinky.s to the project and compile the program. After the program compilation is done, we can then update the debug option to select the required debug interface hardware, and then we download the program to the flash memory and test the application. The program should be able to toggle the LED at 1Hz when the program is running.

As you can see in the listing of LPC11xx.hs in Appendix H, creating an assembly project involves a bit more work to define hardware registers. But once that is done, creating assembly applications for Cortex-M0 is fairly straightforward.

Allocating Data Space for Variables

In the blinky example, the data processing can be handled with just a few registers, so it does not use any stack memory at all. By default, the stack memory allocation is done for us in the default startup code. We could reduce the stack size allocated by modifying the Stack_Size definition from 0x200 to the other stack size required:

```
Stack_Size      EQU     0x00000200

                AREA    STACK, NOINIT, READWRITE, ALIGN=3
Stack_Mem       SPACE   Stack_Size
__initial_sp
```

For most applications, there would be fair number of data variables. For simple applications, we can also allocate memory space in the RAM. For example, we can add a section in our application code to define three data variables: "MyData1" (a word size data variable), "MyData2" (a half word size data variable), and "MyData3" (a byte size data variable):

```
          PRESERVE8 ; Indicate the code here preserve
                    ; 8 byte stack alignment
          THUMB     ; Indicate THUMB code is used
; -------------------------------------------------
; Allocate data variable space
          AREA    | Header Data|, DATA   ; Start of Data definitions
          ALIGN   4

MyData1 DCD     0   ; Word size data
MyData2 DCW     0   ; half Word size data
MyData3 DCB     0   ; byte size data
; -------------------------------------------------
          AREA    |.text|, CODE, READONLY  ; Start of CODE area
          INCLUDE LPC11xx.hs
          EXPORT DataTest
DataTest FUNCTION
          ; Switch Clock to 48MHz
          BL Set48MHzClock

          LDR    R0,=MyData1
          LDR    R1,=0x00001234
          STR    R1,[R0]  ; MyData1 = 0x00001234

          LDR    R0,=MyData2
          LDR    R1,=0x55CC
          STRH   R1,[R0]  ; MyData2 = 0x55CC

          LDR    R0,=MyData3
          LDR    R1,=0xAA
          STRB   R1,[R0]  ; MyData3 = 0xAA

          B      . ; Endless loop
          ENDFUNC
; -------------------------------------------------
Set48MHzClock FUNCTION
          LDR    R0,=LPC_SYSCON_BASE
          ... (details same as previous example)
          ENDFUNC
; -------------------------------------------------
          END
; -------------------------------------------------
```

Once the program is compiled, we can examine the data memory layout by right-clicking on the target name (e.g., "Target 1") in the project window and selecting "open .\data_access. Map" ("data_access" is the name of the project in this case). From the map report file, we can see the address location and size of the variables we allocated:

```
Image Symbol Table
    Local Symbols
    Symbol Name    Value    Ov Type    Size    Object(Section)
    ...
```

```
Header Data    0x10000000    Section    7    data_access.o(Header Data)
MyData1        0x10000000    Data       4    data_access.o(Header Data)
MyData2        0x10000004    Data       2    data_access.o(Header Data)
MyData3        0x10000006    Data       1    data_access.o(Header Data)
STACK          0x10000008    Section   512   startup_lpc11xx.o(STACK)
__initial_sp   0x10000208    Data       0    startup_lpc11xx.o(STACK)
```

Because the RAM in the LPC1114 starts at address 0x10000000 onward, the variables are located starting from this address.

Another way to allocate memory space is to use the stack memory. To allocate memory space for local variables inside a function, we can modify the value of SP at the beginning of a function:

```
MyFunction

        PUSH    {R4, R5}
        SUB     SP, SP , #8  ; Allocate two words for space for local variables
        MOV     R4, SP ; Make a copy of SP to R4
        LDR     R5,=0x00001234
        STR     R5,[R4,#0] ; MyData1 = 0x00001234
        LDR     R5,=0x55CC
        STRH    R5,[R4,#4] ; MyData2 = 0x55CC
        MOVS    R5,#0xAA
        STRB    R5,[R4,#6] ; MyData3 = 0xAA
        ...
        ADD     SP, SP, #8 ; Restore SP back to starting value to free space
        POP     {R4, R5}
        BX      LR
```

The main advantage of using the stack for local variables is that local variables in functions that are not active do not take up any space in RAM. In contrast, many 8-bit microcontroller architectures allocate all data variables in static memory locations, resulting in larger SRAM requirements.

UART Example in Assembly

Once we have prepared the assembly code for system initialization, writing a UART example in assembly is actually not that difficult. To do this, we create an assembly version of UART functions based on the first UART example, as in Chapter 15. We also need to modify the startup code so that the reset handler executes the "UartTest" function in the new assembly program file (Figure 16.2).

Figure 16.3 shows the flowchart of the test code, and the source code listing (uart-test.s) can be found in Appendix H. To simplify the program, we developed a few UART functions for character output (UartPutc), string output (UartPuts), character input

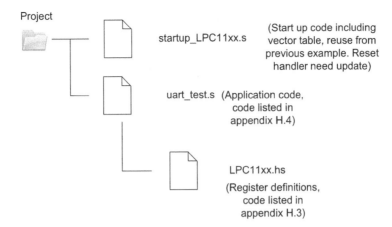

Figure 16.2:
UART test project.

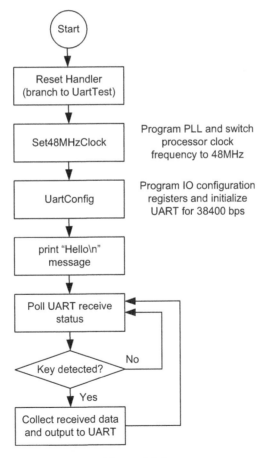

Figure 16.3:
Flowchart for the UART test.

(UartGetRxData), receive buffer checking (UartGetRxDataAvail), and UART initialization (UartConfig).

The program behaves exactly the same as in the C version.

Additional Text Output Functions

In the previous example, we created text output functions for string (UartPuts) and character (UartPutc). We can also create additional UART output functions for other data output.

A function called UartPutHex was developed to send hexadecimal numbers. This function calls the UartPutc function a number of times, which outputs a single ASCII character each time it is executed:

```
UartPutHex FUNCTION
        ; Output register value in hexadecimal format
        ; Input R0 = value to be displayed
        PUSH    {R0, R4-R7, LR} ; Save registers to stack
        MOV     R4, R0      ; Save register value to R3 because R0 is used
                            ; for passing input parameter
        MOVS    R0,#'0'     ; Starting the display with "0x"
        BL      UartPutc
        MOVS    R0,#'x'
        BL      UartPutc
        MOVS    R5, #8      ; Set loop counter
        MOVS    R6, #28     ; Rotate offset
        MOVS    R7, #0xF    ; AND mask
UartPutHex_loop
        RORS    R4, R6      ; Rotate data value left by 4 bits(right 28)
        MOV     R0, R4      ; Copy to R0
        ANDS    R0, R7      ; Extract the lowest 4 bit
        CMP     R0, #0xA    ; Convert to ASCII
        BLT     UartPutHex_Char0to9
        ADDS    R0, #7      ; If larger or equal 10, then convert to A-F
                            ; (R0=R0+7+48)
UartPutHex_Char0to9
        ADDS    R0, #48     ; otherwise convert to 0-9
        BL      UartPutc    ; Output 1 hex character
        SUBS    R5, #1      ; decrement loop counter
        BNE     UartPutHex_loop ; if all 8 hexadecimal characters been displayed
        POP     {R0, R4-R7, PC} ; then return, otherwise process next 4-bit
        ENDFUNC
```

A function called UartPutDec for outputting decimal numbers is also created. Similar to the previous function, it uses the UartPutc function. An array of constant values (referred as masks in the program code) is used in the function to speed up the conversion of the value to a decimal string:

```
UartPutDec FUNCTION
        ; Output register value in decimal format
        ; Input R0 = value to be displayed
        ; For 32-bit value, the maximum number of digits is 10
        PUSH    {R4-R6, LR}  ; Save register values
        MOV     R4, R0       ; Copy input value to R4 because R0 is
                             ; used for character output
        ADR     R6, UartPutDecConst ; Starting address of mask array
UartPutDecCompareLoop1       ; compare until input value is same or
                             ; larger than the current mask (.../100/10/1)
        LDR     R5, [R6]     ; Get Mask value
        CMP     R4, R5       ; Compare input value to mask value
        BHS     UartPutDecStage2 ; Value is same or larger than current mask
        ADDS    R6, #4       ; Next smaller mask address
        CMP     R4, #10      ; Check for zero to 9
        BLO     UartPutDecSmallNumber0to9
        B       UartPutDecCompareLoop1
UartPutDecStage2
        MOVS    R0, #0       ; Initial value for current digit
UartPutDecLoop2
        CMP     R4, R5       ; Compare to mask value
        BLO     UartPutDecLoop2_exit
        SUBS    R4, R5       ; Subtract mask value
        ADDS    R0, #1       ; increment current digit
        B       UartPutDecLoop2
UartPutDecLoop2_exit
        ADDS    R0, #48      ; convert to ascii 0-9
        BL      UartPutc     ; Output 1 character
        ADDS    R6, #4       ; Next smaller mask address
        LDR     R5, [R6]     ; Get Mask value
        CMP     R5, #1       ; Last Mask
        BEQ     UartPutDecSmallNumber0to9
        B       UartPutDecStage2
UartPutDecSmallNumber0to9    ; Remaining value in R4 is from 0 to 9
        ADDS    R4, #48      ; convert to ascii 0-9
        MOV     R0, R4       ; Copy to R0 for display
        BL      UartPutc     ; Output 1 character
        POP     {R4-R6, PC}  ; Restore registers and return
        ALIGN       4
UartPutDecConst              ; array of mask values for conversion
        DCD     1000000000
        DCD     100000000
        DCD     10000000
        DCD     1000000
        DCD     100000
        DCD     10000
        DCD     1000
        DCD     100
        DCD     10
        DCD     1
        ALIGN
        ENDFUNC
```

Using these functions, it is fairly easy to transfer information between your targeted systems to a personal computer running a terminal program, or to output information to a display interface to help software development.

Complex Branch Handling

When a conditional branch operation is based on a combination of input variables, it can take a complex decision sequence to decide if a branch should be taken. In some cases, it is possible to simplify the decision steps using assembly code.

If the branch condition is based on the variable of 5 bits or less, we can encode the branch condition as a 32-bit constant and extract the decision bit using shift or rotate instruction. For example:

$if((x == 0)||(x == 3)||((x > 12)\&\&(x < 19))||(x = 23))$ goto label; // x is a 5-bit data

The decision can be written as follows:

```
    LDR   R0,=x          ; Get address of x
    LDR   R0,[R0]        ; Read x from memory
    LDR   R1,=0x0087E009 ; Encoded branch condition bit 23, 18-13, 3, 0 are set to 1
    ADDS  R0, R0, #1     ; Shift as least one bit
    LSRS  R1, R1, R0     ; Extract branch condition to carry flag
    BCS   label          ; Branch if condition met
```

Alternatively, the branch condition can be encoded into an array of data bytes if the branch condition is more than 5 bits wide:

```
    LDR   R0,=x          ; Get address of x
    LDR   R0,[R0]        ; Read x from memory
    LSRS  R1,R1,R0       ; Get byte offset in look up table
    LDR   R2,=BranchConditionTable
    LDRB  R2,[R2,R1]     ; Get encoded condition
    MOVS  R1, #7
    ANDS  R1, R1, R0     ; Get lowest 3 bit of x
    ADDS  R0, R0, #1     ; Shift as least one bit
    LSRS  R2, R2, R0     ; Extract branch condition to carry flag
    BCS   label          ; Branch if condition met
    ...
BranchConditionTable
    DCB   0x09, 0xE0, 0x87, 0x00, ... ; Byte array of encoded branch condition
```

Mixed-Language Projects

In addition to assembly language projects and C language projects, there are also large number of embedded projects that contain both C language and assembly language. In fact, the

Keil MDK examples in Chapter 14 are already mixed-language projects because the default startup code provided in Keil MDK is written in assembly language.

In mixed-language projects with both C and assembly program files, the AAPCS-compliant requirement is even more important than pure assembly projects. Otherwise the result could be unpredictable: the program might work with one version of the compiler and when switching to a different version, or if the compiler changes, the project might stop working because of conflicts in register usage.

Calling a C Function from Assembly

When calling a C function from an assembly file, we need to be aware of the following areas:

- Register R0 to R3, R12, and LR could be changed. If these registers hold data that are needed for later use, you need to save them to the stack.
- The value of SP should be aligned to a double-word address boundary.
- You need to ensure input parameters are stored in the correct registers (in simple cases of one to four parameters, register R0 to R3 are used).
- The return value (assuming it is 32 bits or smaller) is normally stored in R0.

For example, if you have a C function that adds four values:

```
int my_add_c(int x1, int x2, int x3, int x4)
{
    return (x1 + x2 + x3 + x4);
}
```

In Keil MDK, you can call the C function from assembly by using the following code:

```
        MOVS    R0, #0x1 ; First  parameter (x1)
        MOVS    R1, #0x2 ; Second parameter (x2)
        MOVS    R2, #0x3 ; Third  parameter (x3)
        MOVS    R3, #0x4 ; Fourth parameter (x4)
        IMPORT  my_add_c
        BL      my_add_c ; Call "my_add_c" function. Result store in R0
```

If the assembly code is written as an embedded assembler inside C files, instead of using the IMPORT keyword to import the address symbol, the __CPP keyword should be used:

```
        MOVS    R0, #0x1 ; First  parameter (x1)
        MOVS    R1, #0x2 ; Second parameter (x2)
        MOVS    R2, #0x3 ; Third  parameter (x3)
        MOVS    R3, #0x4 ; Fourth parameter (x4)
        BL      __cpp(my_add_c) ; Call "my_add_c" function. Result store in R0
```

The __cpp keyword is required for Keil MDK in accessing C or C++ compile time constant expressions. For other tool chains, the directive required can be different.

Calling an Assembly Function from C Code

When calling an assembly function from C code, we need to be aware of the following areas when writing the assembly function:

- If we change any values in registers R4 to R11, we need to save the original values on the stack and restore the original values before returning to the C code.
- If we need to call another function inside the assembly function, we need to save the LR on the stack and use it for return.
- The function return value is normally stored in R0.

For example, if we have an assembly function that add four values:

```
        EXPORT my_add_asm
my_add_asm FUNCTION
        ADDS    R0, R0, R1
        ADDS    R0, R0, R2
        ADDS    R0, R0, R3
        BX      LR  ;  Return result in R0
        ENDFUNC
```

In the C code, we need to declare the function as

```
extern int my_add_asm(int x1, int x2, int x3, int x4);
int y;

    ...

    y = my_add_asm(1, 2, 3, 4); // call the my_add_asm function
```

If your assembly code needs to access some data variables in your C code, you can also use the IMPORT keyword. For example, the following code locates the variable "y" in the project; calculate the value of y^2 (square) and put the result back:

```
        EXPORT  CALC_SQUARE_Y
CALC_SQUARE_Y FUNCTION
        IMPORT  y
        LDR     R0,=y  ; Obtain the address value of variable "y"
        LDR     R1, [ R0]
        MULS    R1, R1, R1
        STR     R1, [ R0]
        BX      LR
        ENDFUNC
```

The preceding example assumes the variable "y" is 32 bits (LDR instruction transfers data in 32-bit formats).

Embedded Assembly

In most cases, we might only need one or two simple assembly functions, so we might want to embed the assembly code in the same program file as the C code. In most tool chains, a feature called inline assembler could be used. For ARM tool chains (Keil MDK and ARM RealView Development Suite), an alternative feature called "embedded assembler" is available.

The embedded assembler allows you to develop assembly functions inside C files. For example, the "my_add_e" function that adds four parameters could be written as

```
__asm int my_add_e(int x1, int x2, int x3, int x4)
{
        ADDS    R0, R0, R1
        ADDS    R0, R0, R2
        ADDS    R0, R0, R3
        BX      LR  ; Return result in R0
}
```

You can then call this function in C code just like a normal C function:

```
y = my_add_e(1, 2, 3, 4);
```

Inside embedded assembly functions, you can also import address value or data symbols using the __cpp keyword. For example, a function to increment variable "y" could be

```
__asm void increment_y(void)
{
        LDR     R0, =__cpp(&y)
        LDR     R1, [R0]
        ADDS    R1, R1, #1     ; increment
        STR     R1, [R0]
        BX      LR  ; Return result in R0
}
```

You can also use __cpp to import a function address location. For example,

```
__asm void embedded_asm_call_c(void)
{
        PUSH    {R4, LR}
        ; method 1
        MOVS    R0, #0x1     ; First  parameter (x1)
        MOVS    R1, #0x2     ; Second parameter (x2)
        MOVS    R2, #0x3     ; Third  parameter (x3)
        MOVS    R3, #0x4     ; Fourth parameter (x4)
        BL      __cpp(my_add_c)     ; Call the C function
        ; method 2
        MOVS    R0, #0x1     ; First  parameter (x1)
        MOVS    R1, #0x2     ; Second parameter (x2)
        MOVS    R2, #0x3     ; Third  parameter (x3)
        MOVS    R3, #0x4     ; Fourth parameter (x4)
        LDR     R4, =__cpp(my_add_c) ; Import the address of my_add_c
        BLX     R4                   ; Call the C function
        POP     {R4, PC}     ; Return
}
```

One advantage of an embedded assembler is that it allows you to locate the exception stack frame in exception handlers. Examples of this function can be found in Chapter 12 (the assembly wrapper for the hard fault handler) and Chapter 17 (the assembly wrapper for the SVC handler).

Accessing Special Instructions

In some cases, we might want to access some special instructions that cannot be generated by normal C code. If you are using CMSIS-compliant device drivers, a number of CMSIS functions are available; you can just use these functions to generate the required assembly instructions (Table 16.3).

Table 16.3: CMSIS Functions Support for the Cortex-M0

Instruction	CMSIS Function
ISB	void __ISB(void); // Instruction Synchronization Barrier
DSB	void __DSB(void); // Data Synchronization Barrier
DMB	void __DMB(void); // Data Memory Barrier
NOP	void __NOP(void); // No Operation
WFI	void __WFI(void); // Wait for Interrupt (enter sleep)
WFE	void __WFE(void); // Wait for Event (enter sleep / // clear event latch)
SEV	void __SEV(void); // Send Event
REV	uint32_t __REV(uint32_t value); // Reverse byte order // within a word
REV16	uint32_t __REV16(uint16_t value); // Reverse byte order within // each half word independently
REVSH	int32_t __REVSH(int16_t value); // Reverse byte order in the // lower halfword, and then sign extend // the result in a 32-bit word
CPSIE I	void __enable_irq(void); // Clear PRIMASK
CPSID I	void __disable_irq(void); // Set PRIMASK

The C compiler itself might also provide similar features, which are normally called intrinsic functions. For example, the Keil MDK and the ARM RealView Development Suite provide the intrinsic functions shown in Table 16.4. Beware that some of these functions differ from the CMSIS versions by lowercase characters in the function's names.

To allow your application code to be more portable, you should use CMSIS intrinsic functions if possible.

Idiom Recognitions

Some C compilers also provide a feature called idiom recognition. When the C code is constructed in a particular way, then the C compiler automatically converts the operation into a special instruction. Table 16.5 shows the idiom recognition features available in Keil MDK or ARM RVDS for Cortex-M0.

Table 16.4: Keil MDK or ARM RVDS Intrinsic Functions Support for the Cortex-M0

Instruction	Intrinsic Functions Provided in Keil MDK or ARM RVDS
ISB	void __isb(void); // Instruction Synchronization Barrier
DSB	void __dsb(void); // Data Synchronization Barrier
DMB	void __dmb(void); // Data Memory Barrier
NOP	void __nop(void); // No Operation
WFI	void __wfi(void) ; // Wait for Interrupt (enter sleep)

WFE	void __wfe(void); // Wait for Event (enter sleep /
	 // clear event latch)

SEV	void __sev(void); // Send Event

REV	unsigned int __rev(unsigned int val); // Reverse byte order
	 // within a word

CPSIE I	void __enable_irq(void); // Clear PRIMASK
CPSID I	void __disable_irq(void); // Set PRIMASK

ROR	unsigned int __ror(unsigned int val, unsigned int shift);
	 // rotate a value right by a specific number of bit
	 // "Shift" can be 1 to 31

Table 16.5: Idiom Recognition in Keil MDK or ARM RVDS for the Cortex-M0

Instruction	C Language Code That Can Be Recognized by Keil MDK or ARM RVDS

REV16	```/* recognized REV16 r0,r0 */```
	```int rev16(int x)```
	```{```
	``` return```
	```(((x&0xff)<<8)|((x&0xff00)>>8)|((x&0xff000000)>>8)|((x&0x00ff0000)<<8));```
	```}```

REVSH	```/* recognized REVSH r0,r0 */```
	```int revsh(int i)```
	```{```
	``` return ((i<<24)>>16)|((i>>8)&0xFF);```
	```}```

If the software is ported to a different C compiler without the same idiom recognition feature, the code will still compile because it is using standard C syntax, although the generated instruction sequence might be less efficient than using idiom recognitions.

## CHAPTER 17

Using Low-Power Features in Programming

Overview

In Chapter 11, we covered how the Cortex-M0 processor provides low-power advantages over other processors and looked at an overview of the low-power features in the Cortex-M0. In this chapter, we discuss how these low-power features are used in programming. In the last part of this chapter, we briefly cover the low-power features in a Cortex-M0 microcontroller (NXP LPC111x) and demonstrate how to use sleep modes in this device.

Review of Sleep Modes in the Cortex-M0 Processor

The Cortex-M0 processor supports normal sleep and deep sleep modes. The sleep modes can be entered using WFE or WFI instructions, or using Sleep-on-Exit feature (Figure 17.1).

The actual differences between normal sleep mode and deep sleep mode on a microcontroller depend on the system level design of the chip. For example, normal sleep might result in some of the clock signals being switched off, whereas deep sleep might also reduce voltage supplies to the memory blocks and might switch off additional components in the system.

After entering sleep mode, the processor can be awakened using interrupt requests, debug requests, events, and reset. Figure 17.2 summarizes the wakeup conditions for interrupt requests.

	WFE instruction executed	WFI instruction executed	Enter sleep by Sleep-On-Exit
Normal sleep (SLEEPDEEP bit in System Control Register = 0)	Normal sleep, wake up on events (including interrupts)	Normal sleep, wake up on interrupts	
Deep sleep (SLEEPDEEP bit in System Control Register = 1)	Deep sleep, wake up on events (including interrupts)	Deep sleep, wake up on interrupts	

Figure 17.1:
Normal sleep and deep sleep can both be entered using various methods.

The Definitive Guide to the ARM Cortex-M0. DOI: 10.1016/B978-0-12-385477-3.10017-5

Type	Conditions			Result	
	Priority level (exclude PRIMASK)	SEVONPEND	PRIMASK	Wake up	Execute ISR
WFE	IRQ priority > current level	--	0	Yes	Yes
	IRQ priority > current level	0	1	No	No
	IRQ priority ≤ current level	0	--	No	No
	IRQ priority ≤ current level	1	--	Yes	No
WFI	IRQ priority > current level	--	0	Yes	Yes
	IRQ priority > current level	--	1	Yes	No
	IRQ priority ≤ current level	--	--	No	No

Figure 17.2:
Summaries of wakeup conditions for interrupt requests.

In the System Control Block of the Cortex-M0 processor, there is a programmable register called the System Control Register (SCR; Table 17.1). This register contains several control bits related to sleep features.

Table 17.1: System Control Register (0xE000ED10)

Bits	Field	Type	Reset Value	Descriptions
31:5	Reserved	—	—	Reserved
4	SEVONPEND	R/W	0	Send Event on Pend bit—enable generation of event by a new interrupt pending status, which can wake up the processor from WFE
3	Reserved	—	—	Reserved
2	SLEEPDEEP	R/W	0	Sleep mode type control bit:0: Normal sleep1: Deep sleep
1	SLEEPONEXIT	R/W	0	Sleep-on-Exit bit—when set to 1, enable the Sleep-on-Exit feature
0	Reserved	—	—	Reserved

For the users of CMSIS-compliant device driver library, the System Control Register can be accessed by the register symbol "SCB->SCR." For example, to enable deep sleep mode, you can use

```
SCB->SCR |= 1<<2; /* Enable deep sleep feature */
```

The System Control Register must be accessed using a word-size transfer.

Using WFE and WFI in Programming

In most cases, the device driver libraries from microcontroller vendors contain functions to enter low-power modes that are customized for their microcontrollers. Using these functions will help you to achieve the best level of power optimization for your microcontrollers.

However, if you are developing C code that needs to be portable between multiple Cortex-M microcontrollers, you can use the CMSIS functions shown in Table 17.2 to access the WFE and WFI instructions directly.

Table 17.2: CMSIS Intrinsic Functions for WFE and WFI Instructions

Instruction	CMSIS Functions
WFE	__WFE();
WFI	__WFI();

For users who are not using CMSIS-compliant device drivers, you can use intrinsic functions provided by the C compiler or inline assembly to generate WFE and WFI instructions. In these cases, the software code will be tool chain dependent and less portable. For example, the ARM RealView Development Suite or Keil MDK provides the following C intrinsic functions (unlike the CMSIS version, they are in lowercase letters) (Table 17.3).

Table 17.3: ARM RealView Compiler or Keil MDK Intrinsic Functions for WFI and WFE

Instruction	Built-in Intrinsic Functions Provided in ARM RealView C Compiler or Keil MDK
WFE	__wfe();
WFI	__wfi();

Because the WFE can be awakened by various sources of events, including past events, it is usually used in an idle loop. For example,

```
while (processing_required()==0) {
    __wfe();
    }
```

Users of assembly programming environments can use WFE and WFI directly in their assembly codes.

Using the Send-Event-on-Pend Feature

The Send-Event-on-Pend feature allows any interrupts (including disabled ones) to wake up the processor if the processor entered sleep by executing the WFE instruction. When the

SEVONPEND bit in the System Control Register is set, an interrupt switching from inactive state to pending state generates an event, which wakes up the processor from WFE sleep.

If the pending status of an interrupt was already set before the processor entered the sleep state, a new request from this interrupt during WFE sleep will not wake up the processor.

For users of CMSIS-compliant device driver libraries, the Send-Event-on-Pend feature can be enabled by setting bit 4 in the System Control Register. For example, you can use

```
SCB->SCR |= 1<<4; /* Enable Send-Event-on-Pend feature */
```

If you are not using a CMSIS-compliant device driver library, you can use the following C code to carry out the same operation:

```
#define SCB_SCR (* ((volatile unsigned long *) (0xE000ED10)))
/* Set SEVONPEND bit in System Control Register */
SCB_SCR |= 1<<4;
```

Users of assembly language can enable this feature by using the following assembly code:

```
LDR    r0, =0xE000ED10; System Control Register address
LDR    r1, [r0]
MOVS   r2, #0x10; Set SEVONPEND bit
ORR    r1, r2
STR    r1, [r0]
```

Using the Sleep-on-Exit Feature

The Sleep-on-Exit feature is ideal for interrupt-driven applications. When it is enabled, the processor can enter sleep as soon as it completes an exception handler and returns to Thread mode. It does not cause the processor to enter sleep if the exception handler is returning to another exception handler (nested interrupt). By using Sleep-on-Exit, the microcontroller can stay in sleep mode as much as possible (Figure 17.3).

When the Cortex-M0 enters sleep using the Sleep-on-Exit feature, it is just like executing WFI immediately after the exception exit. However, the unstacking process is not carried out

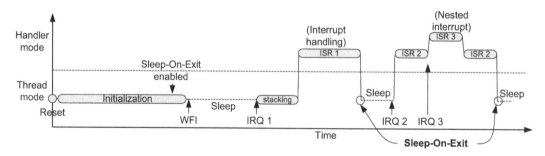

Figure 17.3:
Sleep-on-Exit feature.

because the registers will have to be pushed onto the stack at the next exception entry. The Sleep-on-Exit feature reduces the power consumption of the system (1) by avoiding unnecessary program execution in thread in interrupt-driven applications and (2) by reducing unnecessary stack push and pop operations. When the processor is awakened by a halt debug request, then the unstacking process will be carried out automatically.

When the Sleep-on-Exit feature is used, the WFE or WFI instruction is normally placed in an idle loop:

```
SCB->SCR = SCB->SCR | 0x2; // Enable Sleep-On-Exit feature
while (1) {
  __WFI(); // Execute WFI and enter sleep
  };
```

The loop is required because if the processor is awakened by a halt debug request, the instruction after the WFI (branch back to WFI loop) would be executed when the processor is unhalted after debugging.

If you are not using a CMSIS-compliant device driver, you can use the following C code to enable the Sleep-on-Exit feature:

```
#define SCB_SCR (*((volatile unsigned long *)(0xE000ED10)))
/* Set SLEEPONEXIT bit in System Control Register */
SCB_SCR = SCB_SCR | 0x2;
```

Users of assembly language can enable this feature using the following assembly code:

```
LDR    r0, =0xE000ED10; System Control Register address
LDR    r1, [r0]
MOVS   r2, #0x2
ORR    r1, r2; Set SLEEPONEXIT bit
STR    r1, [r0]
```

In interrupt-driven applications, do not enable the Sleep-on-Exit feature too early during the initialization. Otherwise if the processor receives an interrupt request during the initialization process, it will enter sleep automatically after the interrupt handler is executed, before the rest of the initialization process completes.

Wakeup Interrupt Controller (WIC) Feature

The Wakeup Interrupt Controller (WIC) is an optional component that microcontroller vendors can use to mirror the wakeup decision functionality of the NVIC when all the processor clocks have stopped during deep sleep. This feature also allows the processor to be put into an ultra-low-power state and still be able to be awakened by an interrupt almost instantly. The WIC was first introduced in the Cortex-M3 revision two (r2p0). The same feature was also made available in the Cortex-M0 and Cortex-M4 processors. Some details about the WIC features were introduced in Chapter 11. In this chapter, we will discuss how this feature can be used in embedded applications.

The presence of WIC does not require extra programmable registers. However, the use of the WIC usually requires a system-level power management unit (PMU), which would have device-specific programmable registers. In general, the presence of the WIC feature is usually transparent to the software.

The WIC is used only when the processor enters deep sleep. To use the WIC feature, the following steps are required:

- Enable PMU (device specific)
- Enable deep sleep feature in the System Control Register
- Enter sleep

When WIC is enabled and the processor enters deep sleep, the sequence shown in Figure 17.4 will occur.

Because the task of detecting and masking interrupts for wakeup is offloaded to the WIC, the Cortex-M0 processor can remain in a low-power state and does not require any clocks. To reduce

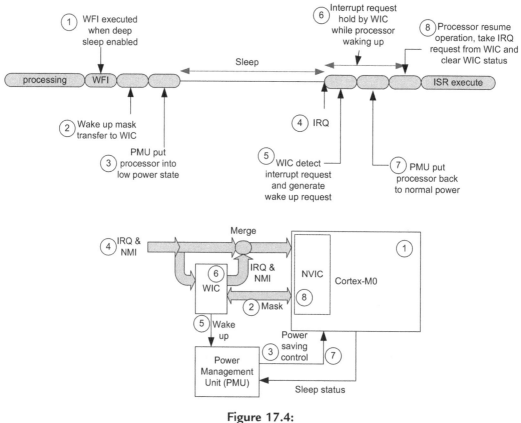

Figure 17.4:
WIC operation sequence.

the power consumption further, microcontrollers can use a special silicon technology called state retention power gating (SRPG) to power down most part of the processor logic, leaving only a small portion of circuit within each register to hold the current status (see Figure 11.11 in Chapter 11). This allows the leakage current of the design to be further lowered. Currently the SRPG is only supported in a limited numbers of silicon technology processes (cell libraries).

The use of WIC does not require a special programming step apart from configuring the device-specific PMU and enabling deep sleep. However, it can result in the SysTick timer being disabled during deep sleep. If your application uses an embedded OS and requires the OS task scheduler to continue to operate during sleep, you might need to do one of the following:

- Enabled a separate timer that is not affected by deep sleep to wake up the processor at a scheduled time.
- Disable the WIC feature.
- Avoid using deep sleep and use normal sleep instead.

Event Communication Interface

One of the wakeup sources for the WFE sleep operation is the external event signal (here the word "external" refers to the processor boundary; the source generating the event can be on chip or off chip). The event signal could be generated by on-chip peripherals or by another processor on the same chip. The event communication and WFE can be used together to reduce power in polling loops.

The Cortex-M0 processor uses two signals for event communication:

- *Transmit Event (TXEV).* A pulse is generated when the SEV instruction is executed.
- *Receive Event (RXEV).* When a pulse is received on this signal, the event latch inside the processor will be set and can cause the processor to wake up from WFE sleep operation.

First, we look at a simple use of the event connection in a single processor system: the event can be generated by a number of peripherals. A DMA controller is used in the example shown in Figure 17.5.

In a microcontroller system, a memory block copying process can be accelerated using a DMA controller. If a polling loop is used to determine the DMA status, this will waste energy, consume memory bandwidth, and might end up slowing down the DMA operation. To save energy, we put the processor into the WFE sleep state. When the DMA operation completes, we can then use a "Done" status signal to wake up the processor and continue program execution.

In the application code, instead of using a simple polling loop that continuously monitors the status of the DMA controller, the polling loop can include a WFE instruction as follows:

```
Enable_DMA_event_mask();  // Write to programmable enable mask register
                          // to enable DMA event
Start_DMA(); // Start DMA operation
do {
    __WFE(); // WFE Sleep operation, wake up when an event is received
} while (check_DMA_completed()==0);
Disable_DMA_event_mask(); // Write to programmable enable mask register
                          // to disable DMA event
```

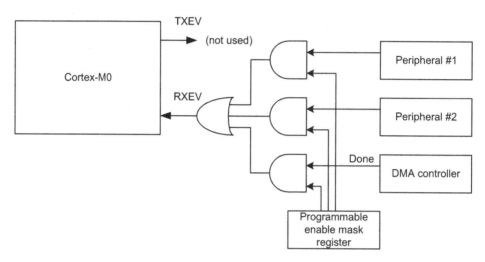

Figure 17.5:
Use of the event interface: example 1—DMA controller.

Because the processor could be awakened by other events, the polling loop must still check the DMA controller status.

For applications using an embedded OS, an OS-specific delay function should be used instead of the WFE to allow the processor to switch to another task. The embedded OS is covered in Chapter 18.

In multiprocessor systems, interprocessor communication such as spin lock often involves polling software flags in shared memory. Similar to the DMA controller example, the WFE sleep operation can be used to reduce power consumption during these activities. In a dual processor system, the event communication interface can be connected in a crossover configuration as shown in Figure 17.6.

In this arrangement, the polling loop for a shared software flag could be written as

```
do {
    __WFE(); // WFE Sleep operation, wake up when an event is received
} while (sw_flag_x==0); // poll software flag
task_X(); // execute task X when software flag for task X is received
```

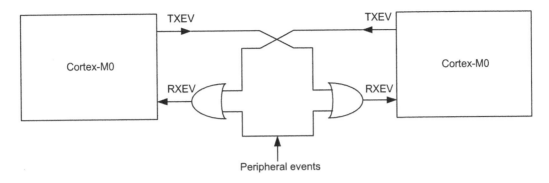

Figure 17.6:
Use of the event interface: example 2—dual processor event crossover.

For the other process that changes "sw_flag_x," it needs to generate an event after the shared variable is updated. This can be done by executing the SEV (Send event) instruction:

```
sw_flag_x = 1; // Set software flag in shared memory
__DSB(); // Data synchronization barrier to ensure the write is completed
         // not essential for Cortex-M0 but is added for software porting
__SEV(); // execute SEV instruction
```

Using this arrangement, the processor running the polling loop can stay in sleep mode until it receives an event. Because the SEV execution sets the internal event latch, this method works even if the polling process and the process that sets the software variable are running at different times on the same processor, as in a single processor multitasking system.

For applications that use an embedded OS, an OS-specific event passing mechanism should be used instead of directly using WFE and SEV.

Developing Low-Power Applications

Most Cortex-M microcontrollers come with various low-power modes to help you reduce the power consumption as low as possible. Although these features are often linked to the low-power features of the processor, each microcontroller product provides different low-power control methods and low-power characteristics. Therefore, it is not possible to cover every method for all the different product types. Here we will only cover some general information about how to reduce power on typical embedded systems and examples on a Cortex-M0 microcontroller: the NXP LPC111x.

In general, various measures can be taken to reduce power consumption:

- Reduction of active power
    1. *Choose the right microcontroller device.* Once the basic system and memory size requirements of the project are clear, you can select a microcontroller with enough memory and peripherals but not too much more.

2. *Run the processor at suitable clock frequency.* Many applications do not require a high clock frequency. When a processor is running at high clock speed, it might require wait states because of flash memory access time and hence reduce the energy efficiency.

3. *Choose the right clock source.* Many low-power microcontrollers provide multiple clock sources including internal ones. Depending on the requirements of your applications, some clock sources might work better than others. There is no general rule of "best choice" for which clock source to use. It entirely depends on the application and the microcontroller you are using.

4. *Do not enable a peripheral unless it is needed.* Some low-power microcontrollers allow you to turn off clock signals to each peripheral. In some cases, you can even turn off the power supply to a certain peripheral to reduce power.

5. *Check out other clock system features.* Some microcontrollers provide various clock dividers for different parts of the system. You can use these dividers to reduce the power—for example, reduce the processor speed when the processing requirement is low.

6. *Select a good power supply design.* A good choice of power supply design can provide optimum voltage for the application.

- Reduction of active cycles

   1. When the processor is idle, the sleep mode can be used to reduce power consumption, even it is going to enter sleep for a short period of time.

   2. Application code can be optimized for speed to reduce active cycles. In some cases (e.g., the C compiler option has been set to speed optimization), it might increase code size, but when there is spare space in the flash memory, then the optimization is worth trying.

   3. Features like Sleep-on-Exit can be used to reduce active cycles in interrupt-driven applications.

- Reduction of power during sleep

   1. *Select the right low-power features.* A low-power microcontroller might support various low-power sleep modes. Using the right sleep modes might help you to reduce the power consumption significantly.

   2. *Turn off unneeded peripherals and clock signals during sleep.* This can reduce the power consumption, but it might also increase the time required to restore the system to an operational state after exiting sleep mode.

   3. *Consider turning off some of the power.* Some microcontrollers can even turn off the power supply to some parts inside the microcontroller like flash memory and oscillators. But doing this usually means it will take longer to wake up the system.

Most microcontroller vendors would provide code library and example code to demonstrate the low-power features of their microcontrollers. Those examples can make the application development much easier.

The first step to take when developing a low-power application is to become familiar with the microcontroller device you are using. There are a few areas to investigate when developing sleep mode support code:

- Determine which sleep mode should be used.
- Determine which clock signals need to remain on.
- Determine if some clock support circuits like crystal oscillators can be switched off.
- Determine if clock source switching is needed.

To demonstrate the process, we will develop a couple of examples to use the low-power features in the LPC111x.

Example of Using Low-Power Features on the LPC111x

The LPC111x supports four power modes (Table 17.4).

Table 17.4: Power Modes in LPC111x

Power Modes	Descriptions
Run mode	The microcontroller system in normal operation: — Clocks to various parts of the microcontroller can be turned on/off using the System AHB clock control register (LPC_SYSCON -> SYSAHBCLKCTRL). — Clocks to several components including the processor can be divided to lower frequency. — Several parts of the system (ADC, oscillator, PLL, etc.) can be powered down using Power-down Configuration Register (LPC_SYSCON -> PDRUNCFG).
Sleep mode	The processor entered sleep mode with the SLEEPDEEP bit in the System Control Register (SCB -> SCR) cleared: — The clock to the processor stopped. — The peripheral clock continued to run (based on LPC_SYSCON -> SYSAHBCLKCTRL).
Deep sleep mode	The processor entered sleep mode with the DEEPSLEEP bit in the System Control Register (SCB -> SCR) set to 1: — The clock to the processor stopped. — Several parts of the system (flash, oscillator, PLL, etc.) can be powered down using the Deep Sleep Configuration Register (LPC_SYSCON -> PDSLEEPCFG). — The microcontroller can be awakened from the "start logic" feature on the I/O port. — When awakened from deep sleep, the value of the Power-down Configuration Register (LPC_SYSCON -> PDRUNCFG) is updated from the Wakeup Configuration Register (LPC_SYSCON -> PDAWAKECFG).
Deep power down mode	In this mode, most parts of the system are powered down. The status of the processor and RAM are lost. However, data in four general-purpose registers inside the power management unit are retained. This mode is entered by entering sleep mode with the following: — The deep sleep mode is enabled (SLEEPDEEP bit in SCB -> SCR set). — The DPDEN bit in the PCON register in the power management unit is set, The processor can be awakened by reset or by the "start logic" feature on the I/O port.

The first example we will be working on is a low-power version of the blinky application. For a simple blinky application, there is no need to use PLL and the oscillator for external crystal. We can just use the 12MHz internal RC oscillator provided in the LPC111x. Therefore, we can edit two define constants in system_LPC11xx.c (full listing in appendix H.1) to disable the oscillator for the external crystal and to disable the PLL setup:

```
#define CLOCK_SETUP      0
#define SYSPLL_SETUP     0
```

In this example we are going to toggle pin 0 of port 2. The processor is put into sleep mode most of the time, and it is awakened only when the 32-bit timer 0 reaches the required value. The Sleep-on-Exit feature is also enabled to obtain the shortest active cycles. The blinky application code can be written as shown:

```
#include "LPC11XX.h"
#define KEIL_MCB1000_BOARD

// Function declarations
void LedOutputCfg(void);       // Set I/O pin connected to LED as output
void Timer0_Intr_Config(void); // Setup timer

int main(void)
{
  SystemInit();   // Switch Clock to 12MHz
                  // (use internal RC oscillator only, PLL disabled)
  // Initialize LED output
  LedOutputCfg();
  Timer0_Intr_Config();      // Program timer interrupt 2 Hz
  SCB->SCR = SCB->SCR | 0x2;// Turn on Sleep-On-Exit feature
  while(1) {
    __WFI(); // Enter sleep mode
    };
} // end main

void Timer0_Intr_Config(void)
{ // Use 32-bit timer 0
  // Enable clock to 32-bit timer 0
  LPC_SYSCON->SYSAHBCLKCTRL = LPC_SYSCON->SYSAHBCLKCTRL | (1<<9);

  LPC_TMR32B0->TCR = 0;        // Disable timer
  LPC_TMR32B0->PR  = 0;        // Prescaler set to 0 (TC increment every cycle)
  LPC_TMR32B0->PC  = 0;        // Prescaler counter current value clear
  LPC_TMR32B0->TC  = 0;        // Timer counter current value clear
  LPC_TMR32B0->MR0 = 5999999;  // Match Register set to "6 million - 1"
  LPC_TMR32B0->MCR = 3;        // When match MR0, generate interrupt and reset
  LPC_TMR32B0->TCR = 1;        // Enable timer
  NVIC_EnableIRQ(TIMER_32_0_IRQn); // Enable 32-bit timer 0 interrupt
  return;
}
  void TIMER32_0_IRQHandler(void)
```

```
{
  LPC_TMR32B0->IR                = LPC_TMR32B0->IR;  // Clear interrupt
#ifdef KEIL_MCB1000_BOARD
  // For Keil MCB1000, use P2.0 for LED output
  LPC_GPIO2->MASKED_ACCESS[1] = ~LPC_GPIO2->MASKED_ACCESS[1]; // Toggle bit 0
#else
  // For LPCXpresso, use P0.7 for LED output
  LPC_GPIO0->MASKED_ACCESS[1<<7] = ~LPC_GPIO0->MASKED_ACCESS[1<<7];
  // Toggle bit 7
#endif
  return;
}

// Switch LED signal (P2_0 ) to output port with no pull up or pulldown
void LedOutputCfg(void)
{
  // Enable clock to IO configuration block (bit[16] of AHBCLOCK Control register)
  // and enable clock to GPIO (bit[6] of AHBCLOCK Control register
  LPC_SYSCON->SYSAHBCLKCTRL = LPC_SYSCON->SYSAHBCLKCTRL | (1<<16) | (1<<6);

#ifdef KEIL_MCB1000_BOARD
  // For Keil MCB1000, use P2.0 for LED output
  // PIO2_0 IO output config
  //  bit[5]   - Hysteresis (0=disable, 1 =enable)
  //  bit[4:3] - MODE(0=inactive, 1 =pulldown, 2=pullup, 3=repeater)
  //  bit[2:0] - Function (0 = IO, 1=DTR, 2=SSEL1)
  LPC_IOCON->PIO2_0 = (0<<5) + (0<<3) + (0x0);

  // Initial bit 0 output is 0
  LPC_GPIO2->MASKED_ACCESS[1] = 0;
  // Set pin 7 to 0 as output
  LPC_GPIO2->DIR = LPC_GPIO2->DIR | 0x1;
#else
  // For LPCXpresso, use P0.7 for LED output
  // PIO0_7 IO output config
  //  bit[5]   - Hysteresis (0=disable, 1 =enable)
  //  bit[4:3] - MODE(0=inactive, 1 =pulldown, 2=pullup, 3=repeater)
  //  bit[2:0] - Function (0 = IO, 1=CTS)
  LPC_IOCON->PIO0_7 = (0x0) + (0<<3) + (0<<5);
  // Initial bit[7] output is 0
  LPC_GPIO0->MASKED_ACCESS[1<<7] = 0;
  // Set pin 7 as output
  LPC_GPIO0->DIR = LPC_GPIO0->DIR | (1<<7);
#endif
  return;
} // end LedOutputCfg
```

The blinky program executes the timer 0 interrupt service routine twice per second, with the LED blinky at a rate of 1 Hz.

To try to push the power consumption lower, we will use the deep sleep feature in the next example. By using the deep sleep mode, we can power down a number of parts in the microcontroller.

Caution

Be careful when developing applications with deep sleep mode or deep power down mode. In these two modes you could lose connectivity between the in-circuit debugger and the micro-controller. If you power down the microcontroller soon after it starts running, you could end up being unable to connect the debugger to the microcontroller to carry out debug operations. This also affects flash programming. There are various solutions to this problem:

1. During software development you could add conditional executed code at the initialization stage so that you can switch a pin at reset to disable the deep sleep or power-down operation. This conditional executed code could be removed from the project later on once you are sure that the power management code is working correctly.

2. Depending on the microcontroller product, there can be a special boot mode to disable the execution of the application programmed in the flash memory. In the NXP LPC111x, port 0 bit 1 can be used in such situation. The NXP111x has an in-system programming (ISP) feature to allow the flash to be programmed using the boot loader and the serial port. By pulling bit 1 of port 0 to low at powerup reset, the ISP program in the boot loader will be executed. You can use the ISP feature to update the flash or to connect the in-circuit debugger to the microcontroller and update the flash.

For example, if you are using Keil MCB1000 board with the NXP LPC1114 microcontroller and accidentally lock up the board because you have used the powerdown feature, you can disconnect power to the board, press and hold the boot button, and connect the power again. Then you should be able to reprogram the flash.

The blinky example for deep sleep is very different from the previous example, for a number of reasons:

- During deep sleep, the internal RC oscillator is stopped, so we use the watchdog oscillator instead.
- To achieve lower power, the prescaler on the watchdog oscillator is used to reduce the clock frequency to a lower speed.
- The microcontroller has to wake up using the start logic on the LPC111x. Therefore, the wakeup exception and wakeup handler are used instead of the timer interrupt.
- Start logic on the NXP LPC111x is triggered by I/O port activities. So we use the timer match event output to drive an I/O port output and then use this signal level to trigger the wakeup as shown in Figure 17.7.

The power management of the LPC111x is controlled by a number of registers (Table 17.5). The details of these registers can be found in the NXP LPC111x User Manual.

Before using the deep sleep mode, we need to configure these registers and then program the System Control Register (SCB -> SCR) to enable the deep sleep mode. We also need to program the NVIC, timer, LED output, watchdog clock, and start logic. The example code for blinky in deep sleep is implemented as follows:

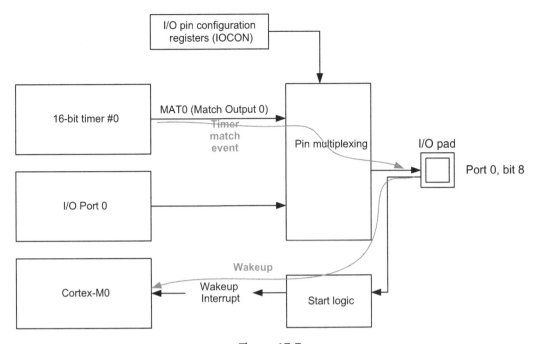

Figure 17.7:
Deep sleep wakeup mechanism used for deep sleep example.

Table 17.5: System Configuration Registers Needed for Deep Sleep Program

Register	Symbol	Descriptions
Power-Down Configuration Register	LPC_SYSCON -> PDRUNCFG	Power-down control for running mode
Deep Sleep Mode Configuration Register	LPC_SYSCON -> PDSLEEPCFG	Power-down configuration to be used when the Cortex-M0 is in deep sleep
Wakeup Configuration Register	LPC_SYSCON -> PDAWAKECFG	Value to be copied to LPC_SYSCON -> PDRUNCFG when the microcontroller wakes up from deep sleep

```
#include "LPC11XX.h"
#define KEIL_MCB1000_BOARD

/* Function declarations */
void LedOutputCfg(void);  // Set I/O pin connected to LED as output
void Timer0_Config(void); // Setup timer

/* Power down control bit definitions */
#define IRC_OUT_PD                      (0x1<<0)
#define IRC_PD                          (0x1<<1)
```

(Continued)

—Cont'd

```
#define FLASH_PD                    (0x1<<2)
#define BOD_PD                      (0x1<<3)
#define ADC_PD                      (0x1<<4)
#define SYS_OSC_PD                  (0x1<<5)
#define WDT_OSC_PD                  (0x1<<6)
#define SYS_PLL_PD                  (0x1<<7)
#define USB_PLL_PD                  (0x1<<8)
#define MAIN_REGUL_PD               (0x1<<9)
#define USB_PHY_PD                  (0x1<<10)
#define RESERVED1_PD                (0x1<<11)
#define LP_REGUL_PD                 (0x1<<12)


int main(void)
{
  /* Use internal OSC for now */
  LPC_SYSCON->SYSAHBCLKDIV = 1; /* AHB clock same as system clock */

  /* Initialize hardware */
  LedOutputCfg();       /* Program LED I/O */
  Timer0_Config();      /* Program timer */

  /* Use port0_8 as wakeup source, i/o pin */
  LPC_IOCON->PIO0_8 = (2<<0); // Function set to MAT0
  /* Only edge trigger. Activation polarity on P0.8 is rising edge. */
  LPC_SYSCON->STARTAPRP0 = LPC_SYSCON->STARTAPRP0 | (1<<8);
  /* Clear all wakeup source */
  LPC_SYSCON->STARTRSRP0CLR = 0xFFFFFFFF;
  /* Enable Port 0.1 as wakeup source. */
  LPC_SYSCON->STARTERP0 = 1<<8;

  NVIC_ClearPendingIRQ(WAKEUP8_IRQn);
  NVIC_EnableIRQ(WAKEUP8_IRQn); // Enable wake up handler

  /* Turn off all other peripheral dividers */
  LPC_SYSCON->SSP0CLKDIV = 0;
  LPC_SYSCON->SSP1CLKDIV = 0;
  LPC_SYSCON->WDTCLKDIV = 0;
  LPC_SYSCON->SYSTICKCLKDIV = 0;

  /* Turn on the watchdog oscillator */
  LPC_SYSCON->PDRUNCFG   &= ~(1<<6);
  LPC_SYSCON->WDTOSCCTRL = (0x1<<5) | 0x1F; // Run watchdog at slow speed,
                                            // with 1/64 prescale

  /* Switch MAINCLKSEL to Watchdog Oscillator */
  LPC_SYSCON->MAINCLKSEL = 2; // Set clock source to watchdog osc
  LPC_SYSCON->MAINCLKUEN = 0; // Enable update to watchdog oscillator
  LPC_SYSCON->MAINCLKUEN = 1;
  while (!(LPC_SYSCON->MAINCLKUEN & 0x01)); // wait to ensure update completed
  /* Enable flash and watchdog oscillator */
```

```
    LPC_SYSCON->PDRUNCFG = ~(WDT_OSC_PD | FLASH_PD | MAIN_REGUL_PD | LP_REGUL_PD);


  /* Copy current run mode power down configuration
     to wake up configuration register so that
     current configuration is restored at wakeup */
  LPC_SYSCON->PDAWAKECFG = LPC_SYSCON->PDRUNCFG;


  /* For deep sleep - retain power to flash, watchdog and reserved */
  LPC_SYSCON->PDSLEEPCFG = ~(FLASH_PD|WDT_OSC_PD|MAIN_REGUL_PD);


  LPC_TMR16B0->TCR = 1;        // Enable timer

  SCB->SCR = SCB->SCR | 0x4; // Turn on deep sleep feature
  while(1) {
    __WFI();     // Enter sleep mode
    };
} // end main

void Timer0_Config(void)
{ // Use 16-bit timer 0
  // Enable clock to 16-bit timer 0
  LPC_SYSCON->SYSAHBCLKCTRL = LPC_SYSCON->SYSAHBCLKCTRL | (1<<7);

  LPC_TMR16B0->TCR = 2;          // Disable and reset timer
  LPC_TMR16B0->TCR = 0;          // Disable timer
  LPC_TMR16B0->PR  = 95;         // Prescaler set to 0 (TC increment every 96 cycle)
   // Watchdog oscillator will be configured as approx 9600 Hz.
   // By having prescale of 96, the timer increment every 10 ms
  LPC_TMR16B0->PC  = 0;          // Prescaler counter current value clear
  LPC_TMR16B0->TC  = 0;          // Timer counter current value clear
  LPC_TMR16B0->MR0 = 199;        // Match Register set to "200 - 1"
                                 // because timer count at 100 Hz,
                                 // match occur once every second
  LPC_TMR16B0->EMR = (0x2<<4);   // Enable match output
  LPC_IOCON->PIO0_8= (2<<0);     // Set PIO0_8 to MAT0 output function
  LPC_TMR16B0->MCR = 2;          // When match MR0, reset counter
  return;
} // end Timer0_Config

void WAKEUP_IRQHandler(void)
{
  unsigned int regVal;
  int i,j;
  regVal = LPC_SYSCON->STARTSRP0;
  if ( regVal != 0 )
  {
      LPC_SYSCON->STARTRSRP0CLR = regVal;
  }
  /* Clear the timer match output to 0 */
  LPC_TMR16B0->EMR = LPC_TMR16B0->EMR & ~(1<<0);
```

(Continued)

—Cont'd

```c
#ifdef KEIL_MCB1000_BOARD
  // For Keil MCB1000, use P2.0 for LED output
  LPC_GPIO2->DIR = LPC_GPIO2->DIR | 0x1; // enable output
  for (i=0; i< 4; i++){
    LPC_GPIO2->MASKED_ACCESS[1] = ~LPC_GPIO2->MASKED_ACCESS[1]; // Toggle bit 0
    for (j=0; j<30;j++) { __ISB(); } // delay
    }
  LPC_GPIO2->DIR = LPC_GPIO2->DIR & ~(0x1); // turn off output to save power
#else
  // For LPCXpresso, use P0.7 for LED output
  LPC_GPIO0->DIR = LPC_GPIO0->DIR | (1<<7);  // enable output
  for (i=0; i< 4; i++){
    LPC_GPIO0->MASKED_ACCESS[1<<7] = ~LPC_GPIO0->MASKED_ACCESS[1<<7];
    //Toggle bit 7
    for (j=0; j<30;j++) { __ISB(); } // delay
    }
  LPC_GPIO0->DIR = LPC_GPIO0->DIR & ~(1<<7);  // turn off output to save power
#endif
  return;
}
void LedOutputCfg(void)
{
  // Enable clock to IO configuration block (bit[16] of AHBCLOCK Control register)
  // and enable clock to GPIO (bit[6] of AHBCLOCK Control register
  LPC_SYSCON->SYSAHBCLKCTRL = LPC_SYSCON->SYSAHBCLKCTRL | (1<<16) | (1<<6);
#ifdef KEIL_MCB1000_BOARD
  // For Keil MCB1000, use P2.0 for LED output
  // PIO2_0 IO output config
  //   bit[5]   - Hysteresis (0=disable, 1 =enable)
  //   bit[4:3] - MODE(0=inactive, 1 =pulldown, 2=pullup, 3=repeater)
  //   bit[2:0] - Function (0 = IO, 1=DTR, 2=SSEL1)
  LPC_IOCON->PIO2_0 = (0<<5) + (0<<3) + (0x0);
  // Initial bit 0 output is 0
  LPC_GPIO2->MASKED_ACCESS[1] = 0;
  // Set pin 7 to 0 as output
  LPC_GPIO2->DIR = LPC_GPIO2->DIR | 0x1;
#else
  // For LPCXpresso, use P0.7 for LED output
  // PIO0_7 IO output config
  //   bit[5]   - Hysteresis (0=disable, 1 =enable)
  //   bit[4:3] - MODE(0=inactive, 1 =pulldown, 2=pullup, 3=repeater)
  //   bit[2:0] - Function (0 = IO, 1=CTS)
  LPC_IOCON->PIO0_7 = (0x0) + (0<<3) + (0<<5);
  // Initial bit[7] output is 0
  LPC_GPIO0->MASKED_ACCESS[1<<7] = 0;
  // Set pin 7 as output
  LPC_GPIO0->DIR = LPC_GPIO0->DIR | (1<<7);
#endif
  return;
} // end LedOutputCfg
```

When the counter timer reaches 99, the counter resets and bit 8 of port 0 is driven high by the timer match output. This triggers the start logic and the wakeup exception. The LED activity is handled within the wakeup exception handler. After the LED stops blinking, the LED output is turned off and the processor returns to sleep.

Note that when a debugger is connected to a system, in some microcontrollers the system design might automatically disable some of the low-power optimizations to allow the debug operation to be performed correctly. Therefore, when trying to measure the power consumption of the system, you might need to disconnect the microcontroller system from the debugger. In some cases, the debugger needs to be disconnected from the system before powering up the microcontroller to minimize the power consumption, because the effect of a debug connection to the power management circuit might retain until the power supply has been disconnected.

## CHAPTER 18

Using SVC, PendSV, and Keil RTX Kernel

Introduction

In Chapter 10 we covered the hardware features in the Cortex-M0 processor related to the operating system. In this chapter we will use the SVC and PendSV features in programming examples. This chapter also introduces the Keil RTX Kernel, which is included in the Keil MDK, including the evaluation version.

In practice, the SVC is rarely used directly without the OS. For applications with an embedded OS, the application programming interface (API) of the OS normally handles these for you. Nevertheless, the information about using SVC and PendSV can still be useful for developers of debugging software.

Using the SVC Exception

SuperVisor Call (SVC) is commonly used in an OS environment for application tasks to access to system services provided by the OS. In general, using the SVC involves the following process:

1. Set up optional input parameters to pass to the SVC handler in registers (e.g., R0 to R3) based on programming practices outlined by AAPCS.
2. Execute the SVC instruction.
3. The SVC exception handler starts execution and can optionally extract the address of the stack frame using SP values.
4. Using the extracted stack frame address, the SVC exception handler can locate and read the input parameters that are stored as stacked registers.
5. Optionally, the SVC exception handler can also track the immediate value in the executed SVC instruction using the stacked PC value in the stack frame.
6. The SVC exception handler then carries out the required processing.
7. If the SVC exception handler needs to return a value back to the application task that made the SVC call, it needs to put the return value back onto the stack frame, usually where the stacked R0 is located.
8. The SVC exception handler executes an exception return, and the contents of the stack frame are restored to the register bank.
9. The modified stacked R0 value in the stack frame, which contains the return value of the SVC handler, is loaded into R0 and can be used by the application task as the return value.

The Definitive Guide to the ARM Cortex-M0. DOI: 10.1016/B978-0-12-385477-3.10018-7

You might wonder why we need to extract the input parameters from the stack frame, instead of just using the values in the register bank. The reason is that if another exception with a priority level higher than the SVC exception occurred during stacking, the other exception handler would be executed first and it could change the values in registers R0 to R3 and R12 before the SVC handler is entered. (In Cortex-M processors, exceptions handlers can be normal C functions; therefore, these registers can be changed.)

Similarly, the return value has to be put into the stack frame. Otherwise, the value stored into R0 will be lost during the unstacking process of returning from the exception.

In the next step, we will see how to do all of this in a programming example. The following example is based in Keil MDK and can also be used on the ARM RealView Development Suite.

First, we need to ensure that the "SVC_Handler" has already been defined in the vector table. If you are using CMSIS-based software packages from microcontroller vendors, the "SVC_Handler" definition should be included in the vector table already. Otherwise, you might need to add this to the vector table.

Second, we need to be able to put the input parameters into the right registers and execute the SVC instruction. With Keil MDK or ARM RVDS, the "__svc" keyword can be used to define the SVC function including the SVC number (the immediate value in the SVC instruction), the input parameters, and the return parameter definitions. You can define multiple SVC functions with different SVC numbers. For example, the following code defined three SVC function prototypes:

```
int __svc(0x00)   svc_service_add(int x, int y);
int __svc(0x01)   svc_service_sub(int x, int y);
int __svc(0x02)   svc_service_incr(int x);
```

Once the SVC functions have been defined, we can use them in our application code. For example,

```
z = svc_service_add(x, y);
```

The code for the SVC handler is separated into two parts in the following example:

- The first part is an assembly wrapper code to extract the starting address of the exception stack frame and put it to register R0 as an input parameter for the second part.
- The second part extracts the SVC number and input parameters from the stack frame and carries out the SVC operation in C. The program code might also need to deal with error conditions if an SVC instruction is executed with an invalid SVC number.

The first half of the SVC handler has to be carried out in assembly because we cannot tell the stack frame starting location from a C-based SVC handler. Even if we can find out the current value of the stack pointers, we do not know how many registers would have been pushed onto the stack at the beginning of C handler.

Using the embedded assembly feature, the first part of the SVC handler can be written as follows:

```
// SVC handler - Assembly wrapper to extract
//                stack frame starting address
__asm void SVC_Handler(void)
{
    MOVS    r0, #4
    MOV     r1, LR
    TST     r0, r1
    BEQ     stacking_used_MSP
    MRS     R0, PSP ; first parameter - stacking was using PSP
    LDR     R1,=__cpp(SVC_Handler_main)
    BX      R1
stacking_used_MSP
    MRS     R0, MSP ; first parameter - stacking was using MSP
    LDR     R1,=__cpp(SVC_Handler_main)
    BX      R1
}
```

We use BX instruction to branch instead of using "B __cpp(SVC_Handler_main)." This is because in case the linker rearranged the positioning of the function order, the BX instruction will still be able to reach the branch destination.

The second part of the SVC handler used the extracted stack frame starting address as the input parameter and used it as a pointer to an integer array to access the stacked register values. The completed example code is listed in the following box:

---

svc_demo.c

```
#include "LPC11XX.h"
#include "uart_io.h"
#include <stdio.h>

// Define SVC function
int __svc(0x00)  svc_service_add(int x, int y);
int __svc(0x01)  svc_service_sub(int x, int y);
int __svc(0x02)  svc_service_incr(int x);

void SVC_Handler_main(unsigned int * svc_args);

// Function declarations
int main(void)
{
    int x, y, z;

    SystemInit();  // System Initialization
    UartConfig();  // Initialize UART

    x = 3; y = 5;
    z = svc_service_add(x, y);
    printf ("3+5 = %d \n", z);

    x = 9; y = 2;
    z = svc_service_sub(x, y);
    printf ("9-2 = %d \n", z);
```

(Continued)

svc_demo.c—Cont'd

```c
  x = 3;
  z = svc_service_incr(x);
  printf ("3++ = %d \n", z);

  while(1);
}
// SVC handler - Assembly wrapper to extract
//                stack frame starting address
__asm void SVC_Handler(void)
{
  MOVS   r0, #4
  MOV    r1, LR
  TST    r0, r1
  BEQ    stacking_used_MSP
  MRS    R0, PSP ; first parameter - stacking was using PSP
  LDR    R1,=__cpp(SVC_Handler_main)
  BX     R1
stacking_used_MSP
  MRS    R0, MSP ; first parameter - stacking was using MSP
  LDR    R1,=__cpp(SVC_Handler_main)
  BX     R1
}

// SVC handler - main code to handle processing
// Input parameter is stack frame starting address
// obtained from assembly wrapper.
void SVC_Handler_main(unsigned int * svc_args)
{
  // Stack frame contains:
  // r0, r1, r2, r3, r12, r14, the return address and xPSR
  // - Stacked R0  = svc_args[0]
  // - Stacked R1  = svc_args[1]
  // - Stacked R2  = svc_args[2]
  // - Stacked R3  = svc_args[3]
  // - Stacked R12 = svc_args[4]
  // - Stacked LR  = svc_args[5]
  // - Stacked PC  = svc_args[6]
  // - Stacked xPSR= svc_args[7]
  unsigned int svc_number;
  svc_number = ((char *)svc_args[6])[-2];
  switch(svc_number)
    {
    case 0: svc_args[0] = svc_args[0] + svc_args[1];
            break;
    case 1: svc_args[0] = svc_args[0] - svc_args[1];
            break;
    case 2: svc_args[0] = svc_args[0] + 1;
            break;
    default: // Unknown SVC request
            break;
    }
  return;
}
```

After the program executes, the UART outputs the expected results generated from the SVC functions.

The priority level of the SVC exception is programmable. To assign a new priority level to the SVC exception, we can use the CMSIS function NVIC_SetPriority. For example, if we want to set the SVC priority level to 0x80, we can use

```
NVIC_SetPriority(SVCall_IRQn, 0x2);
```

The function automatically shifts the priority level value to the implemented bit of the priority level register (0x2<<6 equals 0x80).

Using the PendSV Exception

Unlike the SVC, the PendSV exception is triggered by writing to the Interrupt Control State Register (address 0xE000ED04; see Table 9.6). If the PendSV exception is blocked due to an insufficient priority level, it will wait until the current priority level drops or the blocking (e.g., PRIMASK) is removed.

To put the PendSV exception into pending state, we can use the following C code:

```
SCB->ICSR = SCB->ICSR | (1<<28); // Set PendSV pending status
```

The priority level of the PendSV exception is programmable. To assign a new priority level to the PendSV exception, we can use the CMSIS function NVIC_SetPriority. For example, if we want to set the PendSV priority level to 0xC0, we can use

```
NVIC_SetPriority(PendSV_IRQn, 0x3); // Set PendSV to lowest level
```

The function automatically shifts the priority level value to the implemented bit of the priority level register (0x3<<6 equals 0xC0).

The following code demonstrates the triggering and setup for the PendSV exception. It sets up a timer exception at high priority and the PendSV exception at lower priority. Each time the high-priority timer exception is triggered, the timer handler only executes for a short period of time, carries out essential tasks, and sets the pending status of PendSV. The PendSV is executed after the timer handler completes and reports to the terminal that the timer exception has been executed.

```
pendsv_demo.c

#include "LPC11XX.h"
#include "uart_io.h"
#include <stdio.h>
void Timer0_Intr_Config(void); // declare timer initialization function

int main(void)
```

(Continued)

pendsv_demo.c—Cont'd

```c
{
  SystemInit();   // System Initialization
  UartConfig();   // Initialize UART

  NVIC_SetPriority(TIMER_32_0_IRQn, 0x0); // Set Timer  to highest level
  NVIC_SetPriority(PendSV_IRQn , 0x3);    // Set PendSV to lowest  level

  // Program timer interrupt at 1 Hz.
  // At 48MHz, Timer trigger every 4800000 CPU cycles
  Timer0_Intr_Config();
  while(1);
}

void PendSV_Handler(void)
{
  printf ("[PendSV] Timer interrupt triggered\n");
  return;
}

void TIMER32_0_IRQHandler(void)
{
  LPC_TMR32B0->IR = LPC_TMR32B0->IR;  // Clear interrupt
  SCB->ICSR       = SCB->ICSR | (1<<28); // Set PendSV pending status
  return;
}

void Timer0_Intr_Config(void)
{ // Use 32-bit timer 0
  // Enable clock to 32-bit timer 0
  LPC_SYSCON->SYSAHBCLKCTRL = LPC_SYSCON->SYSAHBCLKCTRL | (1<<9);

  LPC_TMR32B0->TCR = 0;               // Disable timer
  LPC_TMR32B0->PR  = 0;               // Prescaler set to 0 (TC increment every cycle)
  LPC_TMR32B0->PC  = 0;               // Prescaler counter current value clear
  LPC_TMR32B0->TC  = 0;               // Timer counter current value clear
  LPC_TMR32B0->MR0 = 47999999;        // Match Register set to "48 million - 1"
  LPC_TMR32B0->MCR = 3;               // When match MR0, generate interrupt and reset
  LPC_TMR32B0->TCR = 1;               // Enable timer
  NVIC_EnableIRQ(TIMER_32_0_IRQn);    // Enable 32-bit timer 0 interrupt
  return;
}
```

With this arrangement, the processing task required by the timer exception is split into two halves. Because the "printf" process can take a long time, it is executed by the PendSV at a low priority so that other higher or medium priority exceptions can take place while printf is running (Figure 18.1). This type of interrupt processing method can be applied to many applications to help improve the interrupt response of embedded systems.

Another use of the PendSV exception is for context switching in an OS environment; please refer to Chapter 10.

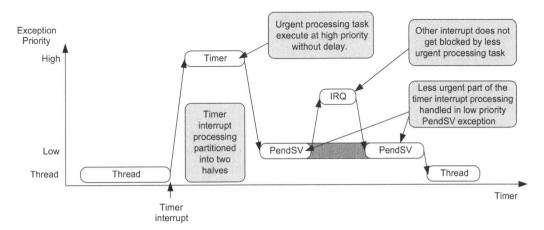

Figure 18.1:
Using PendSV to improve a system-level interrupt response.

Unlike the SVC, the PendSV exception is not precise, which means after the instruction that set the PendSV exception pending status is executed, the processor can still execute a number of instructions before the exception sequence takes place. For this reason, PendSV can only work as a subroutine without any input parameters and output return values.

Using an Embedded OS

When the complexity of applications increases, the application code has to handle more and more tasks in parallel and it is more and more difficult to ensure such applications run smoothly without an embedded OS. An embedded OS divides the available CPU processing time into a number of time slots that carry out different tasks in different time slots. Because the switching of tasks happens a hundred times or more per second, it appears to the application that the tasks are running simultaneously.

Many embedded applications do not require an OS. For example, if the applications do not have to handle many tasks in parallel or if the additional tasks are relatively short so they can be processed inside interrupt handlers, the use of an embedded OS is not required. For simple applications, use of an OS could result in unnecessary overhead. For example, the OS requires extra program size and RAM size, and the OS itself also requires a small amount of processing time. On the other hand, if an application has a number of parallel tasks and requires a good response time for each task, then the use of an embedded OS can be very important.

An embedded OS requires a timer to generate interrupt so that the OS can carry out task scheduling and system management. On the Cortex-M0, the SysTick timer is dedicated for this purpose. An embedded OS might also utilize various OS features on the Cortex-M0, like separate stack pointers for kernel and threads, SVC, and PendSV.

A number of embedded operating systems are available for the Cortex-M0 processor. As an example, we will look at the Keil Real-Time eXecutive (RTX) kernel.

Keil RTX Real-Time Kernel

Features Overview

The Keil RTX Real-Time kernel is a royalty-free, real-time operating system (RTOS) targeted for microcontroller applications. Depending on the version of Keil MDK product you are using, it includes either a precompiled RTX kernel library, or the source code version of the RTX Kernel. You can also get the RTX Kernel source code as a part of the RL-ARM (Real-Time Library) product (Figure 18.2). The precompiled version of RTX kernel library is fully functional and has the same features as the source code version. Even though you are using evaluation versions of the Keil MDK product, the RTX Kernel provided is functionally identical to the source code version, so you can try out all the Keil RTX example projects in this chapter using an evaluation version of the Keil MDK.

The RTX kernel is supported on all Cortex-M processors in addition to traditional ARM processors such as ARM7 and ARM9. It has the following features:

- Flexible scheduler, which supports preemptive, round-robin, and collaborative scheduling schemes
- Support for mailboxes, events (up to 16 per task), semaphores, mutex, and timers
- An unlimited number of defined tasks, with a maximum of 250 active tasks at a time
- Up to 255 task priority levels
- Support for multithreading and thread-safe operations
- Kernel aware debug support in Keil MDK
- Fast context switching time
- Small memory footprint (less than 4 KB for the Cortex-M version, less than 5 KB for the ARM7/9)

In addition, the Cortex-M version of the RTX kernel has the following features:

- SysTick timer support
- No interrupt lockout in the Cortex-M versions (interrupt is not disabled by the OS at any time)

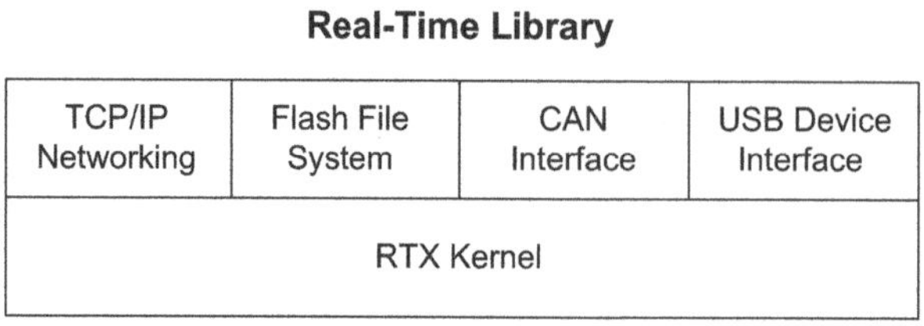

Real-Time Library

TCP/IP Networking	Flash File System	CAN Interface	USB Device Interface
RTX Kernel			

Figure 18.2:
The RL-ARM product.

The RTX kernel can work with or without the other software components in the RL-ARM library. It can also work with third-party software products, such as communication protocol stacks, data processing codecs, and other middleware.

In the RTX kernel, each task has a priority level. Normal tasks can have a priority level from 1 to 254, with 254 being the most important and level 1 the least important. Priority level 0 is reserved for the idle task. If a user task is created with priority level 0, it is automatically changed to level 1 by the task creation function. Priority level 255 is also reserved. Note that the task priority level arrangement is completely separated from interrupt priority.

In the RTX environment, each task can be in one of the states described in Table 18.1.

Table 18.1: Task States in RTX Kernel

State		Description
RUNNING		The task is currently running.
READY		The task in the queue of tasks is ready to run. When the current running task is completed, RTX will select the next highest priority task in the ready queue and start it.
Waiting	WAIT_DLY	The task is waiting for a delay to complete (running os_dly_wait()).
	WAIT_ITV	The task is waiting for an interval to complete (see the period time interval feature discussed in the latter part of this chapter).
	WAIT_OR	The task is waiting for at least one event flag. If ANY of the waiting events occurs, the task is switched to ready state.
	WAIT_AND	The task is waiting for at least one event flag. If ALL of the waiting events occur, the task is switched to ready state.
	WAIT_SEM	The task is waiting for a semaphore.
	WAIT_MUT	The task is waiting for a mutex (mutual exclusive) to become available.
	WAIT_MBX	The task is waiting for a mailbox message.
INACTIVE		The task has not been started or the task has been deleted.

Each task must be declared with __task keyword. For example, a simple task that toggles an LED can be written as

```
__task void blinky(void) {
  while(1) {
    LPC_GPIO2->MASKED_ACCESS[ 1] = ~LPC_GPIO2->MASKED_ACCESS[ 1] ; // Toggle bit 0
    os_dly_wait (50); // delay 50 clock ticks (0.5 second)
    }
  }
```

We need to initialize each task before it can be executed. In addition, the OS kernel also requires initialization steps. Next, we will see how the OS initialization and task initializations are carried out.

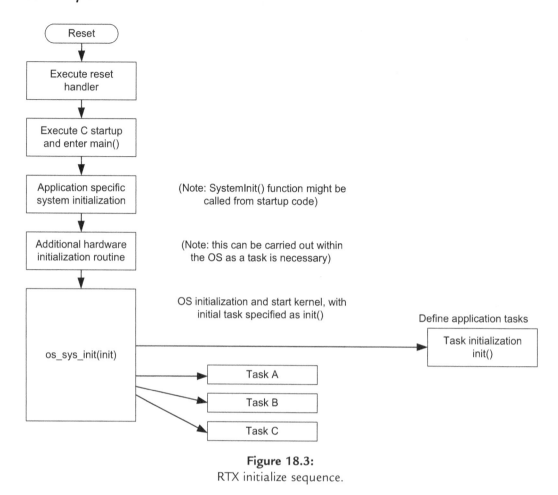

Figure 18.3:
RTX initialize sequence.

OS Startup Sequence

An application using the RTX kernel has the startup sequence shown in (Figure 18.3).

The init () task is the first task executed by the OS and can be used to create additional tasks. The name "init ()" used in the flowchart is just an example, other names can be used. The os_sys_init () function initializes and starts the OS. It must be called from main() and does not return.

Simple OS Example

In the first example of using the RTX kernel, we will create a simple task that toggles an LED. The steps for creating the project are the same as they are for creating a project without an OS. In addition to the usual project options, we also need to enable the RTX kernel option (Figure 18.4).

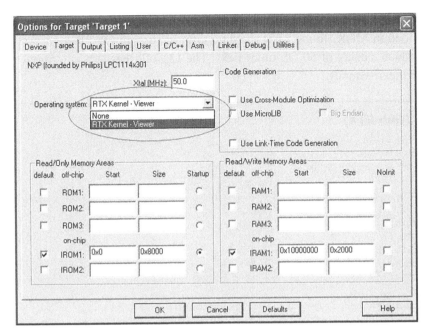

Figure 18.4:
RTX kernel option in Keil MDK.

The project also requires a configuration file called RTX_config.c. This file defines a number of parameters used by the RTX kernel, including clock frequency settings and stack settings. An example of the RTX_config.c is provided in Appendix H.5. You can also find this file in the examples in the Keil MDK installations. You can reuse the RTX_config.c from RTX examples for Cortex-M0 or Cortex-M3 in your Cortex-M0 project.

You need to edit a few parameters in the RTX_config.c to ensure the setting matches your project settings. Some of these are described in Table 18.2.

The descriptions for the remaining parameters can be found in the comments in RTX_config.c in the appendix H.

Table 18.2: Several Example Parameters in RTX_config.c

	Value in this Project	**Descriptions**
OS_CLOCK	48000000	Clock frequency
OS_TICK	10000	Time interval between the OS timer tick in us. 10000 = 10 ms.
OS_TASKCNT	6	Maximum number of concurrent running tasks
OS_PRIVCNT	0	Number of tasks with a user-provided stack
OS_STKSIZE	50	Stack size for tasks (in bytes)
OS_STKCHECK	1	Enables the stack stacking code to detect stack overflow
OS_TIMERCNT	0	Number of user timers
OS_ROBIN	1	Enables round-robin task switching
OS_ROBINTOUT	5	Identifies how long a task will execute before a task switch

The actual code for the application is very simple. To use the OS, we need to use the "RTL.h" header provided in MDK. In the LED toggling task, we use an OS function `os_dly_wait` `(50)` to produce a delay of 50 OS timer ticks. The LED flashes at a rate of once per second (toggles twice per second).

```c
blinky.c (with RTX)
#include <RTL.h>
#include "LPC11XX.h"
#define KEIL_MCB1000_BOARD

OS_TID t_blinky; // Declare a task ID for blink

__task void blinky(void) {
  while(1) {
#ifdef KEIL_MCB1000_BOARD
    // For Keil MCB1000, use P2.0 for LED output
    LPC_GPIO2->MASKED_ACCESS[1] = ~LPC_GPIO2->MASKED_ACCESS[1]; // Toggle bit 0
#else
    // For LPCXpresso, use P0.7 for LED output
    LPC_GPIO0->MASKED_ACCESS[1<<7] = ~LPC_GPIO0->MASKED_ACCESS[1<<7];//Toggle bit 7
#endif
    os_dly_wait (50);  // delay 50 clock ticks
    }
  }

__task void init (void) {
  t_blinky = os_tsk_create (blinky, 1); // Create a task "blinky" with priority 1
  os_tsk_delete_self ();
}

// Switch LED signal (P2_0 ) to output port with no pull up or pulldown
void LedOutputCfg(void)
{
  // Enable clock to IO configuration block (bit[16] of AHBCLOCK Control register)
  // and enable clock to GPIO (bit[6] of AHBCLOCK Control register
  LPC_SYSCON->SYSAHBCLKCTRL = LPC_SYSCON->SYSAHBCLKCTRL | (1<<16) | (1<<6);

#ifdef KEIL_MCB1000_BOARD
  // For Keil MCB1000, use P2.0 for LED output
  // PIO2_0 IO output config
  //   bit[5]   - Hysteresis (0=disable, 1 =enable)
  //   bit[4:3] - MODE(0=inactive, 1 =pulldown, 2=pullup, 3=repeater)
  //   bit[2:0] - Function (0 = IO, 1=DTR, 2=SSEL1)
  LPC_IOCON->PIO2_0 = (0<<5) + (0<<3) + (0x0);

  // Initial bit 0 output is 0
  LPC_GPIO2->MASKED_ACCESS[1] = 0;
  // Set pin 7 to 0 as output
  LPC_GPIO2->DIR = LPC_GPIO2->DIR | 0x1;
#else
  // For LPCXpresso, use P0.7 for LED output
```

```
// PIO0_7 IO output config
//  bit[5]   - Hysteresis (0=disable, 1 =enable)
//  bit[4:3] - MODE(0=inactive, 1 =pulldown, 2=pullup, 3=repeater)
//  bit[2:0] - Function (0 = IO, 1=CTS)
LPC_IOCON->PIO0_7 = (0x0) + (0<<3) + (0<<5);

// Initial bit[7] output is 0
LPC_GPIO0->MASKED_ACCESS[1<<7] = 0;
// Set pin 7 as output
LPC_GPIO0->DIR = LPC_GPIO0->DIR | (1<<7);
#endif
  return;
} // end LedOutputCfg

int main(void)
{
  SystemInit();     // Switch Clock to 48MHz
  LedOutputCfg();   // Initialize LED output
  os_sys_init(init); // Initialize OS
} // end main
```

For each task (apart from the initial task), a task identifier value is required, and this is defined with data type `OS_TID`. This task ID value is assigned when the task is created and is required for intertask communications, which will be demonstrated later.

Apart from `os_tsk_create()`, various other functions can be used to create tasks (Table 18.3).

You can also detect the task ID value using the function shown in Table 18.4.

Table 18.3: Functions to Create New Tasks

Task Creation Functions	Description
os_tsk_create	Create a new task.
os_tsk_create_ex	Create a new task with an argument passing to the new task.
os_tsk_create_user	Create a new task with separate stack.
os_tsk_create_user_ex	Create a new task with a separate stack and with an argument passing to the new task.

Table 18.4: Functions to Determine Task ID

Task ID Function	Description
os_tsk_self	Return the task ID of the task.

After a task has been created, it can also be deleted. For example, the initial task (init) deletes itself after all the required tasks have been created (Table 18.5).

After a task has been created, you can change the priority of a task (Table 18.6).

Table 18.5: Functions to Delete Task

Task Delete Functions	Description
os_tsk_delete	Delete a task.
os_tsk_delete_self	Delete the task itself.

Table 18.6: Functions to Manage Task Priority Level

Task Priority Functions	Description
os_tsk_prio	Change the priority level of a task.
os_tsk_prio_self	Change the priority level of a current task.

Details of each OS function can be found on the Keil web site: RL-ARM User's Guide.

Intertask Communications

In most complex applications, there can be various interaction between tasks. Instead of using polling loops to check the status of shared variables, we should use the intertask communication features provided in the OS. Otherwise, a task waiting for input from another task will stay in the ready task queue and could be executed when a time slot is available. This can end up wasting the processing time of the processor.

Most embedded OSs provide a number of methods to handle intertask communications. For example, a simple handheld device might have the tasks and interactions shown in (Figure 18.5).

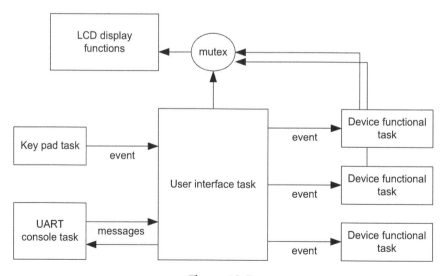

Figure 18.5:
Interactions between tasks in an example embedded project.

In Keil RTX kernel, the communication between tasks can be handled by the following:

- Events
- Mailbox (messages)
- Semaphore
- Mutual exclusive (MUTEX)

You can also combine these communication channels and use shared data to handle data transfers between tasks. Using these OS-provided functions correctly allows the task schedule in the OS kernel to understand the task processing activities and allow the tasks to be scheduled efficiently.

Event Communications

In RTX kernel, each task can have up to 16 event inputs. Each event input is represented by a bit in a 16-bit event pattern. The following example demonstrates the most simple event communication—two tasks are created, one (eventgen) generates events regularly to the blinky task. The blinky task then toggles the LED when an event is received (Figure 18.6).

The code for the event communication demonstration is shown next. (The LED I/O configuration code is the same as in the previous example and is omitted from the listing.)

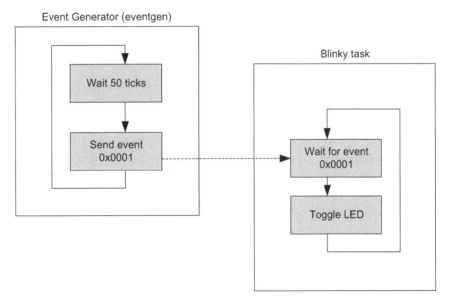

Figure 18.6:
Simple event communication between two tasks.

---

blinky_event.c (the LED I/O configuration code omitted as it is same as previous)

```
#include <RTL.h>
#include "LPC11XX.h"

OS_TID t_eventgen; // Declare a task ID for event generator
OS_TID t_blinky;   // Declare a task ID for blink

__task void blinky(void) {
  while(1) {
    os_evt_wait_and (0x0001, 0xffff); // wait for an event flag 0x0001, no timeout
    LPC_GPIO2->MASKED_ACCESS[1] = ~LPC_GPIO2->MASKED_ACCESS[1]; // Toggle bit 0
  }
}

__task void eventgen(void) { // Event generator
  while (1) {
    os_dly_wait(50);
    os_evt_set (0x0001, t_blinky);  // Send a event 0x0001 to blinky task
  }
}

__task void init (void) {
  t_blinky   = os_tsk_create (blinky,  1); // Create a task "blinky" with priority
1
  t_eventgen = os_tsk_create (eventgen, 1); // Create a task "eventgen" with
priority 1
  os_tsk_delete_self ();
}

int main(void)
{
  SystemInit();    // Switch Clock to 48MHz
  LedOutputCfg();  // Initialize LED (GPIO #2, bit0) output
  os_sys_init(init); // Initialize OS
} // end main
```

The event functions in the Keil RTX kernel allow a task to wait for multiple events and continue if any one of the events is asserted (OR arrangement) or continue if all of the required events are asserted (AND arrangement) (Table 18.7).

Table 18.7: Functions for Event Communications

Event Functions	Description
os_evt_set	Send an event pattern to a task
os_evt_clr	Clear an event from a task.
os_evt_wait_and	Wait until all the required flags are received.
os_evt_wait_or	Wait until any of the required flags are received.
os_evt_get	Obtained the bit pattern value of the event received.
isr_evt_set	Set the event to a task from interrupt service routine.

For example, a task that was waiting for an event might need to respond differently based on the source of the received event. In this case, you can use each bit of the event signal for each event source, and then detect which event it received using the `os_evt_get` function (Figure 18.7).

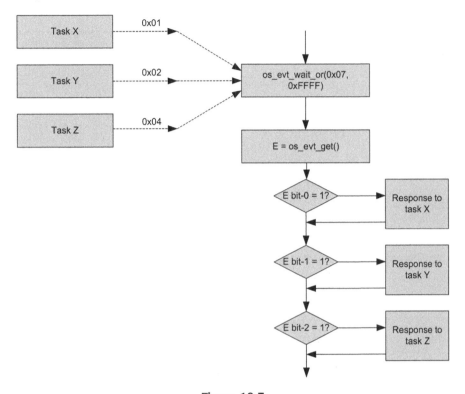

Figure 18.7:
Using the os_evt_get() function to detect which task generates the event.

Mutual Exclusive

Very often, multiple tasks need access to the same resource such as hardware peripherals. In such cases, we can define mutual exclusive (MUTEX) to ensure that only one task can access to a hardware resource at one time (Figure 18.8). The following example shows two

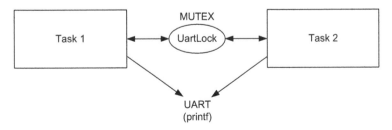

Figure 18.8:
Using the MUTEX to control hardware resource sharing.

tasks both using the UART interface. To prevent error, the MUTEX is used so that only one printf can be executed at a time.

The program code to demonstrate MUTEX is as follows:

```
mutex_demo.c
#include <RTL.h>
#include "LPC11XX.h"
#include "uart_io.h"
#include <stdio.h>

OS_TID t_task1; // Declare a task ID for task1
OS_TID t_task2; // Declare a task ID for task2
OS_MUT UartLock; // Declare a MUTEX for Uart access

__task void task1(void) {
  while(1) {
    os_dly_wait(50);
    os_mut_wait(UartLock, 0xFFFF); // Wait for UART access lock,
                                   // with indefinite timeout period
    printf ("Task 1 running.\n");
    os_mut_release(UartLock);      // Finished using UART, MUTEX release
    }
  }

__task void task2(void) {
  while (1) {
    os_dly_wait(50);
    os_mut_wait(UartLock, 0xFFFF); // Wait for UART access lock,
                                   // with indefinite timeout period
    printf ("Task 2 running.\n");
    os_mut_release(UartLock);      // Finished using UART, MUTEX release
    }
  }

__task void init (void) {os_mut_init(UartLock);
  t_task1 = os_tsk_create (task1, 1); // Create a task "task1" with priority 1
  t_task2 = os_tsk_create (task2, 1); // Create a task "task2" with priority 1

  os_tsk_delete_self ();
}

int main(void)
{
  SystemInit();    // Switch Clock to 48MHz
  UartConfig();    // Configure UART
  os_sys_init(init); // Initialize OS
} // end main
```

In this example we did not define a timeout value while waiting for the lock. You can define a time out value when using the `os_mut_wait` function. The `os_mut_wait` function return value can be used to determine if the mutex lock has been gained successfully (Table 18.8).

Table 18.8: Functions for the Mutual Exclusive Operation

Mutex Functions	Description
os_mut_init	Initialize a mutex
os_mut_wait	Attempt to obtain a mutex lock, if the mutex is locked by another task, wait until the mutex is free.
os_mut_release	Release a locked mutex.

Semaphore

The semaphore feature is similar to MUTEX. Whereas MUTEX limits just one task access to a shared resource, semaphore can limit a fixed number of tasks to access a pool of shared resources. Imagine that only a limited number of toys are available in a playroom, and several children want to play with these toys. A token system can be set up so that only the children who get the token can enter the playroom. Once a child finishes playing, he or she returns the token, leaves the room, and another child can then take the token and enter the playroom. This is how a semaphore works. A semaphore object needs to be initialized to the maximum number of tasks (i.e., number of toys available) that can use the shared resource, and before each task can use the resource, it needs to request a token. If no tokens are left, the task must wait. The token must be returned when the task finishes with the shared resource.

The mutex is a special case of semaphore for which the maximum number of available tokens is 1. To illustrate, we can modify the last MUTEX example to use the semaphore feature instead:

```
semaphore_demo.c
#include <RTL.h>
#include "LPC11XX.h"
#include "uart_io.h"
#include <stdio.h>

OS_TID t_task1; // Declare a task ID for task1
OS_TID t_task2; // Declare a task ID for task2
OS_SEM UartLock; // Declare a Semaphore for Uart access
```

(Continued)

semaphore_demo.c—Cont'd

```
__task void task1(void) {
  while(1) {
  os_dly_wait(50);
  os_sem_wait(UartLock, 0xFFFF); // Wait for UART access lock,
                                 // with indefinite timeout period
  printf ("Task 1 running.\n");
  os_sem_send(UartLock);       // Finished using UART, increment semaphore
  }
}


__task void task2(void) {
  while (1) {
  os_dly_wait(50);
  os_sem_wait(UartLock, 0xFFFF); // Wait for UART access lock,
                                 // with indefinite timeout period
  printf ("Task 2 running.\n");
  os_sem_send(UartLock);       // Finished using UART, increment semaphore
  }
}


__task void init (void) {
  os_sem_init(UartLock, 1);   // Create a semaphore called UartLock with
                              // initial count of 1
  t_task1 = os_tsk_create (task1, 1); // Create a task "task1" with priority 1
  t_task2 = os_tsk_create (task2, 1); // Create a task "task2" with priority 1
  os_tsk_delete_self ();
}


int main(void)
{
  SystemInit();    // Switch Clock to 48MHz
  UartConfig();    // Configure UART
  os_sys_init(init); // Initialize OS
} // end main
```

Table 18.9 lists the semaphore functions available in RTX.

Table 18.9: Functions for Mutual Exclusive Operation

Semaphore Functions	Description
os_sem_init	Initializes a semaphore with a value.
os_sem_wait	Attempts to obtain a semaphore token. If the semaphore value is 0, waits until another task releases a token.
os_sem_send	Releases a semaphore token (value increases by 1).
isr_sem_send	Releases a semaphore token (increases the number of tokens in a semaphore) from an interrupt service routine.

Mailbox Messages

For transferring of more complex information between tasks, we can use the mailbox feature in the Keil RTX kernel. This feature allows us to define a data set in the memory used by one task and then transfer the pointer of the data set to another task. A mailbox in Keil RTX supports multiple messages, so several tasks can send messages to the same mailbox at the same time without losing any information. The maximum number of messages a mailbox can hold is defined when the mailbox is declared. In most cases, a capacity of 20 messages should be sufficient.

In the following example, three tasks are defined: two for creating and sending messages and one for receiving and printing out the messages. The message contents are created in a memory pool for fixed block allocation. The pointers to the messages' contents are transferred using the mailbox (Figure 18.9).

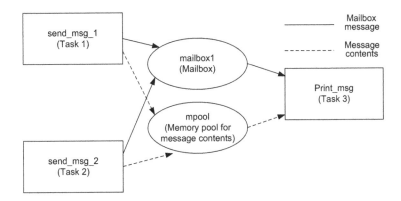

Figure 18.9:
A simple demonstration of using the mailbox feature to transfer messages.

The two tasks that create messages generate messages regularly, with each message containing two words. The messages are stored in a memory buffer called "mpool", a fixed block-size memory pool. Each time a message is created, the task needs to allocate space in the memory pool to hold the message content. The pointer to the location of the message content is then sent to the third task by the mailbox.

```
mailbox_demo.c
#include <RTL.h>
#include "LPC11XX.h"
#include "uart_io.h"
#include <stdio.h>


OS_TID t_sendmsg_1; // Declare a task ID for task1
OS_TID t_sendmsg_2; // Declare a task ID for task2
OS_TID t_printmsg;  // Declare a task ID for printmsg
```

(Continued)

mailbox_demo.c—Cont'd

```c
// define a memory pool for message box
os_mbx_declare (mailbox1, 16);

// Declare 16 blocks of 20 bytes for message contents
// each message contains two integers
_declare_box(mpool, 20, 16);

__task void send_msg_1(void) { // sender of message
  unsigned int *msg;
  int         counter=0;
  while(1) {
    os_dly_wait(100);
    msg = _alloc_box(mpool); // allocate block for message content
    if (*msg == 0) {
     printf ("_alloc_box failed\n");
     while (1);}
    msg[0] = counter;// Message
    msg[1] = 0x1234; // Message
    os_mbx_send(mailbox1, (void *) msg, 0xFFFF); // Send to mail box
    counter ++;
    }
  }

__task void send_msg_2(void) { // sender of message
  unsigned int *msg;
  int         counter=0x100;
  while(1) {
    os_dly_wait(100);
    msg = _alloc_box(mpool); // allocate block for message content
    if (*msg == 0) {
     printf ("_alloc_box failed\n");
     while (1);}
    msg[0] = counter;// Message
    msg[1] = 0x4567; // Message
    os_mbx_send(mailbox1, (void *) msg, 0xFFFF); // Send to mail box
    counter ++;
    }
  }

__task void print_msg(void) { // receiver of message
  unsigned int *msg;
  unsigned int received_values[2];
  while (1) {
    os_mbx_wait(mailbox1, (void *)&msg, 0xFFFF);
    received_values[0] = msg[0];
    received_values[1] = msg[1];
    _free_box(mpool, msg);
    printf ("Received values %x, %x\n",received_values[0],received_values[1]);
    }

  }
```

```
__task void init (void) {
 int status;

 // Initial fixed block size memory pool,  // block size is 20
 status = _init_box (mpool, size of (mpool), 20);
 if (status != 0) {
 printf ("_init_box failed\n");
 while (1);}

 os_mbx_init(mailbox1, size of(mailbox1)); // Initialize mailbox1

 // Create a task "send_msg_1" with priority 1
 t_sendmsg_1= os_tsk_create (send_msg_1,1);
 // Create a task "send_msg_2" with priority 1
 t_sendmsg_2= os_tsk_create (send_msg_2,1);
 // Create a task "print_msg"  with priority 1
 t_printmsg = os_tsk_create (print_msg, 1);
 os_tsk_delete_self ();
}

int main(void)
{
 SystemInit();     // Switch Clock to 48MHz
 UartConfig();     // Configure UART
 os_sys_init(init); // Initialize OS
} // end main
```

The third task (`print_msg`) receives the message, and then from the pointer received, it can extract the message contents from mpool and free the allocated space after the message content is read.

A number of RTX functions were used in this example. These functions are briefly described in Table 18.10.

Table 18.10: Functions for Mailbox and Messaging Operation

Mailbox Functions	Description
os_mbx_declare	Create a macro to define a mailbox object.
os_mbx_init	Initialize a mailbox object.
os_mbx_send	Send a message pointer to a mailbox object.
os_mbx_wait	Wait for a message from a mailbox object. If a message is available, get the pointer of the message.
os_mbx_check	Check how many messages can still be added to a mailbox object (it is not used in the example).
_declare_box	Declare a memory pool for fixed block size allocation.
_init_box	Initialize a fixed block size memory pool.
_alloc_box	Allocate a block of memory from the memory pool.
_free_box	Return the allocated memory block to the memory pool.

To run this example, OS_STKSIZE setting in RTX_config.c might need to be adjusted (a value of 100 was used, and it worked successfully).

A number of additional RTX functions are available for accessing the mailbox from interrupt service routines (Table 18.11).

Table 18.11: Functions for Mailbox and Messaging Operation from an Interrupt Service Routine

Mailbox Functions	Description
isr_mbx_check	Check available space in the mailbox from the interrupt service routine.
isr_mbx_receive	Receive a mailbox message in an interrupt service routine.
isr_mbx_send	Send a mailbox message in an interrupt service routine.

Periodic Time interval

In addition to the delay function `os_dly_wait`, you can also set up a task so that it wakes up periodically using the periodic time interval feature (Table 18.12). Different from the `os_dly_wait`, the periodic time interval feature can ensure the task is awake between a fixed number of ticks, even if the task's running time is more than one tick (Figure 18.10).

Table 18.12: Periodic Time Interval Functions

Mailbox Functions	Description
os_itv_set	Set the periodic time interval value.
os_itv_wait	Wait until the periodic time interval reaches its value.

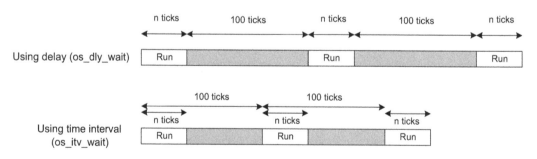

Figure 18.10:
The difference between a delay function and a time interval function.

We can modify the previous blinky example to create an example for the periodic time interval version fairly easily. The only change is the delay setup code inside the blinky task:

blinky.c (using time interval feature)

```c
#include <RTL.h>
#include "LPC11XX.h"

OS_TID t_blinky; // Declare a task ID for blink

__task void blinky(void) {
  os_itv_set(50); // set time interval to 50
  while(1) {
    LPC_GPIO2->MASKED_ACCESS[1] = ~LPC_GPIO2->MASKED_ACCESS[1]; // Toggle bit 0
    os_itv_wait();  // wait until time interval reach 50
    }
  }

__task void init (void) {
  t_blinky = os_tsk_create (blinky, 1); // Create a task "blinky" with priority 1
  os_tsk_delete_self ();
}

// Switch LED signal (P2_0 ) to output port with no pull up or pulldown
void LedOutputCfg(void)
{
  // Enable clock to IO configuration block (bit[16] of AHBCLOCK Control register)
  // and enable clock to GPIO (bit[6] of AHBCLOCK Control register
  LPC_SYSCON->SYSAHBCLKCTRL = LPC_SYSCON->SYSAHBCLKCTRL | (1<<16) | (1<<6);

  // PIO2_0 IO output config
  //   bit[5]  - Hysteresis (0=disable, 1 =enable)
  //   bit[4:3] - MODE(0=inactive, 1 =pulldown, 2=pullup, 3=repeater)
  //   bit[2:0] - Function (0 = IO, 1=DTR, 2=SSEL1)
  LPC_IOCON->PIO2_0 = (0<<5) + (0<<3) + (0x0);

  // Initial bit 0 output is 0
  LPC_GPIO2->MASKED_ACCESS[1] = 0;
  // Set pin 7 to 0 as output
  LPC_GPIO2->DIR = LPC_GPIO2->DIR | 0x1;
  return;
} // end LedOutputCfg

int main(void)
{
  SystemInit();    // Switch Clock to 48MHz
  LedOutputCfg();  // Initialize LED (GPIO #2, bit0) output
  os_sys_init(init); // Initialize OS
} // end main
```

Note: You cannot mix periodic time interval functions and `os_dly_wait()` functions in a single task.

Other RTX Features

In addition to the mentioned functions, the RTX kernel library has a number of other features and functions that are not covered in the previous examples. Table 18.13 shows some of the additional functions in the RTX kernel.

Table 18.13: Some Additional Functions Provided in the RTX Kernel

Functions	Description
tsk_lock	Disable RTX kernel task switching (for a program sequence that is not thread safe).
tsk_unlock	Enable RTX kernel task switching.
os_tmr_create	Set up and start a timer.
os_tmr_call	Call this user-defined function when a timer created by os_tmr_create is reached.
os_tmr_kill	Delete a timer created by os_tmr_create.

The details of the available OS functions are presented on the Keil web site within the RL-ARM User's Guide.

Application Example

Using the RTX kernel, it is simple to develop applications that have to deal with several concurrent tasks. For example, the dial control interface covered in Chapter 15 can be modified to use RTX and add LED toggling with variable speed (controlled by the dial) as a separate task (Figure 18.11).

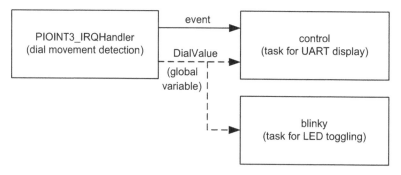

Figure 18.11:
Dial control interface example using RTX.

In this example, the event is generated from an interrupt handler; therefore, the function `isr_evt_set` is used instead of `os_evt_set`.

dial_ctrl.c (using RTX)

```c
#include <RTL.h>
#include <LPC11XX.h>
#include <stdio.h>
#include "uart_io.h"
//#define KEIL_MCB1000_BOARD


OS_TID t_blinky; // Declare a task ID for LED blinky
OS_TID t_ctrl;   // Declare a task ID for main control


// Function declarations
void DialIOcfg(void);  // Configure Port 3 for dial interface
void LedOutputCfg(void);  // Set I/O pin connected to LED as output


// Global variable for communicating between main program and ISR
volatile int  DialValue;   // Dial value (0 to 0xFF)
short    int  last_state;  // Last state of I/O port signals

__task void blinky(void) {
  while(1) {
#ifdef KEIL_MCB1000_BOARD
    // For Keil MCB1000, use P2.0 for LED output
    LPC_GPIO2->MASKED_ACCESS[1] = ~LPC_GPIO2->MASKED_ACCESS[1]; // Toggle bit 0
#else
    // For LPCXpresso, use P0.7 for LED output
    LPC_GPIO0->MASKED_ACCESS[1<<7] = ~LPC_GPIO0->MASKED_ACCESS[1<<7];//Toggle bit 7
#endif
    os_dly_wait (3 + (DialValue>>2));  // delay
    }
  }

__task void control(void) {
  DialValue=0;
  last_state=(LPC_GPIO3->DATA & 0xC)>>2;
     // capture and save signal levels for next compare
  while(1) {
    os_evt_wait_or(0x0001, 0xFFFF);
    printf("%d\n",DialValue);
    }
  }

__task void init (void) {
  t_blinky = os_tsk_create (blinky, 1); // Create a task "blinky" with priority 1
  t_ctrl   = os_tsk_create (control, 1); // Create a task "blinky" with priority 1
  os_tsk_delete_self ();
}

int main(void)
{
```

(Continued)

dial_ctrl.c (using RTX)—Cont'd

```c
  SystemInit();  // Switch Clock to 48MHz
  UartConfig();  // Initialize UART
  DialIOcfg();    // IO port and interrupt setup
  LedOutputCfg();// Set I/O pin connected to LED
  printf ("\nDial test\n"); // Test message
  os_sys_init(init); // Initialize OS
} // end main

void DialIOcfg(void)
{ // The inputs are P3.2 and P3.3
  // Enable clock to GPIO block (bit[6] of AHBCLOCK Control register
  LPC_SYSCON->SYSAHBCLKCTRL = LPC_SYSCON->SYSAHBCLKCTRL | (1<<6);
  // PIO1_7 IO output config
  //  bit[5]   - Hysteresis (0=disable, 1 =enable)
  //  bit[4:3] - MODE(0=inactive, 1 =pulldown, 2=pullup, 3=repeater)
  //  bit[2:0] - Function (0 = IO, 1=nDCD)
  LPC_IOCON->PIO3_2 = (0x0) + (0<<3) + (1<<5);
  // PIO1_6 IO output config
  //  bit[5]   - Hysteresis (0=disable, 1 =enable)
  //  bit[4:3] - MODE(0=inactive, 1 =pulldown, 2=pullup, 3=repeater)
  //  bit[2:0] - Function (0 = IO, 1=RI)
  LPC_IOCON->PIO3_3 = (0x0) + (0<<3) + (1<<5);
  // Set direction of P3.2 and P3.3 as input
  LPC_GPIO3->DIR    = LPC_GPIO3->DIR & ~(0x0C); // Clear bit [3:2] to 0
  // Set interrupt of P3.2 and P3.3 as edge sensitive
  LPC_GPIO3->IS     = LPC_GPIO3->IS  & ~(0x0C); // Clear bit [3:2] to 0

  // Set interrupt of P3.2 and P3.3 for both rising and falling edge
  LPC_GPIO3->IBE    = LPC_GPIO3->IBE |  (0x0C); // Set bit [3:2] to 1
  // Set interrupt mask of P3.2 and P3.3
  LPC_GPIO3->IE     = LPC_GPIO3->IE  |  (0x0C); // Set bit [3:2] to 1
  // Clear any previous interrupt of P3.2 and P3.3
  LPC_GPIO3->IC     = LPC_GPIO3->IC  |  (0x0C); // write bit [3:2] to 1
  // Clear any previous occurred interrupt for port 3
  NVIC_ClearPendingIRQ(EINT3_IRQn);
  // Set priority of port 3 interrupt
  NVIC_SetPriority(EINT3_IRQn, 0);
  // Enable interrupt at NVIC
  NVIC_EnableIRQ(EINT3_IRQn);
  return;
}
// Switch LED signal (P0_7) to output port with no pull up or pulldown
void LedOutputCfg(void)
{
  // Enable clock to IO configuration block (bit[16] of AHBCLOCK Control register)
  // and enable clock to GPIO (bit[6] of AHBCLOCK Control register
  LPC_SYSCON->SYSAHBCLKCTRL = LPC_SYSCON->SYSAHBCLKCTRL | (1<<16) | (1<<6);
#ifdef KEIL_MCB1000_BOARD
```

```
  // For Keil MCB1000, use P2.0 for LED output
  // PIO2_0 IO output config
  //  bit[5]   - Hysteresis (0=disable, 1 =enable)
  //  bit[4:3] - MODE(0=inactive, 1 =pulldown, 2=pullup, 3=repeater)
  //  bit[2:0] - Function (0 = IO, 1=DTR, 2=SSEL1)
  LPC_IOCON->PIO2_0 = (0<<5) + (0<<3) + (0x0);

  // Initial bit 0 output is 0
  LPC_GPIO2->MASKED_ACCESS[1] = 0;
  // Set pin 7 to 0 as output
  LPC_GPIO2->DIR = LPC_GPIO2->DIR | 0x1;
#else
  // For LPCXpresso, use P0.7 for LED output
  // PIO0_7 IO output config
  //  bit[5]   - Hysteresis (0=disable, 1 =enable)
  //  bit[4:3] - MODE(0=inactive, 1 =pulldown, 2=pullup, 3=repeater)
  //  bit[2:0] - Function (0 = IO, 1=CTS)
  LPC_IOCON->PIO0_7 = (0x0) + (0<<3) + (0<<5);
  // Initial bit[7] output is 0
  LPC_GPIO0->MASKED_ACCESS[1<<7] = 0;
  // Set pin 7 as output
  LPC_GPIO0->DIR = LPC_GPIO0->DIR | (1<<7);

#endif
    return;
} // end LedOutputCfg
// Interrupt handler for port 3
void PIOINT3_IRQHandler(void)
{
  short int new_state;
  // Pattern for determine the direction of changes
  // Clock wise       pattern is 00 -> 01 -> 11 -> 10 -> 00 -> ...
  // Anti Clock wise pattern is 00 -> 10 -> 11 -> 01 -> 00 -> ...
  // After merging the new_state and last_state, clock wise can be
  // pattern b0100(4), b1101(13), b1011(11) and b0010(2)
  // anti-clockwise can be pattern b1000(8), b1110(14),b0111(7) and b0001(1)
  const signed char Pattern[] = { 0,-1,1,0,  1,0,0,-1,  -1,0,0,1,  0,1,-1,0};
  // Clear asserted interrupt P3.2 or P3.3
  LPC_GPIO3->IC = LPC_GPIO3->MIS & (0x0C); // write bit [3:2] to 1
  // Extract bit 3 and 2 and combine with last state
  new_state = (LPC_GPIO3->DATA & 0xC) | last_state;
  // Obtain increment/decrement info from new_state and calculate new DialValue
  DialValue = (DialValue + Pattern[new_state]) & 0xFF;
  // Save the current state for next time
  last_state  = (new_state & 0xC) >> 2;
  isr_evt_set(0x1, t_ctrl);// Send event
  return;
}
```

## CHAPTER 19

Getting Started with the ARM RealView Development Suite

Overview

In addition to the ARM Keil Microcontroller Development Kit (MDK), ARM also provides another development suite called the RealView Development Suite (RVDS). Although it is based on the same C compiler, RVDS is targeted at the higher end of the market and has the following additional features:

*   Supports all modern ARM processors including Cortex-A9/A8/A5
*   RealView debugger provides multiprocessor debug support and full CoreSight debug support
*   Offers profiler-driven compilation
*   Supports Windows, Linux, and Unix
*   Offers instruction set simulation (ISS) models
*   Offers real-time system models (fast models)

When using RVDS, instead of targeting a microcontroller device, you target the compilation for a processor or ARM architecture version. This results in a number of differences between command line options in RVDS and Keil MDK. In this chapter, we will cover the basic steps of using RVDS to create your programs.

Typically, the software compilation flow using RVDS can be summarized as shown in Figure 19.1.

An integrated development environment (IDE) called the ARM Workbench IDE is included in RVDS. It is based in the open-source Eclipse IDE. You can also use RVDS either with the IDE or on the command line. In this chapter, we will focus mainly on the command line operation. Details of using the ARM Workbench IDE are covered in the ARM Workbench IDE User Guide (document DUI0330, reference 5). It can be downloaded from the ARM web site (http://informcenter.arm.com).

Simple Application Example

Using RVDS on the command line is straightforward. Based on the CMSIS version of the blinky example, we can compile the project with a few commands. First, we collect all the files necessary to build the application into a project folder. The folder named "Project" shown in Figure 19.2 is only an example; you can use other folder names.

The Definitive Guide to the ARM Cortex-M0. DOI: 10.1016/B978-0-12-385477-3.10019-9

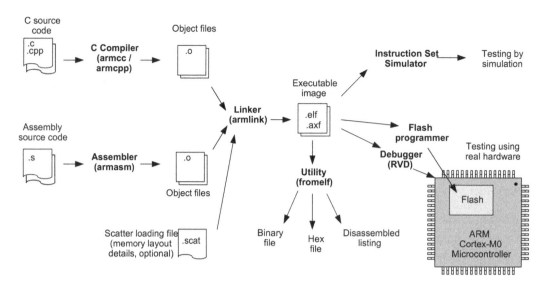

Figure 19.1:
Example of the software generation flow in RVDS.

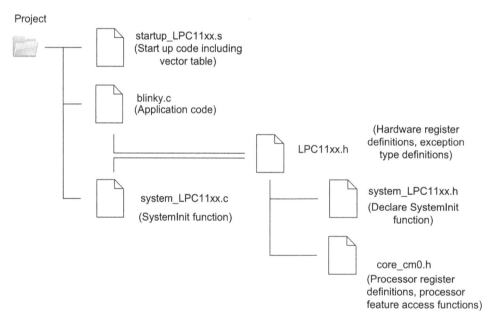

Figure 19.2:
Blinky project in RVDS.

In this example, we are going to reuse the program code from the previous Keil MDK projects for NXP LPC1114. If you have Keil MDK installed, you can find these files in the locations identified in Table 19.1.

Table 19.1: Locations of LPC1114 Support Files in Keil MDK Installation

File	Location (assuming you have Keil MDK installed in C:\KEIL)
startup_LPC11xx.s	C:\Keil\ARM\Startup\NXP\LPC11xx
system_LPC11xx.c	C:\Keil\ARM\Startup\NXP\LPC11xx, or Appendix H of this book
system_LPC11xx.h	C:\Keil\ARM\INC\NXP\LPC11xx, or Appendix H of this book
LPC11xx.h	C:\Keil\ARM\INC\NXP\LPC11xx
Blinky.c	See example code in Chapter 15
core_cm0.h	C:\Keil\ARM\INC

If you do not have Keil MDK software installed, you can download the required files from the NXP web site at http://ics.nxp.com/support/documents/microcontrollers/zip/code.bundle. lpc11xx.keil.zip. In addition, you can also download the generic CMSIS support files from www.onarm.com.

Once you have copied the files into the project (see the note on the include path that follows), we can compile the project with the following commands:

```
# Assemble startup code
armasm -g --cpu Cortex-M0 startup_LPC11xx.s -o startup_LPC11xx.o

# Compile application code
armcc -c -g --cpu Cortex-M0 blinky.c -o blinky.o -I c:\keil\arm\inc -I
    c:\keil\arm\inc\nxp\LPC11xx

# Compile system initialization code
armcc -c -g --cpu Cortex-M0 system_LPC11xx.c -o system_LPC11xx.o -I c:\keil\arm\inc -I
    c:\keil\arm\inc\nxp\LPC11xx

# Linking stage
armlink startup_LPC11xx.o system_LPC11xx.o blinky.o --ro-base 0x0 --rw_base
    0x10000000 "--keep=startup_LPC11xx.o(RESET)" "--first=startup_LPC11xx.o(RESET)"
    --map --entry=Reset_Handler -o blinky.elf

# Generate disassembled listing for checking
fromelf -c blinky.elf —output=list.txt
```

In the preceding example, we used the files from Keil MDK directly by adding an include path using the "-I" option. In this way, we do not have to copy the files across to the local directory.

The commonly used options for the program generation flows in RVDS include those described in Tables 19.2, 19.3, 19.4, and 19.5.

Table 19.2: Commonly Used Options for armasm

armasm Options	Descriptions
--cpu Cortex-M0	Define the processor type.
-I *<directory>*	Define the search path for include files.
-g	Generate an object with debug information (DWARF debug table).
-o *<file>*	Output an object file name.
--list *<file>*	Output a detailed listing of the assembly language produced.
--thumb	Use the UAL Thumb assembly syntax (equivalent to the use of the THUMB directive in the assembly header).

Table 19.3: Commonly Used Options for armcc

armcc Options	Descriptions
--cpu Cortex-M0	Define the processor type.
-I *<directory>*	Define the search path for include files.
-c	Perform the compile step, but not the link step.
-g	Generate the object with debug information (DWARF debug table).
-o *<file>*	Output an object file name.
--list *<file>*	Output a detailed listing of the assembly language produced.
--thumb	Target the Thumb instruction set. It is not required if you have specified the processor type to Cortex-M0, as it only supports Thumb.
-Ospace/-Otime	Optimize for code size or execution speed.
-O0/-O1/-O2/-O3	Set the optimization level. If this option is not used, the compiler uses level 2 by default.
--multiply_latency=*<n>*	If you are using a Cortex-M0 processor with a small multiplier (e.g., Cortex-M0 in minimum size configuration), the multiply instruction take 32 cycle. By using "--multiply_latency = 32", option the C compiler can optimize the code better.

Table 19.4: Commonly Used Options for armlink

armlink options	Descriptions
--keep=<object>(<section>)	Specify that the section inside the specific object file must not be removed by an unused section elimination.
--first=<object>(<section>)	Specify the section to be put in the beginning of the execution region.
--scatter <file>	Specify the scatter loading file (memory layout information).
--map	Enable printing of a memory map.
--entry=<location>	Specify the entry point of the image (program starting point).
--ro_base=<address>	Specify the starting address of the execution region.
--rw_base=<address>	Specify the starting address of the RW region (e.g., RAM).
-o <file>	Output an image file name.

Table 19.5: Commonly Used Options for fromelf

fromelf Options	Descriptions
-c / --text -c	Create a disassembled listing (output from this operation cannot be fed back to armasm).
-d / --text -d	Print the contents of the data sections.
-s / --text -s	Print the symbol table and versioning table.
-e / --text -e	Decode the exception table information for the object. Use with -c when disassembling the image.
--disassemble	Create a disassembled version of the image. (You can reassemble the output using armasm.)
--bin	Produce a binary file (for example, **fromelf** --bin --output=outfile.bin infile.axf).
--i32	Generate a hex file in the Intel Hex32 format.
--m32	Generate the Motorola 32-bit format (32-bit S-record).
--vhx	Generate the Verilog Hex format.
--output=<*file*>	Specify an output file.

Additional details of the available command line options can be found in the following documents: (Note: These document names are for RealView Development Suite 4.0. other versions of RVDS might have different document names.)

- armasm: RealView Compilation Tools Assembler Guide
- armcc: RealView Compilation Tools Compiler User Guide
- armlink: RealView Compilation Tools Linker User Guide
- fromelf: RealView Compilation Tools Utilities Guide

All of these documents are automatically installed on your machine when you install RVDS, or alternatively you can find them on the ARM web site at http://informcenter.arm.com.

Apart from using a batch file (for Windows platforms) or shell scripts (for Linux/Unix platforms), users of Linux, UNIX, or Windows with Cygwin (or similar) environments can also create files to handle the compile process.

Using the Scatter Loading File

In the previous example, we used command line options to specify the address range of the read-only region (flash/ROM) and read-write region (RAM). This is fine for many simple projects. However, for applications with a more complex memory layout, or if you want to arrange the memory map in specific ways, a scatter loading file should be used to describe the layout of the memory system.

To use the scatter-loading feature, we use the --scatter option in armlink. Using the previous blinky example, we can convert the program generation flow to use scatter loading:

```
# Assemble startup code
    armasm -g --cpu Cortex-M0 startup_LPC11xx.s -o startup_LPC11xx.o

# Compile application code
```

```
armcc -c -g --cpu Cortex-M0 blinky.c -o blinky.o -I c:\keil\arm\inc -I
    c:\keil\arm\inc\nxp\LPC11xx

# Compile system initialization code
armcc -c -g --cpu Cortex-M0 system_LPC11xx.c -o system_LPC11xx.o -I c:\keil\arm\inc -I
    c:\keil\arm\inc\nxp\LPC11xx ,

# Linking stage
armlink startup_LPC11xx.o system_LPC11xx.o blinky.o --scatter scatter.scat "--
    keep=startup_LPC11xx.o(RESET)" --entry Reset_Handler --map -o blinky.elf

# Generate disassembled listing for checking
fromelf -c -e -d -s blinky.elf --output=list.txt
```

In this example, the scatter loading file we used is called scatter.scat:

```
scatter.scat
LOAD_REGION 0x00000000 0x00200000
{ ; flash memory start at 0x00000000
  ;; Maximum of 48 exceptions (48*4 bytes == 0xC0)
  VECTORS 0x0 0xC0
  {
    ; Provided by the user in startup_LPC11xx.s
    * (RESET,+FIRST)
  }

  CODE 0xC0 FIXED
  { ; The rest of the program code start from 0xC0
    * (+RO)
  }

  DATA 0x10000000 0x2000
  { ; In LPC1114, the SRAM start at 0x10000000
    * (+RW, +ZI)
  }
}
```

Using the scatter loading file, the "RESET" section in the startup code is allocated to the beginning of the memory. As a result, we do not need to use the --first option when running armlink. For systems with multiple ROM regions, you can add memory sections and assign different objects to different memory sections.

The scatter loading file can also be used to define stack memory and heap memory. This will be demonstrated in the next example when the vector table is coded in C. For this example, this step is not required because the stack and heap are defined in the startup code.

Details of scatter loading syntax can be found in the RealView Compilation Tools Developer Guide (reference 9).

Example with Vector Table in C

In the Keil MDK, the default startup codes for ARM microcontrollers are written in assembly. However, you can also create the vector table in C. In the following example, we will convert the blinky example to use a C vector table. Based on the exception vectors definition for NXP LPC1114, the following file "exceptions.c" is created:

```
exceptions.c
//*************************************************************************
// Function definitions
//*************************************************************************

static void Default_Handler(void);
// The following functions are declared with weak attributes.
// If another handler with the same name is presented it will be overridden.
void __attribute__ ((weak)) NMI_Handler(void);
void __attribute__ ((weak)) HardFault_Handler(void);
void __attribute__ ((weak)) SVC_Handler(void);
void __attribute__ ((weak)) PendSV_Handler(void);
void __attribute__ ((weak)) SysTick_Handler(void);
void __attribute__ ((weak)) WAKEUP_IRQHandler(void);
void __attribute__ ((weak)) SSP1_IRQHandler(void);
void __attribute__ ((weak)) I2C_IRQHandler(void);
void __attribute__ ((weak)) TIMER16_0_IRQHandler(void);
void __attribute__ ((weak)) TIMER16_1_IRQHandler(void);
void __attribute__ ((weak)) TIMER32_0_IRQHandler(void);
void __attribute__ ((weak)) TIMER32_1_IRQHandler(void);
void __attribute__ ((weak)) SSP0_IRQHandler(void);
void __attribute__ ((weak)) UART_IRQHandler(void);
void __attribute__ ((weak)) ADC_IRQHandler(void);
void __attribute__ ((weak)) WDT_IRQHandler(void);
void __attribute__ ((weak)) BOD_IRQHandler(void);
void __attribute__ ((weak)) PIOINT3_IRQHandler(void);
void __attribute__ ((weak)) PIOINT2_IRQHandler(void);
void __attribute__ ((weak)) PIOINT1_IRQHandler(void);
void __attribute__ ((weak)) PIOINT0_IRQHandler(void);
#pragma weak NMI_Handler = Default_Handler
#pragma weak HardFault_Handler = Default_Handler
#pragma weak SVC_Handler = Default_Handler
#pragma weak PendSV_Handler = Default_Handler
#pragma weak SysTick_Handler = Default_Handler
#pragma weak WAKEUP_IRQHandler = Default_Handler
#pragma weak SSP1_IRQHandler = Default_Handler
#pragma weak I2C_IRQHandler = Default_Handler
#pragma weak TIMER16_0_IRQHandler = Default_Handler
#pragma weak TIMER16_1_IRQHandler = Default_Handler
#pragma weak TIMER32_0_IRQHandler = Default_Handler
#pragma weak TIMER32_1_IRQHandler = Default_Handler
#pragma weak SSP0_IRQHandler = Default_Handler
#pragma weak UART_IRQHandler = Default_Handler
#pragma weak ADC_IRQHandler = Default_Handler
#pragma weak WDT_IRQHandler = Default_Handler
#pragma weak BOD_IRQHandler = Default_Handler
#pragma weak PIOINT3_IRQHandler = Default_Handler
#pragma weak PIOINT2_IRQHandler = Default_Handler
#pragma weak PIOINT1_IRQHandler = Default_Handler
#pragma weak PIOINT0_IRQHandler = Default_Handler
```

(Continued)

exceptions.c—Cont'd

```c
//*******************************************************************************
// Default handler
//*******************************************************************************
static void Default_Handler(void)
{
    while(1); // infinite loop
}

//*******************************************************************************
// Vector table
//*******************************************************************************
#pragma arm section rodata="RESET"

typedef void(* const ExecFuncPtr)(void) __irq;

//ExecFuncPtr exception_table[] = {
void (* const exception_table[])(void) = {
    /* (ExecFuncPtr)&Image$$ARM_LIB_STACK$$ZI$$Limit, */
        /* Initial SP, already provided by library */
    /* (ExecFuncPtr)&__main, */
        /* Initial PC, already provided by library */
    NMI_Handler,
    HardFault_Handler,
    0, 0, 0, 0, 0, 0, 0,    /* Reserved */
    SVC_Handler,
    0, 0,                   /* Reserved */
    PendSV_Handler,
    SysTick_Handler,
    /* Configurable interrupts start here...*/
    WAKEUP_IRQHandler,      /* 16+ 0: Wakeup PIO0.0 */
    WAKEUP_IRQHandler,      /* 16+ 1: Wakeup PIO0.1 */
    WAKEUP_IRQHandler,      /* 16+ 2: Wakeup PIO0.2 */
    WAKEUP_IRQHandler,      /* 16+ 3: Wakeup PIO0.3 */
    WAKEUP_IRQHandler,      /* 16+ 4: Wakeup PIO0.4 */
    WAKEUP_IRQHandler,      /* 16+ 5: Wakeup PIO0.5 */
    WAKEUP_IRQHandler,      /* 16+ 6: Wakeup PIO0.6 */
    WAKEUP_IRQHandler,      /* 16+ 7: Wakeup PIO0.7 */
    WAKEUP_IRQHandler,      /* 16+ 8: Wakeup PIO0.8 */
    WAKEUP_IRQHandler,      /* 16+ 9: Wakeup PIO0.9 */
    WAKEUP_IRQHandler,      /* 16+10: Wakeup PIO0.10 */
    WAKEUP_IRQHandler,      /* 16+11: Wakeup PIO0.11 */
    WAKEUP_IRQHandler,      /* 16+12: Wakeup PIO1.0 */
    0,                      /* 16+13: Reserved */
    SSP1_IRQHandler,        /* 16+14: SSP1 */
    I2C_IRQHandler,         /* 16+15: I2C */
    TIMER16_0_IRQHandler,   /* 16+16: 16-bit Counter-Timer 0 */
    TIMER16_1_IRQHandler,   /* 16+17: 16-bit Counter-Timer 1 */
    TIMER32_0_IRQHandler,   /* 16+18: 32-bit Counter-Timer 0 */
    TIMER32_1_IRQHandler,   /* 16+19: 32-bit Counter-Timer 1 */
    SSP0_IRQHandler,        /* 16+20: SSP0 */
    UART_IRQHandler,        /* 16+21: UART */
    0,                      /* 16+22: Reserved */
    0,                      /* 16+23: Reserved */
```

```
    ADC_IRQHandler,          /* 16+24: A/D Converter */
    WDT_IRQHandler,          /* 16+25: Watchdog Timer */
    BOD_IRQHandler,          /* 16+26: Brown Out Detect */
    0,                       /* 16+27: Reserved */
    PIOINT3_IRQHandler,      /* 16+28: PIO INT3 */
    PIOINT2_IRQHandler,      /* 16+29: PIO INT2 */
    PIOINT1_IRQHandler,      /* 16+30: PIO INT1 */
    PIOINT0_IRQHandler       /* 16+31: PIO INT0 */
};

#pragma arm section
```

The C-based vector table in "exceptions.c" does not contain the initial stack pointer value and the reset vector. This will be inserted at the linking stage.

In the previous examples, the assembly startup code contains definitions for stack and heap memory. Because the C-based vector table does not contain such information, the scatter loading file is modified to include stack and heap memory definitions. The reset vector and initial stack pointer values are also defined in the scatter loading file.

scatter.scat (with heap and stack definitions)

```
LOAD_REGION 0x00000000 0x00200000
{
  ;; Maximum of 48 exceptions (48*4 bytes == 0xC0)
  VECTORS 0x0 0xC0
  {
    ; First two entries provided by library
    ; Remaining entries provided by the user in exceptions.c

    * (:gdef:__vectab_stack_and_reset, +FIRST)
    * (RESET)
  }

  CODE 0xC0 FIXED
  {
    * (+RO)
  }

  DATA 0x10000000 0x2000
  {
    * (+RW, +ZI)
  }

  ;; Heap starts at 4KB and grows upwards
  ARM_LIB_HEAP 0x10001000 EMPTY 0x1800-0x1000
  {
  }

  ;; Stack starts at the end of the 8KB of RAM
```

(Continued)

---

scatter.scat (with heap and stack definitions)—Cont'd

```
;; And grows downwards for 2KB
ARM_LIB_STACK 0x10002000 EMPTY -0x800
{
}
}
```

With the vector table and the scatter loading file ready, we can generate the program image using the following command lines:

```
# Compile vector table and default handler
armcc -c -g --cpu Cortex-M0 exceptions.c -o exceptions.o

# Compile application code
armcc -c -g --cpu Cortex-M0 blinky.c -o blinky.o -I c:\keil\arm\inc -I
    c:\keil\arm\inc\nxp\LPC11xx

# Compile system initialization code
armcc -c -g --cpu Cortex-M0 system_LPC11xx.c -o system_LPC11xx.o -I c:\keil\arm\inc -I
    c:\keil\arm\inc\nxp\LPC11xx

# Linking stage
armlink exceptions.o system_LPC11xx.o blinky.o --scatter scatter.scat "--
    keep=exceptions.o(RESET)" --map -o blinky.elf

# Generate disassembled listing for checking
fromelf -c -e -d -s blinky.elf --output=list.txt
```

Using MicroLIB in RVDS

In Keil MDK, one of the project options uses MicroLIB to reduce code size. MicroLIB is an implementation of the C library targeted specially for microcontroller applications where available program memory size could be limited. Because MicroLIB is optimized for small code size, the performance of the library functions is less than those in the standard C library.

MicroLIB can also be used with RVDS. To demonstrate the use of MicroLIB, we use the RVDS blinky example and modify the program generation script to include the "--library_type=microlib" option:

```
# Assemble start up code
armasm -g --cpu Cortex-M0 startup_LPC11xx.s -o startup_LPC11xx.o
    --library_type=microlib --pd "__MICROLIB SETA 1"

# Compile application code
armcc -c -g --cpu Cortex-M0 blinky.c -o blinky.o -I c:\keil\arm\inc -I
    c:\keil\arm\inc\nxp\LPC11xx --library_type=microlib

# Compile system initialization code
armcc -c -g --cpu Cortex-M0 system_LPC11xx.c -o system_LPC11xx.o -I
    c:\keil\arm\inc -I c:\keil\arm\inc\nxp\LPC11xx --library_type=microlib

# Linking stage
```

```
armlink startup_LPC11xx.o system_LPC11xx.o blinky.o "--
    keep=startup_LPC11xx.o(RESET)" "--first=startup_LPC11xx.o(RESET)" --entry
    Reset_Handler --rw_base 0x10000000 --map --ro-base 0x0 -o blinky.elf
    --library_type=microlib

# Generate disassembled listing for checking
    fromelf -c blinky.elf --output=list.txt
```

In the assembly stage for the startup code, we added the additional option of --pd
"__MICROLIB SETA 1." This is because MicroLIB has different stack and heap definitions
compared to the standard C library; as a result, the startup code contains the conditional
assembly directive, controlled by the __MICROLIB option.

Using Assembly for Application Development in RVDS

For small projects, it is possible to develop the entire project using assembler. For example,
a program that calculates the sum of 1 to 10 can be as simple as that shown here:

---

simple_prog.s (calculate sum of 1 to 10)

```
Stack_Size      EQU      0x00000200

                AREA     STACK, NOINIT, READWRITE, ALIGN=3
Stack_Mem       SPACE    Stack_Size
__initial_sp

                PRESERVE8
                THUMB

; Vector Table Mapped to Address 0 at Reset

                AREA     RESET, DATA, READONLY
                EXPORT   __Vectors

__Vectors       DCD      __initial_sp         ; Top of Stack
                DCD      Reset_Handler        ; Reset Handler
                DCD      NMI_Handler          ; NMI Handler
                DCD      HardFault_Handler    ; Hard Fault Handler
                DCD      0                    ; Reserved
                DCD      0                    ; Reserved
                DCD      0                    ; Reserved
                DCD      0                    ; Reserved
                DCD      0                    ; Reserved
                DCD      0                    ; Reserved
                DCD      0                    ; Reserved
                DCD      SVC_Handler          ; SVCall Handler
                DCD      0                    ; Reserved
                DCD      0                    ; Reserved
                DCD      PendSV_Handler       ; PendSV Handler
                DCD      SysTick_Handler      ; SysTick Handler

                AREA     |.text|, CODE, READONLY

; Reset Handler
```

(Continued)

simple_prog.s (calculate sum of 1 to 10)—Cont'd

```
Reset_Handler    PROC
                 EXPORT   Reset_Handler                  [WEAK]
; --- Start of main program code ---
; Calculate the sum of 1 to 10
                 MOVS     R0, #10
                 MOVS     R1, #0
                 ; Calculate 10+9+8...+1
loop
                 ADDS     R1, R1, R0
                 SUBS     R0, R0, #1
                 BNE      loop
                 ; Result is now in R1
                 B        .      ; Branch to self (infinite loop)
; --- End  of main program code  ---

                 ENDP

; Dummy Exception Handlers (infinite loops which can be modified)

NMI_Handler      PROC
                 EXPORT   NMI_Handler                    [WEAK]
                 B        .
                 ENDP
HardFault_Handler PROC
                 EXPORT   HardFault_Handler              [WEAK]
                 B        .
                 ENDP
SVC_Handler      PROC
                 EXPORT   SVC_Handler                    [WEAK]
                 B        .
                 ENDP
PendSV_Handler   PROC
                 EXPORT   PendSV_Handler                 [WEAK]
                 B        .
                 ENDP
SysTick_Handler PROC
                 EXPORT   SysTick_Handler                [WEAK]
                 B        .
                 ENDP

                 ALIGN
                 END
```

The program combined the vector table, the main application code, some dummy exception handlers, and the stack definition. Because there is only one object file, there is no need to carry out a separate link stage. All you need to assemble the program and generate the execution image can be carried out in just one `armasm` command:

```
# Generate executable image
armasm -g --cpu Cortex-M0 simple_prog.s -o simple_prog.elf

# Generate disassembled code for checking
fromelf -c simple_prog.elf --output=list.txt
```

For most applications, it is common to separate the vector table and the application code. For example, we can reuse the assembly version of the blinky example in Chapter 16 and assemble it using RVDS as follows:

```
# Assemble the vector table and startup code
armasm -g --cpu Cortex-M0 startup_LPC11xx.s -o startup_LPC11xx.o

# Assemble the application code
armasm -g --cpu Cortex-M0 blinky.s -o blinky.o

# Linking stage
armlink startup_LPC11xx.o blinky.o "--keep=startup_LPC11xx.o(RESET)" "--
   first=startup_LPC11xx.o(RESET)"--entry Reset_Handler --rw_base 0x10000000 --map
   --ro-base 0x0 -o blinky.elf

# Create disassembled list for checking
fromelf -c blinky.elf --output=list.txt
```

The general assembly programming techniques for RVDS are the same as in Keil MDK, which is covered in Chapter 16.

Flash Programming

After generating the compiled image, we often need to program the image into the flash memory of the microcontroller for testing. RVDS contains flash programming features for a number of ARM microcontroller devices. The details of using flash programming with RVDS are described in the ARM Workbench IDE User Guide. At the time of this writing, the RVDS installation (4.0-SP3) does not have flash programming support on the NXP LPC11xx product. However, you can create your own flash programming configuration files, and this topic is also covered in the ARM Workbench IDE User Guide (reference 5) and ARM Application Note 190, "Creating Flash Algorithms with Eclipse."

Alternatively, there are a number of other solutions:

1. Using Keil MDK
   - If you have access to Keil MDK and a supported in-circuit debugger (e.g., ULINK 2), you can use the flash programming feature in Keil MDK to program the image created in RVDS into the flash memory.
   - To use Keil MDK to program on your image, you need to change the file extension from .elf to .axf.
   - The next step is to create a μVision project in the same directory with the same name (e.g., blinky). In the project creation wizard, select the microcontroller

device you use, and when asked to copy the default startup code, click "no" to prevent the existing file (if there is one) from being overwritten.

- Now you need to set the debug options to use your in-circuit debugger (e.g., ULINK 2) and flash programming option if necessary. By default the flash programming option should have been set up for you automatically.
- Once the execution image (with .AXF file extension) has been built, click the flash programming button on the toolbar ⚞. The compiled image will then be programmed into the flash memory.
- After the image is programmed in to the flash memory, you can start a debug session using the μVision debugger to debug your program.

2. Using other flash programming utilities

- Most ARM microcontrollers have flash programming utilities provided by the microcontroller vendors or other third parties. For example, the NXP LPC11xx microcontroller devices can be programmed using a tool called "Flash Magic," provided by Embedded Systems Academy (www.flashmagictool.com). This tool works with the built-in flash memory programming firmware on the LPC11XX microcontroller devices and allows the device to be programmed with a serial communication connection.

Note that with the NXP LPC11xx series, the address 0x1C-0x1F in the flash memory is used as a checksum for the on-chip boot loader. Although flash memory programmers in Keil ULINK products handle this address automatically, third-party flash memory programmers may not be aware of this and might report an error when the programmed value and the read back value do not match. In version 4.10 and later of the Keil MDK, a utility called ELFDWT is included that can insert the checksum value in the generated AXF executable image. This allows all flash programmers supporting LPC11xx to use the created image. More information on the ELFDWT utility is covered in Chapter 14.

Debugging Using RealView Debugger

RVDS includes the RealView Debugger (RVD). To use RVD with a target system, you need to have a run-time control unit called RealView-ICE (Figure 19.3). The RealView-ICE ships with a standard 20-pin JTAG connector and can be connected to the debug host using either a USB or an Ethernet connection.

By default, RealView-ICE supports the JTAG debug protocol. To demonstrate the use of RVD with the Cortex-M0 with the JTAG debug protocol, here we use the Keil Microcontroller Prototyping System (MPS), a FPGA platform for prototyping Cortex-M systems, commonly used for system-on-chip or ASIC prototyping (Figure 19.4). (The existing NXP LPC11xx microcontrollers do not support JTAG debug.)

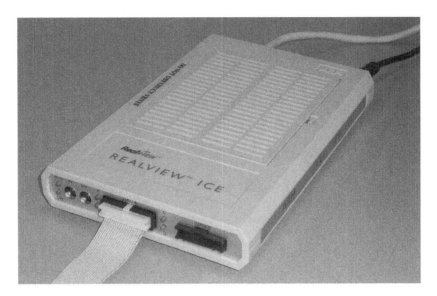

Figure 19.3:
RealView-ICE.

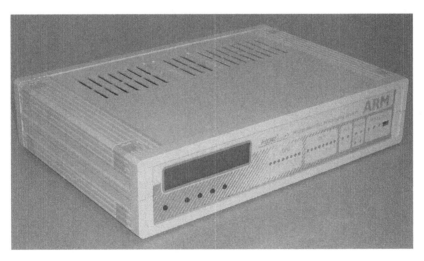

Figure 19.4:
Microcontroller Prototyping System (MPS).

The MPS system contains two main FPGAs. One is for the processor and memory interface, and the other is a peripheral FPGA, which users can modify for SoC development. The processor FPGA can be switched between various Cortex-M processors including the Cortex-M0.

The MPS memory system contains 64 MB of flash memory and two SRAMs of 4 MB each. By default, the boot loader remaps one of the 4MB SRAMs to address 0x0 after it has been booted up so that users can download test code to SRAM and execute it from SRAM at high speed (zero wait state at 50 MHz).

After launching RVD, you need to create a connection to the RealView ICE. Do this by accessing the Connect-to-Target function in the pull-down menu (Figure 19.5).

You can then add a connection configuration in the Connect-to-Target window (Figure 19.6).

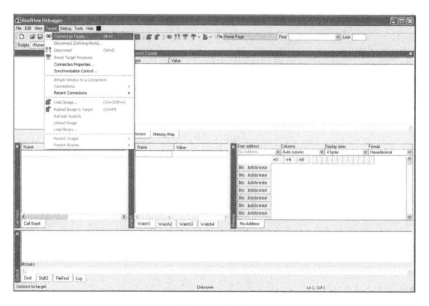

Figure 19.5:
Connect-to-target function in the RealView debugger.

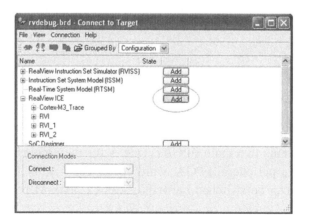

Figure 19.6:
Connect-to-Target window.

With RealView-ICE connected to your host personal computer (either by Ethernet or by USB), it will be detected in the RVconfig window (Figure 19.7).

After the RealView-ICE is connected, you can then use the Auto Configure function to detect the Cortex-M0 on the JTAG scan chain (Figure 19.8). We can then close the RVconfig window and save the configuration.

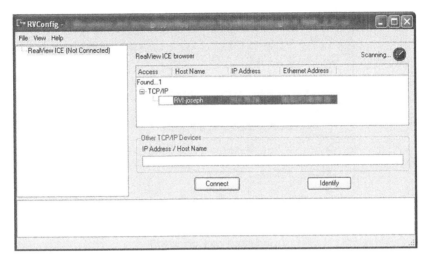

Figure 19.7:
Connect to Target window.

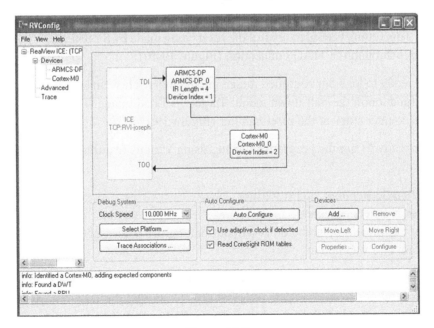

Figure 19.8:
Cortex-M0 detected by auto configure.

Back at the Connect-to-Target window, we can now expand the RealView ICE, locate the Cortex-M0 configuration, and connect to it by double-clicking on the Cortex-M0 connection (Figure 19.9). After the connection is made, the processor is put into a debug state and halts.

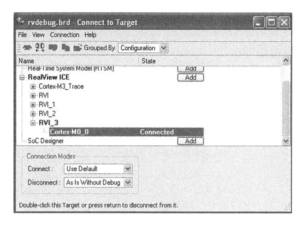

Figure 19.9:
Connected to Cortex-M0.

Before we load a compiled image and start debugging, we need to make the following adjustment: Click on the debug tab of the "Registers" windows, and change the reset type to "Ctrl_Reg." This setting is used to control the reset being made by SYSRESETREQ rather than using the reset through the JTAG connection (Figure 19.10).

Now we are ready to load our compiled image into memory. This can be done by accessing the load image function in the pull-down menu: Target → Load image. The image is loaded and the program counter stops at the reset handler (Figure 19.11).

Now we are ready to run the program or debug using various features in RVD, including the following:

- Halting, single stepping, and restarting
- Processor registers accesses
- Memory examination (can be done without halting the processor)
- Breakpoint and watchpoint

In RVD, most of the debug operations can also be automated using scripts. For example, we can put the operations from changing the reset type to starting the program execution into a short script file called "blinky.inc" (script files in RVD normally have an ".inc" file extension).

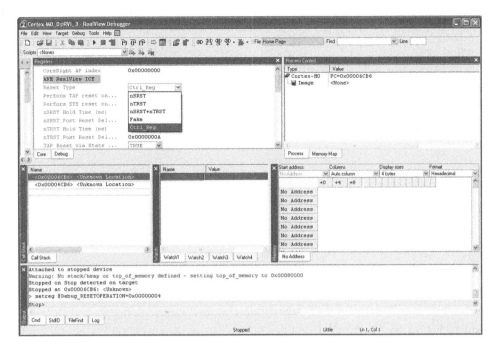

Figure 19.10:
Change the reset-type setting before continuing.

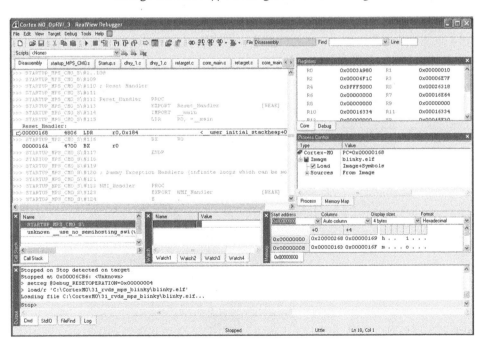

Figure 19.11:
Image loaded.

```
// Change reset type to control register
setreg @Debug_RESETOPERATION=0x00000004

// Load file
load/r 'C:\CortexM0\31_rvds_mps_blinky\blinky.elf'

// Reset
reset

// start program execution
go
```

By creating this file, we can set up the system and get the program running by starting the script from the pull-down menu: choose Tools → Include commands from file, and then select the RVD script that we created.

Using Serial Wire Debug with the RealView Debugger

A serial wire debug interface can be used with the RVD and the RealView-ICE. However, this can only be achieved using an LVDS probe v2 (Figure 19.12). The 20-pin IDC connection cannot be used for serial wire debug operations. You also need RealView-ICE version 3.4 or later for serial wire debug operations.

To enable the serial wire debug operation, follow these steps:

- Connect the LVDS probe v2 to the JTAG B connector at the front of the RealView-ICE.
- Start the RealView ICE Update utility, and connect to the RealView-ICE. Then check if you have updated firmware in the RealView-ICE (e.g., version 3.4 or later). If not, update the firmware to version 3.4.

Figure 19.12:
LVDS probe v2.

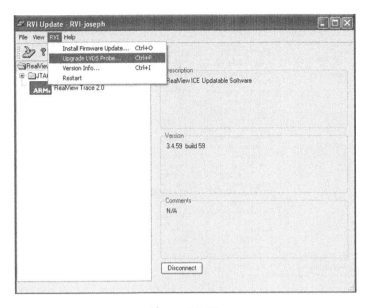

Figure 19.13:
Update the LVDS probe.

- You also need to update the LVDS probe by selecting "Update LVDS probe" in the Real-View ICE Update utility (Figure 19.13).
- Reboot the RealView-ICE when the update has completed.
- Disconnect the RealView ICE Update utility from the RealView-ICE, and then start the RealView debugger.
- Create a new connection and connect to the RealView-ICE as shown in Figure 19.6
- In the RVConfig window, select "Advanced" settings; you will then find the serial wire debug option (as shown in Figure 19.14).
- Select SWD for LVDS Debug Interface mode (Figure 19.14). The use of the SWJ switching sequence is not essential for most Cortex-M0 devices. However, it is essential for most Cortex-M3 devices because the debug access port in most of the Cortex-M3 devices supports both JTAG and serial wire debug protocol, and JTAG protocol is used by default.
- Now you can connect to the Cortex-M0 using the serial wire debug interface (Figure 19.15).

Note that the first version of the LVDS probe cannot support serial wire debug.

Retargeting in RVDS

One of the advanced features available in RealView Debugger RVD is semihosting support. This allows I/O functions like "printf," "scanf," and even file operations like "fopen" and "fread" to be carried out via the debugger. For example, a small program that requests user inputs ("scanf") and generates output messages ("printf") can be compiled with RVDS and then tested with RVD though the StdIO console window.

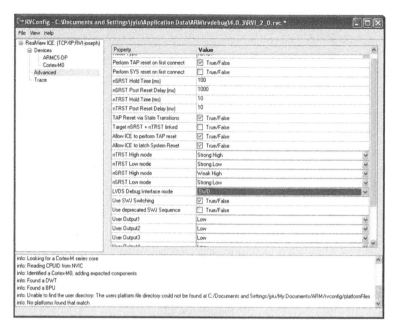

Figure 19.14:
Serial wire debug option in RVI advance settings.

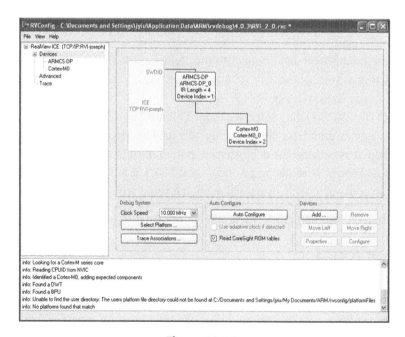

Figure 19.15:
RealView-ICE connected to the Cortex-M0 using the serial wire debug interface.

To demonstrate, a simple "hello" program is created:

```
hello.c
#include "MPS_CM0.h"
#include "stdio.h"
int main(void)
{
  char name[20];
  SystemInit();

  while(1){
  printf("Please enter your name :");
  scanf ("%s", &name[0]);
  printf("Hello %s, nice to meet you.\n\n", name);
  }

} // end main
```

This program is then compiled with RVDS, without retargeting I/O through the UART.

When the program is executed in RVD, the StdIO console window will display the output message and allow us to input information (bottom of Figure 19.16).

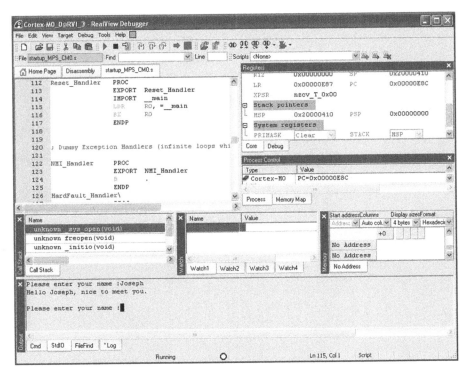

Figure 19.16:
StdIO console provides input and output functions.

Although it is possible to handle input output functions by setting up retargeting using UART, in some system-on-chip designs a UART interface might not be available. The semihosting support also allows data used for software testing to be stored on the debug host (personal computer) and accessed by the application running on the target system to access these files.

## CHAPTER 20

Getting Started with the GNU C Compiler

Overview

Apart from development tool chains produced by ARM, you can also develop software for Cortex-M0 microcontrollers using various development tools from other vendors. These include a number of development suites based on the GNU tool chain. Some of these tool chains are available free of charge, and others are available at a low cost and provide various additional features to assist software development.

The full release of the GNU C compiler is available from GNU Compiler home page (http://gcc.gnu.org).In this chapter, we will demonstrate compiling programs for Cortex-M0 using Sourcery G++ Lite, available from CodeSourcery (www.codesourcery.com). This product is available free of charge and is available as various precompiled packages including prebuilt packages for Windows and Linux. Using a prebuilt package is generally much easier than building a tool chain from the source. Prebuilt packages usually also include linker script examples and startup code. On the CodeSourcery web site, you can find various versions of Sourcery G++ Lite (Table 20.1).

For Cortex-M0 software development, in most cases we should be using the EABI version of the Sourcery G++. If your application is going to be running on a Cortex-M0 system with a µClinux operation system, then you should use the µClinux version.

The Sourcery G++ Lite edition supports software development tools in command line versions only. Apart from the Lite edition, CodeSourcery also provides a number of other editions of Sourcery G++ (Table 20.2).

Table 20.1: Available Sourcery G++ Lite Package

Target OS	Development Platform	Descriptions
EABI	Windows/Linux	For development without a targeted operating system
µClinux	Windows/Linux	For development of applications running on the µClinux operating system
Linux	Windows/Linux	For development of applications running on the Linux operating system
SymbianOS	Windows/Linux	For development of applications running on the Symbian operating system

The Definitive Guide to the ARM Cortex-M0. DOI: 10.1016/B978-0-12-385477-3.10020-5

Table 20.2: Sourcery G++ Editions

Edition	Features
Lite edition	Free command lines tools only, unsupported.
Personal edition	Low-cost development suite with limited supports. Features included the following: — Integration of the Eclipse integrated development environment (IDE) — Simulator and JTAG/BDM support — Supports for various ARM microcontrollers including ready-to-use linker scripts debug configurations and peripheral registers browsing — Preconfigured board supports and Board Builder Wizard to set up support for customer boards — Debug interface supports, including the following: Keil ULINK 2 ARMUSB SEGGER J-Link — Design examples
Academic edition	Same as the personal edition, for academic institute noncommercial use only
Standard edition	Includes all features in the personal edition, with additional libraries and unlimited technical support
Professional edition	Includes all features in standard edition, with priority support and critical defect correction

Apart from CodeSourcery, a number of other tool vendors provide GNU-based development packages. Because the examples in this chapter cover the development of software using the command line tools, most of the information is also applicable to other GNU-based tool chains. Information about the use of IDE, project management, and the debug environment is typically tool chain dependent; please refer to the documentation available from tool vendors.

Typical Development Flow

The GNU tool chain contains the C compiler, assembler, linker, libraries, debugger, and additional utilities. You can develop applications using C, assembly, or a mix of both languages (Table 20.3).

Table 20.3: Command Names

Tools	Generic Command Name	Command Name in CodeSourcery ARM EABI Package
C compiler	gcc	arm-none-eabi-gcc
Assembler	as	arm-none-eabi-as
Linker	ld	arm-none-eabi-ld
Binary file generation	objcopy	arm-none-eabi-objcopy
Disassembler	objdump	arm-none-eabi-objdump

The prefix of commands reflects the type of the prebuilt tool chain. In this case, the tool chain is prebuilt for the ARM EABI version without targeted OS.

The typical development flow of software development using gcc is shown in Figure 20.1. Unlike ARM RVDS or Keil MDK-ARM, the linking stage is usually carried out by the C compiler rather than as a separate step. This ensures that the details of the required parameters and libraries are passed on to the linker correctly. Using the linker as a separate step can be error prone and is not recommended.

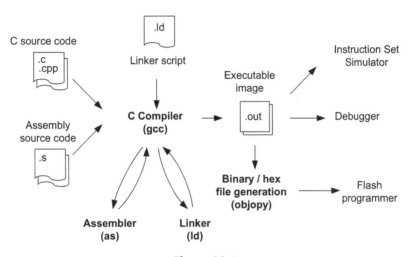

Figure 20.1:
Typical program generation flow.

Simple C Application Development

Based on the blinky example developed previously (Figure 20.2), we can compile the same program using gcc.

When compiling applications for Cortex-M processors, Sourcery G++ provides a vector table. The vector table use by Sourcery G++ is a part of the CodeSourcery Common Startup Code Sequence (CS3) feature. As a result, it is not necessary to have a file for the vector table and startup code.

The source code for blinky.c is as follows:

---

blinky.c for Sourcery G++

```
#include "LPC11XX.h"
#pragma weak __cs3_isr_systick = SysTick_Handler

#define KEIL_MCB1000_BOARD
```

(*Continued*)

blinky.c for Sourcery G++—Cont'd

```c
// Function declarations
void LedOutputCfg(void);  // Set I/O pin connected to LED as output

int main(void)
{
  // Switch Clock to 48MHz
  SystemInit();
  // Initialize LED output
  LedOutputCfg();
  // Program SysTick timer interrupt at 1KHz.
  // At 48MHz, SysTick trigger every 48000 CPU cycles
  SysTick_Config(48000);

  while(1);
} // end main

// SysTick handler to toggle LED every 500 ticks
void SysTick_Handler(void)
{
static short int TickCount = 0;
if ((TickCount++) == 500) { // for every 500 counts, toggle LED
  TickCount = 0; // reset counter to 0
#ifdef KEIL_MCB1000_BOARD
  // For Keil MCB1000, use P2.0 for LED output
  LPC_GPIO2->MASKED_ACCESS[1] = ~LPC_GPIO2->MASKED_ACCESS[1];
  // Toggle bit 0
#else
  // For LPCXpresso, use P0.7 for LED output
  LPC_GPIO0->MASKED_ACCESS[1<<7] = ~LPC_GPIO0->MASKED_ACCESS[1<<7]; // Toggle bit 7
#endif
  }
return;
}
// Switch LED signal (P0_7) to output port with no pull up or pulldown
void LedOutputCfg(void)
{
  // Enable clock to IO configuration block (bit[16] of AHBCLOCK Control register)
  // and enable clock to GPIO (bit[6] of AHBCLOCK Control register
  LPC_SYSCON->SYSAHBCLKCTRL = LPC_SYSCON->SYSAHBCLKCTRL | (1<<16) | (1<<6);

#ifdef KEIL_MCB1000_BOARD
  // For Keil MCB1000, use P2.0 for LED output
  // PIO2_0 IO output config
  // bit[5]  - Hysteresis (0=disable, 1 =enable)
  // bit[4:3] - MODE(0=inactive, 1 =pulldown, 2=pullup, 3=repeater)
  // bit[2:0] - Function (0 = IO, 1=DTR, 2=SSEL1)
  LPC_IOCON->PIO2_0 = (0<<5) + (0<<3) + (0x0);

  // Initial bit 0 output is 0
  LPC_GPIO2->MASKED_ACCESS[1] = 0;
  // Set pin 7 to 0 as output
  LPC_GPIO2->DIR = LPC_GPIO2->DIR | 0x1;
```

```
#else
  // For LPCXpresso, use P0.7 for LED output
  // PIO0_7 IO output config
  //   bit[5]   - Hysteresis (0=disable, 1 =enable)
  //   bit[4:3] - MODE(0=inactive, 1 =pulldown, 2=pullup, 3=repeater)
  //   bit[2:0] - Function (0 = IO, 1=CTS)
  LPC_IOCON->PIO0_7 = (0x0) + (0<<3) + (0<<5);
  // Initial bit[7] output is 0
  LPC_GPIO0->MASKED_ACCESS[1<<7] = 0;
  // Set pin 7 as output
  LPC_GPIO0->DIR = LPC_GPIO0->DIR | (1<<7);
#endif
  return;
} // end LedOutputCfg
```

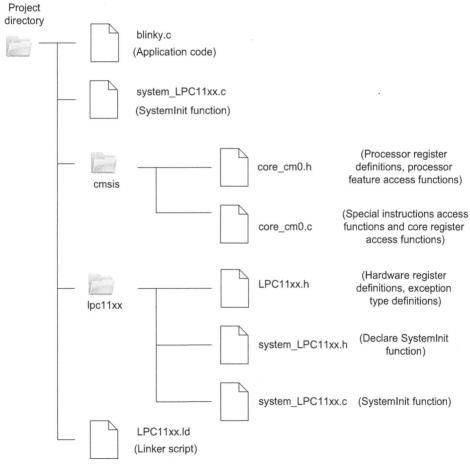

Figure 20.2:
Simple blinky project.

The line "#pragma weak __cs3_isr_systick = SysTick_Handler" is used to map the exception handler name into the CS3 exception vector name. The CS3 feature is specific to the Sourcery G++ tool chain. Other tool chains might require you to provide your own vector table definition. This will be covered in another example later.

Other files used in the project can be found in the previous examples or in the LPC11xx example software package from the NXP web site.

To compile the program and generate the executable image, we can use the following commands:

```
# Compile and link the application
arm-none-eabi-gcc -g -O3 -mcpu=cortex-m0 -mthumb blinky.c lpc11xx\system_LPC11xx.c
   -I cmsis -I lpc11xx -T LPC1114.ld -o blinky.o
# Generate disassembled listing for checking
arm-none-eabi-objdump -S blinky.o > list.txt
# Generate binary image file
arm-none-eabi-objcopy -O binary  blinky.o  blinky.bin
# Generate hex file (Intel hex format)
arm-none-eabi-objcopy -O ihex    blinky.o  blinky.hex
```

The generation of the listing file, binary file, and hex files are all optional. In most cases, the debugger or flash programmer can take the generated executable image directly.

The command line of above the gcc compilation contains the options shown in Table 20.4.

Table 20.4: Common Line Option for gcc Used in the Example

Option	Descriptions
-g	Include debug information
-O3	Optimization level (0 to 3)
-mcpu=cortex-m0	Processor choice
-mthumb	Specify Thumb instruction set
-lm	Link with math library (not used in the preceding example); this option is required when you use math functions like sin, cos, sinf, cosf, and so on
-I <directory>	Include directory
-T <linker script>	Specify linker script
-o <output file>	Specify output file name

The linker script LPC11xx.ld can be found in Appendix G. The linker script is tool chain specific because the symbol names being used are tool chain specific. Users of Sourcery G++ can find example linker scripts in the directory arm-none-eabi/lib of the Sourcery G++ installation. Users of other tool chains may find linker script examples from the tool installation. Alternatively, you can use the linker script in Appendix G as a starting point and modify it.

After the program has been compiled, the generated program image blinky.o is ready to be programmed on the microcontroller for testing.

CodeSourcery Common Startup Code Sequence (CS3)

The Sourcery G++ uses the CS3 for the C startup code and vector table. The CS3 vector table for the Cortex-M processors is called "`__cs3_interrupt_vector_micro`" in the linker script and is predefined with the following vector names shown in Table 20.5.

Table 20.5: Vector Table Symbol in the CS3 Vector Table for the Cortex-M Processor

Exception Number	CS3 Vector Symbol	Descriptions
0	__cs3_stack	Initial main stack pointer
1	__cs_reset	Reset vector
2	__cs_isr_nmi	Nonmaskable interrupt
3	__cs_isr_hard_fault	Hard fault
4	__cs_isr_mpu_fault	Memory management fault (not available in the Cortex-M0)
5	__cs_isr_bus_fault	Bus fault (not available in the Cortex-M0)
6	__cs_isr_usage_fault	Usage fault (not available in the Cortex-M0)
7 ... 10	__cs_isr_reserved_7 ... __cs_isr_reserved_10	Reserved exception type
11	__cs_isr_svcall	SuperVisor Call
12	__cs_isr_debug	Debug monitor (not available in the Cortex-M0)
13	__cs_isr_reserved_13	Reserved exception type
14	__cs_isr_pendsv	PendSV exception
15	__cs_isr_systick	System Tick Timer exception
16 ... 47	__cs_isr_external_0 ... __cs_isr_external_31	External interrupt

Because the symbol names do not match the exception handler names we used in the application code, we used the line "`#pragma weak __cs3_isr_systick = SysTick_Handler`" in the blinky.c so that the linker can insert the correct vector value to the vector table. Alternatively, we can handle the mapping by modifying the linker script.

In other GNU-based tool chains, other startup code and vector table handling mechanisms are available. Because of the different of symbol names, the linker scripts used for Sourcery G++ cannot be used directly on other GNU tool chains.

Using a User-Defined Vector Table

You can replace the CS3 vector table with your own vector table implementation. In the CMSIS software package from ARM (you can download it from www.onarm.com), you can find examples of the CMSIS vector table in assembly targeted for Sourcery G++. This can be modified to be used with LPC11xx.

The following assembler code, "startup_LPC11xx.s," provides the CMSIS version of the vector table for LPC11xx:

```
startup_LPC11xx.s

/******************************************************************************/
/* startup_LPC11xx.s: Startup file for LPC11xx device series                  */
/******************************************************************************/
/* Version: CodeSourcery Sourcery G++ Lite (with CS3)                         */
/******************************************************************************/


/*
//*** <<< Use Configuration Wizard in Context Menu >>> ***
*/

/*
// <h> Stack Configuration
//   <o> Stack Size (in Bytes) <0x0-0xFFFFFFFF:8>
// </h>
*/

    .equ    Stack_Size, 0x00000100
    .section ".stack", "w"
    .align  3
    .globl  __cs3_stack_mem
    .globl  __cs3_stack_size
__cs3_stack_mem:
    .if     Stack_Size
    .space  Stack_Size
    .endif
    .size   __cs3_stack_mem,  . - __cs3_stack_mem
    .set    __cs3_stack_size, . - __cs3_stack_mem

/*
// <h> Heap Configuration
//   <o>  Heap Size (in Bytes) <0x0-0xFFFFFFFF:8>
// </h>
*/

    .equ    Heap_Size,  0x00001000

    .section ".heap", "w"
    .align  3
    .globl  __cs3_heap_start
    .globl  __cs3_heap_end
__cs3_heap_start:
    .if     Heap_Size
    .space  Heap_Size
    .endif
__cs3_heap_end:


/* Vector Table */
```

```
    .section ".cs3.interrupt_vector"
    .globl   __cs3_interrupt_vector_cortex_m
    .type    __cs3_interrupt_vector_cortex_m, %object

__cs3_interrupt_vector_cortex_m:
    .long   __cs3_stack              /* Top of Stack              */
    .long   __cs3_reset              /* Reset Handler             */
    .long   NMI_Handler              /* NMI Handler               */
    .long   HardFault_Handler        /* Hard Fault Handler        */
    .long   0                        /* Reserved                  */
    .long   0                        /* Reserved                  */
    .long   0                        /* Reserved                  */
    .long   0                        /* Reserved                  */
    .long   0                        /* Reserved                  */
    .long   0                        /* Reserved                  */
    .long   0                        /* Reserved                  */
    .long   SVC_Handler              /* SVCall Handler            */
    .long   0                        /* Reserved                  */
    .long   0                        /* Reserved                  */
    .long   PendSV_Handler           /* PendSV Handler            */
    .long   SysTick_Handler          /* SysTick Handler           */

    /* External Interrupts */
    .long   WAKEUP_IRQHandler        /* 16+ 0: Wakeup PIO0.0      */
    .long   WAKEUP_IRQHandler        /* 16+ 1: Wakeup PIO0.1      */
    .long   WAKEUP_IRQHandler        /* 16+ 2: Wakeup PIO0.2      */
    .long   WAKEUP_IRQHandler        /* 16+ 3: Wakeup PIO0.3      */
    .long   WAKEUP_IRQHandler        /* 16+ 4: Wakeup PIO0.4      */
    .long   WAKEUP_IRQHandler        /* 16+ 5: Wakeup PIO0.5      */
    .long   WAKEUP_IRQHandler        /* 16+ 6: Wakeup PIO0.6      */
    .long   WAKEUP_IRQHandler        /* 16+ 7: Wakeup PIO0.7      */
    .long   WAKEUP_IRQHandler        /* 16+ 8: Wakeup PIO0.8      */
    .long   WAKEUP_IRQHandler        /* 16+ 9: Wakeup PIO0.9      */
    .long   WAKEUP_IRQHandler        /* 16+10: Wakeup PIO0.10     */
    .long   WAKEUP_IRQHandler        /* 16+11: Wakeup PIO0.11     */
    .long   WAKEUP_IRQHandler        /* 16+12: Wakeup PIO1.0      */
    .long   0                        /* 16+13: Reserved           */
    .long   SSP1_IRQHandler          /* 16+14: SSP1               */
    .long   I2C_IRQHandler           /* 16+15: I2C                */
    .long   TIMER16_0_IRQHandler     /* 16+16: 16-bit Counter-Timer 0 */
    .long   TIMER16_1_IRQHandler     /* 16+17: 16-bit Counter-Timer 1 */
    .long   TIMER32_0_IRQHandler     /* 16+18: 32-bit Counter-Timer 0 */
    .long   TIMER32_1_IRQHandler     /* 16+19: 32-bit Counter-Timer 1 */
    .long   SSP0_IRQHandler          /* 16+20: SSP                */
    .long   UART_IRQHandler          /* 16+21: UART               */
    .long   0                        /* 16+22: Reserved           */
    .long   0                        /* 16+23: Reserved           */
    .long   ADC_IRQHandler           /* 16+24: A/D Converter      */
    .long   WDT_IRQHandler           /* 16+25: Watchdog Timer     */
```

(Continued)

startup_LPC11xx.s—Cont'd

```
    .long    BOD_IRQHandler              /* 16+26: Brown Out Detect      */
    .long    0                           /* 16+27: Reserved              */
    .long    PIOINT3_IRQHandler          /* 16+28: PIO INT3              */
    .long    PIOINT2_IRQHandler          /* 16+29: PIO INT2              */
    .long    PIOINT1_IRQHandler          /* 16+30: PIO INT1              */
    .long    PIOINT0_IRQHandler          /* 16+31: PIO INT0              */

    .size    __cs3_interrupt_vector_cortex_m, . - __cs3_interrupt_vector_cortex_m

    .thumb

/* Reset Handler */

    .section .cs3.reset,"x",%progbits
    .thumb_func
    .globl   __cs3_reset_cortex_m
    .type    __cs3_reset_cortex_m, %function
__cs3_reset_cortex_m:
    .fnstart

    LDR      R0, =SystemInit
    BLX      R0
    LDR      R0,=_start
    BX       R0
    .pool
    .cantunwind
    .fnend
    .size    __cs3_reset_cortex_m,.-__cs3_reset_cortex_m

    .section ".text"

/* Exception Handlers */

    .weak    NMI_Handler
    .type    NMI_Handler, %function
NMI_Handler:
    B        .
    .size    NMI_Handler, . - NMI_Handler

    .weak    HardFault_Handler
    .type    HardFault_Handler, %function
HardFault_Handler:
    B        .
    .size    HardFault_Handler, . - HardFault_Handler

    .weak    SVC_Handler
    .type    SVC_Handler, %function
SVC_Handler:
    B        .
    .size    SVC_Handler, . - SVC_Handler

    .weak    PendSV_Handler
    .type    PendSV_Handler, %function
```

```
PendSV_Handler:
    B       .
    .size   PendSV_Handler, . - PendSV_Handler
    .weak   SysTick_Handler
    .type   SysTick_Handler, %function
SysTick_Handler:
    B       .
    .size   SysTick_Handler, . - SysTick_Handler


/* IRQ Handlers */

    .globl  Default_Handler
    .type   Default_Handler, %function
Default_Handler:
    B       .
    .size   Default_Handler, . - Default_Handler

    .macro  IRQ handler
    .weak   \handler
    .set    \handler, Default_Handler
    .endm

    IRQ     WAKEUP_IRQHandler
    IRQ     SSP1_IRQHandler
    IRQ     I2C_IRQHandler
    IRQ     TIMER16_0_IRQHandler
    IRQ     TIMER16_1_IRQHandler
    IRQ     TIMER32_0_IRQHandler
    IRQ     TIMER32_1_IRQHandler
    IRQ     SSP0_IRQHandler
    IRQ     UART_IRQHandler
    IRQ     ADC_IRQHandler
    IRQ     WDT_IRQHandler
    IRQ     BOD_IRQHandler
    IRQ     PIOINT3_IRQHandler
    IRQ     PIOINT2_IRQHandler
    IRQ     PIOINT1_IRQHandler
    IRQ     PIOINT0_IRQHandler

    .end
```

Based on the previous blinky example, the user-defined vector table can be included in the compilation stage:

```
arm-none-eabi-gcc -g -O3 -mcpu=cortex-m0 -mthumb blinky.c lpc11xx\system_LPC11xx.c
    startup_LPC11xx.s -I cmsis -I lpc11xx -T LPC1114.ld -o blinky.o
```

Using Printf in gcc

Gcc supports retargeting (e.g., `printf`). The retargeting implementation in Gcc is different from ARM development tools like Keil MDK-ARM or RVDS. The following example demonstrates the retargeting of the text I/O function (`printf`).

A simple hello world program "hello.c" is created as follows:

```
hello.c
#include "LPC11XX.h"
#include <stdio.h>
#include "uart_io.h"

int main(void)
{
  SystemInit();  // Switch Clock to 48MHz
  UartConfig();
  printf ("Hello world\n");
  while(1);
} // end main

// Retarget function
int _write_r(void *reent, int fd, char *ptr, size_t len)
{
  size_t i;
  for (i=0; i<len;i++) {
    UartPutc(ptr[i]); // call character output function in uart_io.c
    }
  return len;
}
```

The printf retargeting is handled by the "_write_r" function. This function calls the UART function "UartPutc" to output a character. We also include a UART program file called "uart_io.c" in this example, which contains the "UartConfig" function for initialization of the UART interface.

To compile and link this example, the following command line is used:

```
arm-none-eabi-gcc -g -O3 -mcpu=cortex-m0 -mthumb hello.c uart_io.c
   lpc11xx\system_LPC11xx.c -I cmsis -I lpc11xx -T LPC1114.ld -o hello.o
```

When the program is executed, the UART is initialized and the message "Hello world" is output through the UART interface.

Inline Assembler

The GNU C compiler supports inline assembler. The general syntax is as follows:

```
__asm ("     inst1  op1, op2, ... \n"
       "     inst2  op1, op2, ... \n"
       ...
       "     instN  op1, op2, ... \n"
       : output_operands    /* optional */
       : input_operands     /* optional */
       : clobbered_operands /* optional */
       );
```

In simple cases where the assembly instruction does not require parameters, it can be as simple as

```
void Sleep(void)
{ // Enter sleep using WFI instruction
  __asm ("    WFI\n");
  return;
}
```

If the assembly code requires input and output parameters, then you might need to define the input and output operands and the clobbered register lists if any other register is modified by the inline assembly operation. For example, the inline assembly code to multiply a value by 10 can be written as

```
unsigned int DataIn, DataOut;
...
__asm("    movs  r0, %0\n"
      "    movs  r3, #10\n"
      "    muls  r0, r0, r3\n"
      "    movs  %1, r0\n"
     :"=r (DataOut) : "r" (DataIn) : "cc", "r0", "r3");
```

In the code example, %0 is the first input parameter and %1 is the first output parameter. Because the operand order is output_operands, input_operands, and clobbered_operands, "DataOut" is assigned to %0, and "DataIn" is assigned to %1. The code changes register R3, so it needs to be added to the clobbered operand list.

More details of the inline assembly in GNU C compiler can be found online in the GNU tool chain documentation, GCC-Inline-Assembly-HOWTO.

SVC Example in gcc

You can mix C program and assembly code in a single gcc compilation step. The following SVC example demonstrates the use of inline assembler in C code, as well as compiling C code files and assembly language file in a single gcc compilation step.

An assembly file is used for the wrapper function for the SVC handler. Details of this arrangement are covered in Chapter 18. To generate the SVC instruction and to set up input parameters, inline assembly is used. A C language–based SVC handler is also included to display the input parameters and the SVC number. The C SVC handler requires an input parameter that indicates the starting address of the SVC exception stack frame.

Instead of using "printf" for display, we used our own UART functions to display the text and values so as to reduce code size. The following is the program listing for "svc_demo.c":

---

svc_demo.c

```
#include "LPC11XX.h"
#include <stdio.h>
#include "uart_io.h"
```

(Continued)

```
svc_demo.c—Cont'd
int main(void)
{
  unsigned int DataIn1, DataIn2;
  SystemInit();   // Switch Clock to 48MHz
  UartConfig();
  UartPuts("SVC demo\n");
  DataIn1 = 0x12;
  DataIn2 = 0x34;

    __asm (
    "    movs   r0,%0\n"
    "    movs   r1,%1\n"
    "    svc    0x3\n"
    : : "r" (DataIn1), "r" (DataIn2) : "cc", "r0", "r1" );

  while(1);
} // end main
void SVC_Handler_c(unsigned int * svc_args)
{
  // Stack frame contains:
  // r0, r1, r2, r3, r12, r14, the return address and xPSR
  // - Stacked R0  = svc_args[0]
  // - Stacked R1  = svc_args[1]
  // - Stacked R2  = svc_args[2]
  // - Stacked R3  = svc_args[3]
  // - Stacked R12 = svc_args[4]
  // - Stacked LR  = svc_args[5]
  // - Stacked PC  = svc_args[6]
  // - Stacked xPSR= svc_args[7]

  unsigned int svc_number;
  svc_number = ((char *)svc_args[6])[-2];
  UartPuts("SVC Handler:\n");
  UartPuts("- R0 = 0x");
  UartPutHex(svc_args[0]);
  UartPutc('\n');

  UartPuts("- R1 = 0x");
  UartPutHex(svc_args[1]);
  UartPutc('\n');

  UartPuts("- SVC number = 0x");
  UartPutHex(svc_number);
  UartPutc('\n');
  return;
}
```

A separate assembly file called "handlers.s" is created, which contains the assembly wrapper function for the SVC handler. This wrapper extracts the starting address of the SVC exception stack frame and passes it to the C-based SVC handler to display the results.

handlers.s

```
    .text
    .syntax unified
    .thumb
    .type   SVC_Handler, %function
    .global SVC_Handler
    .global SVC_Handler_c

SVC_Handler:
    movs   r0, #4
    mov    r1, lr
    tst    r0, r1
    beq    svc_stacking_used_MSP
    mrs    r0, psp /* first parameter - stacking was using PSP */
    ldr    r1,=SVC_Handler_c
    bx     r1
svc_stacking_used_MSP:
    mrs    r0, msp /* first parameter - stacking was using MSP */
    ldr    r1,=SVC_Handler_c
    bx     r1
    .end
```

When the SVC instruction in the svc_demo.c is executed, the SVC exception starts the SVC_Handler in "handlers.s." The starting address of the stack frame is extracted from the current value of LR and stores it in R0 as an input parameter for the SVC_Handler_c function in "svc_demo.c." The C-based SVC handler can then extract the input parameters (stacked register values) from the stack frame, and it can extract the SVC number used when the SVC instruction is executed.

To build this example, the following command is used:

```
arm-none-eabi-gcc -g -O2 -mcpu=cortex-m0 -mthumb svc_demo.c uart_io.c handlers.s
   startup_LPC11xx.s lpc11xx\ system_LPC11xx.c -I cmsis -I lpc11xx -T LPC1114.ld -o
   svc_demo.o
```

This assembly code "handlers.s" and the assembly startup code are handled by the C compiler automatically in a single step compilation.

Alternatively, you can avoid using a separate assembly file by creating the SVC handler wrapper using an attribute naked C function, which contains inline assembly of the SVC_Handler that extracts the stack frame location and branch to the C handler "SVC_Handler_c." In this way, all can be done in just one C file.

```
void SVC_Handler(void) __attribute__((naked));
void SVC_Handler(void)
{
    __asm("    movs   r0, #4\n"
```

(Continued)

```
—Cont'd
    "      mov     r1, lr \n"
    "      tst     r0, r1\n"
    "      beq     svc_stacking_used_MSP\n"
    "      mrs     r0, psp \n"
    "      ldr     r1,=SVC_Handler_c \n"
    "      bx      r1\n"
    "svc_stacking_used_MSP:  \n"
    "      mrs     r0, msp\n"
    "      ldr     r1,=SVC_Handler_c\n"
    "      bx      r1\n");
}
```

Hard Fault Handler Example

Using the same techniques, we can create a hard fault handler that reports the occurrence of the hard fault, and we can extract stacked register values including the stacked PC value. The stacked PC value is very useful for identifying locations of problems in software. By creating a disassembly listing of the compiled program image, we can use the stacked program counter value to identify where the hard fault occurred. The hard fault handler demonstrated here also reports other stacked register values. For example, the IPSR value in the stacked xPSR indicates if the fault occurred within an exception handler, and other registers might indicate address values being used for an invalid memory access.

The program listing to demonstrate the hard fault handler is as follows:

```
hardfault_handler_demo.c
#include "LPC11XX.h"
#include <stdio.h>
#include "uart_io.h"

#define INVALID_ADDRESS (*((volatile unsigned long *)(0xFFFF0000)))

int main(void)
{
  SystemInit();   // Switch Clock to 48MHz
  UartConfig();
  UartPuts("Hardfault handler demo\n");

  /* Generate fault */
  INVALID_ADDRESS = INVALID_ADDRESS + 1;

  while(1);
} // end main
```

```c
void HardFault_Handler_c(unsigned int * hf_args)
{
  // Stack frame contains:
  // r0, r1, r2, r3, r12, r14, the return address and xPSR
  // - Stacked R0   = hf_args[0]
  // - Stacked R1   = hf_args[1]
  // - Stacked R2   = hf_args[2]
  // - Stacked R3   = hf_args[3]
  // - Stacked R12  = hf_args[4]
  // - Stacked LR   = hf_args[5]
  // - Stacked PC   = hf_args[6]
  // - Stacked xPSR = hf_args[7]

  UartPuts("HardFault Handler:\n");

  UartPuts("-  R0 = 0x");
  UartPutHex(hf_args[0]);
  UartPutc('\n');

  UartPuts("-  R1 = 0x");
  UartPutHex(hf_args[1]);
  UartPutc('\n');

  UartPuts("-  R2 = 0x");
  UartPutHex(hf_args[2]);
  UartPutc('\n');

  UartPuts("-  R3 = 0x");
  UartPutHex(hf_args[3]);
  UartPutc('\n');

  UartPuts("- R12 = 0x");
  UartPutHex(hf_args[4]);
  UartPutc('\n');

  UartPuts("-  LR = 0x");
  UartPutHex(hf_args[5]);
  UartPutc('\n');

  UartPuts("-  PC = 0x");
  UartPutHex(hf_args[6]);
  UartPutc('\n');

  UartPuts("- xPSR= 0x");
  UartPutHex(hf_args[7]);
  UartPutc('\n');

  while (1);
  return;
}
```

The program requires an assembly wrapper, "handlers.s," to extract the exception stack frame starting address. The program code for this wrapper function is as follows:

```
handlers.s
    .text
    .syntax unified
    .thumb
    .type   HardFault_Handler, %function
    .global HardFault_Handler
    .global HardFault_Handler_c

HardFault_Handler:
    movs    r0, #4
    mov     r1, lr
    tst     r0, r1
    beq     hf_stacking_used_MSP
    mrs     r0, psp /* first parameter - stacking was using PSP */
    ldr     r1,=HardFault_Handler_c
    bx      r1
hf_stacking_used_MSP:
    mrs     r0, msp /* first parameter - stacking was using MSP */
    ldr     r1,=HardFault_Handler_c
    bx      r1
    .end
```

We can compile this example in the same way as the SVC demonstration example:

```
arm-none-eabi-gcc -g -O2 -mcpu=cortex-m0 -mthumb hardfault_handler_demo.c uart_io.c
   handlers.s startup_LPC11xx.s lpc11xx\ system_LPC11xx.c -I cmsis -I lpc11xx -T
   LPC1114.ld -o hardfault_handler_demo.o
```

Note that the C-based HardFault handler can only work if the main stack pointer is still pointing at valid memory location.

Again, just like the SVC example, you can create the HardFault handler wrapper using an attribute naked C function with inline assembly code to locate the stack frame and jump to the C code. This could all be done in just one C file.

```
void HardFault_Handler(void) __attribute__((naked));
void HardFault_Handler(void)
{
    __asm("      movs    r0, #4\n"
          "      mov     r1, lr \n"
          "      tst     r0, r1\n"
          "      beq     hf_stacking_used_MSP\n"
          "      mrs     r0, psp \n"
          "      ldr     r1,=HardFault_Handler_c \n"
          "      bx      r1\n"
```

```
        "hf_stacking_used_MSP:  \n"
        "    mrs    r0, msp\n"
        "    ldr    r1,=HardFault_Handler_c\n"
        "    bx     r1\n");
}
```

Flash Programming and Debug

After the program image is generated, we need to program the image onto the flash memory of the microcontroller and test/debug the application. If you are using the Sourcery G++ personal edition, academic edition, or professional edition, the development suite already includes flash programming and debug interface support for a number of ARM microcontrollers. Other development suites based on the GNU C compiler might also include device support for the Cortex-M0 microcontroller you use.

If you are using the Sourcery G++ Lite (the free version), then you will need a third-party tool for flash programming and debug. Chapter 19 outlined a number of possible solutions, including third-party flash programming tools and alternative debug arrangements. Because almost all development tools for ARM support the DWARF file format (which is used by gcc), compiled images from gcc can often be imported to other debug environments.

## CHAPTER 21

Software Porting

Overview

As software reuse becomes more common, software porting is becoming a more common task for embedded software developers. In this chapter, we will look into differences between various common ARM processors for microcontrollers and what areas in a program need to be modified when porting software between them.

This chapter also covers software porting of software from 8-bit and 16-bit architectures.

ARM Processors

A number of ARM processors are used in microcontroller products (Table 21.1).

Table 21.1: Commonly Used ARM Processors on Microcontrollers

Processor	Descriptions
ARM7TDMI	A very popular 32-bit processor and widely supported by development tools. It is based on ARM architecture version 4T and supports both ARM and Thumb instruction set. Upward compatible to ARM9, ARM11, and Cortex-A/R processors.
ARM920T/922T/ 940T	Microcontrollers based on these processors are less common nowadays. They are based on ARM architecture version 4T but with Harvard bus architecture. They also support cache, MMU, or MPU features.
ARM9E processor family	Most of the ARM9 microcontrollers are based on the ARM9E processor family. They are based on ARM architecture version v5TE (with Enhanced DSP instructions) and various memory system features (cache, TCM, MMU, MPU, DMA, etc.) depending on processor model. Usually they are targeted at higher end of microcontroller application space with high operating frequency and larger memory system support.
Cortex-M3	The first ARM Cortex processor designed specifically for microcontroller applications. It combines high-performance, high-energy efficiency, low interrupt latency and ease of use. It is based on ARM Architecture v7-M and supports the Thumb instruction set only. Upward compatible to Cortex-M4.
Cortex-M1	A processor design specifically for FPGA application. Based in ARM architecture v6-M, a subset of ARMv7-M, the Cortex-M1 supports a smaller instruction set compared to Cortex-M3. It uses the same exception processing model and shares the same benefits—C friendly and easy to use—as in Cortex-M3.

(Continued)

The Definitive Guide to the ARM Cortex-M0. DOI: 10.1016/B978-0-12-385477-3.10021-7

Table 21.1: Commonly Used ARM Processors on Microcontrollers—Cont'd

Processor	Descriptions
Cortex-M0	Using the ARMv6-M architecture, the Cortex-M0 is developed for ultra low-power designs and is target for general microcontroller applications where good performance, high energy efficiency, and deterministic behavior are required.
Cortex-M4	The latest edition of the ARM Cortex-M processor family targeted at the digital signal controller applications. Based on ARMv7-ME architecture, the Cortex-M4 provides all the features of the Cortex-M3 and also single precision floating point (optional) and SIMD instructions.

The main differences between the Cortex-M processors are illustrated in Figure 21.1.

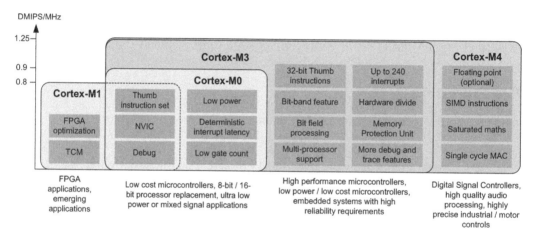

Figure 21.1:
The Cortex-M processor family.

In this chapter we will cover the detailed differences between the Cortex-M0 and some of these processors.

Differences between the ARM7TDMI and the Cortex-M0

There are a large number of differences between the ARM7TDMI and the Cortex-M0.

Operation Mode

The ARM7TDMI has a number of operation modes, whereas the Cortex-M0 only has two modes, as described in Table 21.2.

Some of the exception models from the ARM7TDMI are combined in Handler mode in the Cortex-M0 with different exception types. Consider the example presented in Table 21.3.

The reduction of operation modes simplifies Cortex-M0 programming.

Table 21.2: Operation Modes Comparison between the ARM7TDMI and the Cortex-M0

Operation Modes in ARM7TDMI	Operation Modes in Cortex-M0
System	Thread
Supervisor	Handler
IRQ	
FIQ	
Undefined (Undef)	
Abort	
User	

Table 21.3: Exception Comparison between the ARM7TDMI and the Cortex-M0

Exceptions in the ARM7TDMI	Exception in the Cortex-M0
IRQ	Interrupts
FIQ	Interrupts
Undefined (Undef)	Hard fault
Abort	Hard fault
Supervisor	SVC

Registers

The ARM7TDMI has a register bank with banked registers based on current operation mode. In Cortex-M0, only the SP is banked (Figure 21.2). And in most simple applications without an OS, only the MSP is required.

There are some differences between the CPSR (Current Program Status Register) in the ARM7TDMI and the xPSR in the Cortex-M0. For instance, the mode bits in CPSR are removed, replaced by IPSR, and interrupt masking bit I-bit is replaced by the PRIMASK register, which is separate from the xPSR.

Despite the differences between the register banks, the programmer's model or R0 to R15 remains the same. As a result, Thumb instruction codes on the ARM7TDMI can be reused on the Cortex-M0, simplifying software porting.

Instruction Set

The ARM7TDMI supports the ARM instructions (32-bit) and Thumb instructions (16-bit) in ARM architecture v4T. The Cortex-M0 supports Thumb instructions in ARMv6-M, which is a superset of the Thumb instructions supported by the ARM7TDMI. However, the Cortex-M0

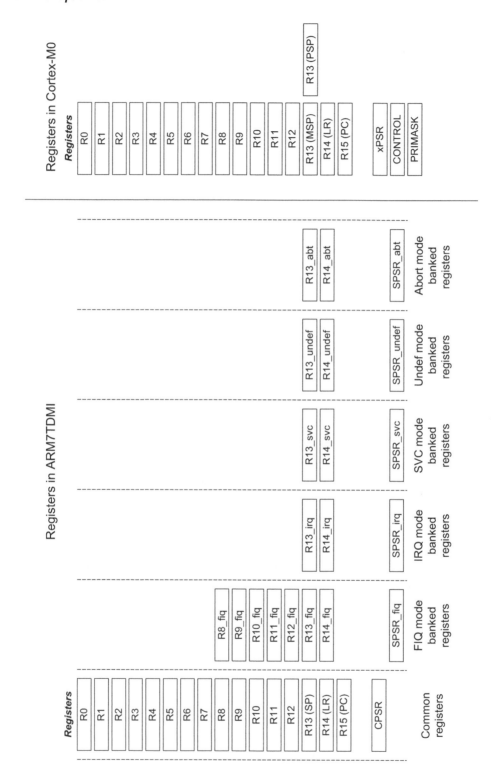

Figure 21.2:
Register bank differences between the ARM7TDMI and the Cortex-M0.

does not support ARM instructions. Therefore, applications for the ARM7TDMI must be modified when porting to Cortex-M0.

Interrupts

The ARM7TDMI supports an IRQ interrupt input and a Fast Interrupt (FIQ) input. Normally a separate interrupt controller is required in an ARM7TDMI microcontroller to allow multiple interrupt sources to share the IRQ and FIQ inputs. Because the FIQ has more banked registers and its vector is located at the end of the vector table, it can work faster by reducing the register stacking required, and the FIQ handler can be placed at the end of vector table to avoid branch penalty.

Unlike the ARM7TDMI, the Cortex-M0 has a built-in interrupt controller called NVIC with up to 32 interrupt inputs. Each interrupt can be programmed at one of the four available priority levels. There is no need to separate interrupts into IRQ and FIQ, because the stacking of registers is handled automatically by hardware. In addition, the vector table in the Cortex-M0 stores the starting address of each interrupt service routine, while in the ARM7TDMI the vector table holds instructions (usually branch instructions that branch to interrupt service routines).

When the ARM7TDMI receives an interrupt request, the interrupt service routine starts in ARM state (using ARM instruction). Additional assembly wrapper code is also required to support nested interrupts. In the Cortex-M0, there is no need to use assembly wrappers for normal interrupt processing.

Porting Software from the ARM7TDMI to the Cortex-M0

Application code for the ARM7TDMI must be modified and recompiled to be used on the Cortex-M0.

Startup Code and Vector Table

Because the vector table and the initialization sequence are different between the ARM7TDMI and the Cortex-M0, the startup code and the vector table must be replaced (Table 21.4).

Example of startup code for the Cortex-M0 can be found in various examples in this book.

Interrupt

Because the interrupt controller used in microcontrollers with the ARM7TDMI would be different from the NVIC in the Cortex-M0, all the interrupt control code needs to be updated. It is recommended to use the NVIC access functions defined in CMSIS for portability reason.

The interrupt wrapper function for nested interrupt support in the ARM7TDMI must be removed. If the interrupt service routine was written in assembly, the handler code will probably require

Table 21.4: Vector Table Differences between the ARM7TDMI and the Cortex-M0

Vector Table in the Arm7TDMI	Vector Table in the Cortex-M0
Vectors	Vectors
B Reset_Handler	IMPORT __main
B Undef_Handler	DCD _stack_top ; Main SP starting value
B SWI_Handler	DCD __main ; Enter C startup
B PrefetchAbort_Handler	DCD NMI_Handler
B DataAbort_Handler	DCD HardFault_Handler
B IRQ_Handler	DCD 0, 0, 0, 0, 0, 0, 0
B FIQ_Handler	DCD SVC_Handler
Reset_Handler ; Setup Stack for each mode	DCD 0, 0
LDR R0,=Stack_Top	DCD PendSV_Handler
MSR CPSR_c, #Mode_IRQ:OR:I_Bit:OR:F_Bit	DCD SysTick_Handler
MOV SP, R0	... ; vectors for other interrupt handlers
... ; setup stack for other modes	
IMPORT __main	
LDR R0, =__main ; Enter C startup	
BX R0	

rewriting because many ARM instructions cannot be directly mapped to Thumb instructions. For example, the exception handler in the ARM7TDMI can be terminated by "MOVS PC, LR" (ARM instruction). This is not valid for the Cortex-M0 and must be replaced by "BX LR".

FIQ handlers for the ARM7TDMI might rely on the banked registers R8 to R14 in the ARM7TDMI to save execution time. For example, constants used by the FIQ handler might be preloaded into these banked registers before the FIQ is enabled so that the FIQ handler can be simplified. When porting such handlers to the Cortex-M0 processor, the banked registers are not available and therefore these constants must be loaded into the registers within the handler.

In some cases you might find assembly code being used to enable or disable interrupts by modifying the I-bit in CPSR. In the Cortex-M0, this is replaced by the PRIMASK interrupt masking register. Note that in the ARM7TDMI you can carry out the exception return and change the I-bit in a single exception return instruction. In the Cortex-M0 processor, PRIMASK and xPSR are separate registers, so if the PRIMASK is set during the exception handler, it must be cleared before the exception exit. Otherwise the PRIMASK will remain set and no other interrupt can be accepted.

C Program Code

Apart from the usual changes caused by peripherals, memory map, and system-level feature differences, the C applications might require changes in the following areas:

- Compile directives like "#pragma arm" and "#pragma thumb" are no longer required because the Cortex-M0 supports Thumb instructions only.

- For ARM RVDS or Keil MDK, all inline assembly has to be rewritten, either using embedded assembler, separate assembly code, or as C functions. Inline assembly in these tools only supports ARM instructions. Users of the GNU C compiler might also need to modify their inline assembly code.
- Exception handlers can be simplified because in the Cortex-M0, each interrupt has its own interrupt vector. There is no need to use software to determine which interrupt service is required, and there is no software overhead in supporting nested interrupts.
- Although the "__irq" directive is not essential in the Cortex-M0 exception handlers, this directive for interrupt handlers can be retained in ARM RVDS or Keil MDK projects for clarity. It might also help software porting if the application has to be ported to other ARM processors in the future.

The C code should be recompiled to ensure that only Thumb instructions are used and no attempt to switch to ARM state should be contained in the compiled code. Similarly, library files must also be updated to ensure they will work with the Cortex-M0.

Assembly Code

Because the Cortex-M0 does not support the ARM instruction set, assembly code that uses ARM instructions has to be rewritten.

Be careful with legacy Thumb programs that use the CODE16 directive. When the CODE16 directive is used, the instructions are interpreted as traditional Thumb syntax. For example, data processing op-codes without S suffixes are converted to instructions that update APSR when the CODE16 directive is used. However, you can reuse assembly files with the CODE16 directive because it is still supported by existing ARM development tools. For new assembly code, the THUMB directive is recommended, which indicates to the assembly that the Unified Assembly Language (UAL) is used. With UAL syntax, data processing instructions updating the APSR require the S suffix.

Fault handlers and system exception handlers like SWI must also be updated to work with the Cortex-M0.

Atomic Access

Because Thumb instructions do not support swap (SWP and SWPB instructions), the code for handling atomic access must be changed. For single processor systems without other bus masters, you can use either the exception mechanism or PRIMASK to achieve atomic operations. For example, because there can only be one instance of the SVC exception running (when an exception handler is running, other exceptions of the same or lower priority levels are blocked), you can use SVC as a gateway to handle atomic operations.

Optimizations

After getting the software working on the Cortex-M0, there are various areas you can look into to optimize your application code.

For assembly code migrated from the ARM7TDMI, the data type conversion operation is one of the potential areas for improvement because of new instructions available in the ARMv6-M architecture.

If the interrupt handlers were written in assembly, there might be chance that the stacking operations can be reduced because the exception sequence automatically stacks R0-R3 and R12.

More sleep modes features are available in the Cortex-M0 that can be used to reduce power consumption. To take the full advantages of the low-power features on a Cortex-M0 microcontroller, you will need to modify your application code to make use of the power management features in the microcontroller. These features are dependent on the microcontroller product, and the information in this area can usually be found in user manuals or application notes provided by the microcontroller vendors.

With the nested interrupts being automatically handled by processor hardware and the availability of programmable priority levels in the NVIC, the priority level of the exceptions can be rearranged for best system performance.

Differences between the Cortex-M1 and the Cortex-M0

Both the Cortex-M1 and the Cortex-M0 are based on the ARM architecture v6-M, so the differences between the Cortex-M1 and the Cortex-M0 are relatively small.

Instruction Set

In the Cortex-M1 processor, WFI, WFE and SEV instructions are executed as NOPs. There is no sleep feature on current implementations of the Cortex-M1 processor.

SVC instruction support is optional in the Cortex-M1 (based on the design configuration parameter defined by an FPGA designer), whereas in the Cortex-M0 processor, SVC instruction is always available.

NVIC

SVC and PendSV exceptions are optional in the Cortex-M1 processor. They are always present in the Cortex-M0. Interrupt latency are also different between the two processors. Some optimizations related to interrupt latency (e.g. zero jitter) are not available on the current implementations of Cortex-M1 processor.

System-Level Features

The Cortex-M1 has Tightly Coupled Memory (TCM) support to allow memory blocks in the FPGA to connect to the Cortex-M1 directly for high-speed access, whereas the Cortex-M0 processor has various low-power support features like WIC (Wakeup Interrupt Controller).

There are also a number of differences in the configuration options between the two processors. These options are only available for FPGA designers (for Cortex-M1 users) or ASIC designers (for Cortex-M0 microcontroller vendors). For example, with the Cortex-M1 processor you can include both the serial wire debug and the JTAG debug interface, whereas Cortex-M0 microcontrollers normally only support either the serial wire or the JTAG debug interface.

Porting Software between the Cortex-M0 and the Cortex-M1

In general, software porting between Cortex-M0 and Cortex-M1 is extremely easy. Apart from peripheral programming model differences, there are few required changes.

Because both processors are based on the same instruction set, and the architecture version is the same, the same software code can often be used directly when porting from one processor to another. The only exception is when the software code uses sleep features. Because the Cortex-M1 does not support sleep mode, application code using WFI and WFE might need to be modified.

There is also a small chance that the software needs minor adjustment because of execution timing differences.

At the time of writing, no CMSIS software package is available for the Cortex-M1. However, you can use the same CMSIS files for the Cortex-M0 on Cortex-M1 programming, because they are based on the same version of the ARMv6-M architecture.

Differences between the Cortex-M3 and the Cortex-M0

The Cortex-M3 processor is based on the ARMv7-M architecture. It supports many more 32-bit Thumb instructions and a number of extra system features. The performance of the Cortex-M3 is also higher than that for the Cortex-M0. These factors make the Cortex-M3 very attractive to demanding applications in the automotive and industrial control areas.

Programmer's Model

The ARMv7-M architecture is a superset of the ARMv6-M architecture. So it provides all the features available in the ARMv6-M. The Cortex-M3 processor also provides various additional features. For the programmer's model, it has an extra nonprivileged mode (User Thread) when the processor is not executing exception handlers. The user Thread mode access

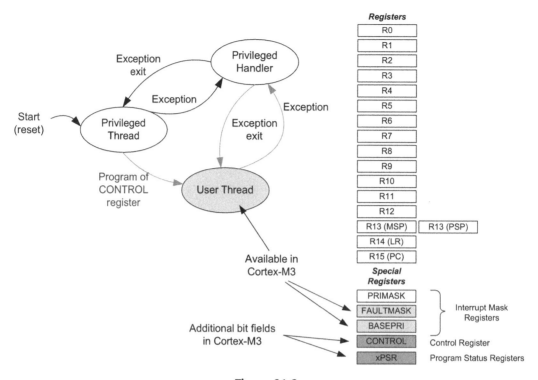

Figure 21.3:
Programmer's model differences between the Cortex-M0 and the Cortex-M3.

to the processor configuration registers (e.g., NVIC, SysTick) is restricted, and an optional memory protection unit (MPU) can be used to block programs running in user threads from accessing certain memory regions (Figure 21.3).

Apart from the extra operation mode, the Cortex-M3 also has additional interrupt masking registers. The BASEPRI register allows interrupts to of certain priority level or lower to be blocked, and the FAULTMASK provides additional fault management features.

The CONTROL register in the Cortex-M3 also has an additional bit (bit[0]) to select whether the thread should be in privileged or user Thread mode.

The xPSR in the Cortex-M3 also has a number of additional bits to allow an interrupted multiple load/store instruction to be resumed from the interrupted transfer and to allow an instruction sequence (up to four instructions) to be conditionally executed.

NVIC and Exceptions

The NVIC in the Cortex-M3 supports up to 240 interrupts. The number of priority levels is also configurable by the chip designers, from 8 levels to 256 levels (in most cases 8 levels to

32 levels). The priority level settings can also be configured into preemption priority (for nested interrupt) and subpriority (used when multiple interrupts of the same preempt priority are happening at the same time) by software.

One of the major differences between the NVIC in the Cortex-M3 and Cortex-M0 is that most of the NVIC registers in the Cortex-M3 can be accessed using word, half word, or byte transfers. With the Cortex-M0, the NVIC must be accessed using a word transfer. For example, if an interrupt priority register needs to be updated, you need to read the whole word (which consists of priority-level settings for four interrupts), modify 1 byte, and then write it back. In the Cortex-M3, this can be carried out using just a single byte-size write to the priority-level register. For users of the CMSIS device driver library, this difference does not cause a software porting issue, as the CMSIS NVIC access function names are the same and the functions use the correct access method for the processor.

The NVIC in the Cortex-M3 also supports dynamic changing of priority levels—in contrast to the Cortex-M0, where the priority level of an interrupt should not be changed after it is enabled.

The Cortex-M3 has additional fault handlers with programmable priority levels. It allows the embedded systems to be protected by two levels of fault exception handlers (Figure 21.4).

When used together with the memory protection unit in the Cortex-M3, robust systems can be build for embedded systems that require high reliability.

The NVIC in the Cortex-M3 also supports the following features:

- *Vector Table Offset Register.* The vector table can be relocated to another address in the CODE memory region or the SRAM memory region.
- *Software Trigger Interrupt Register.* Apart from using NVIC Interrupt Pending Set Register, the pending status of interrupts can be set using this register.

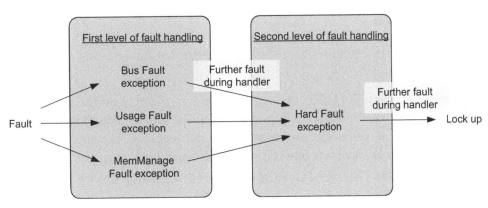

Figure 21.4:
Multiple levels of fault handling in the Cortex-M3.

- *Interrupt Active Status Register.* The active status of each interrupt can be determined by software.
- Additional fault status registers for indicating causes of fault exceptions and fault address
- An additional exception called the debug monitor for debug purposes.

Instruction Set

In addition to the Thumb instructions supported in the Cortex-M0 processor, the Cortex-M3 also supports a number of additional 16-bit and 32-bit Thumb instructions. These include the following:

- Signed and unsigned divide instructions (SDIV and UDIV)
- Compare and branch if zero (CBZ), compare and branch if not zero (CBNZ)
- IF-THEN (IT) instruction, allowing up to four subsequence instructions to be conditionally executed based on the status in APSR.
- Multiply and accumulate instructions for 32-bit and 64-bit results.
- Count leading zero (CLZ)
- Bit field processing instructions for bit order reversing, bit field insert, bit field clear, and bit field extract
- Table branch instructions (commonly used for the switch statement in C)
- Saturation operation instructions
- Exclusive accesses for multiprocessor environments
- Additional instructions that allows high registers (R8 and above) to be used in data processing, memory accesses, and branches

These additional instructions allow faster processing of complex data like floating point values. They also allow the Cortex-M3 to be used in audio signal processing applications, real time control systems.

System-Level Features

The Cortex-M3 includes a number of system-level features that are not available on the Cortex-M0. These include the following:

- *Memory protection unit (MPU).* A memory access monitoring unit that provides eight memory regions. Each memory region can be defined with different locations and size, as well as different memory access permissions and access behavior. If an access violation is found, the access is blocked and a fault exception is triggered. The OS can use the MPU to ensure each task can only access permitted memory space to increase system reliability.
- *Unaligned memory accesses.* In the Cortex-M0, all the data transfer operations must be aligned. This means a word-size data transfer must have an address value divisible by 4,

and half-word data transfer must occur at even addresses. The Cortex-M3 processor allows many memory access instructions to generate unaligned transfers. On the Cortex-M0 processor, access of unaligned data has to be carried out by multiple instructions.

- *Bit band regions.* The Cortex-M3 has two bit addressable memory regions called the bit-band regions. The first bit-band region is in the first 1 MB of the SRAM region, and the second one is the first 1 MB of the peripheral region. Using another memory address range called bit-band alias, the bit data in the bit band region can be individually accessed and modified.

- *Exclusive accesses.* The Cortex-M3 supports exclusive accesses, which are used to handle shared data in multiprocessor systems such as semaphores. The processor bus interface supports additional signals for connecting to an exclusive access monitor unit on the bus system.

Debug Features

The Cortex-M3 provides additional breakpoints and data watchpoints in its debug system. The breakpoint unit can also be used to remap instruction or literal data accesses from the original address (e.g., mask ROM) to a different location in the SRAM region. This allows nonerasable program memories to be patched with a small programmable memory (Table 21.5).

Table 21.5: Debug and Trace Feature Comparison

	Cortex-M0	Cortex-M3
Breakpoints	Up to 4	Up to 8
Watchpoints	Up to 2	Up to 4
Instruction trace	—	Optional
Data trace	—	Yes
Event trace	—	Yes
Software trace	—	Yes

In addition to the standard debug features, the Cortex-M3 also has trace features. The optional Embedded Trace Macrocell (ETM) allows information about instruction execution to be captured so that the instruction execution sequence can be reconstructed on debugging hosts. The Data Watch-point and Trace (DWT) unit can be used to generate trace for watched data variables or access to memory ranges. The DWT can also be used to generate event trace, which shows information of exception entrance and exit. The trace data can be captured using a trace port analyzer such as the ARM RealView-Trace unit or an in-circuit debugger such as the Keil ULINK*Pro*.

The Cortex-M3 processor also supports software-generated trace though a unit called the Instrumentation Trace Macrocell (ITM). The ITM provides 32 message channels and allows software to generate text messages or data output.

Porting Software between the Cortex-M0 and the Cortex-M3

Although there are a number of differences between the Cortex-M0 (ARMv6-M) and the Cortex-M3 (ARMv7-M), porting software between the two processors is usually easy. Because the ARMv7-M supports all features in the ARMv6-M, applications developed for the Cortex-M0 can work on the Cortex-M3 directly, apart from changes that result from their peripheral differences (Figure 21.5).

Normally, when porting an application from the Cortex-M0 to the Cortex-M3, you only need to change the device driver library, change the peripheral access code, and update the software for system features like clock speed, sleep modes, and the like.

Porting software from the Cortex-M3 to the Cortex-M0 might require more effort. Apart from switching the device driver library, you also need to consider the following areas:

• NVIC and SCB (System Control Block) registers in the Cortex-M0 can only be accessed in word-size transfers. If any program code accesses these registers in byte-size transfers or half-word transfers, they need to be modified. If the NVIC and SCB are accessed by using CMSIS functions, switching the CMSIS-compliant device driver to use the Cortex-M0 should automatically handle these differences.
• Some registers in the NVIC and the SCB in the Cortex-M3 are not available in the Cortex-M0. These include the Interrupt Active Status Register, the Software Trigger Interrupt Register, the Vector Table Offset Register, and some of the fault status registers.
• The bit-band feature in the Cortex-M3 is not available in the Cortex-M0. If the bit-band alias access is used, it needs to be converted to use normal memory accesses and handle bit extract or bit modification by software.
• If the application contains assembly code or embedded assembly code, the assembly code might require modification because some of the instructions are not available on the Cortex-M0. For C application code, some instructions such as hardware divide are not

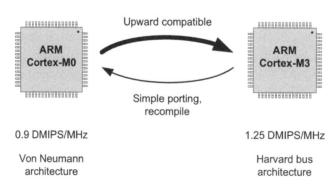

Figure 21.5:
Compatibility between the Cortex-M0 processor and the Cortex-M3 processor.

available in the Cortex-M0. In this case, the compiler will automatically call the C library to handle the divide operation.

*   Unaligned data transfer is not available on the Cortex-M0.
*   Some instructions available in the Cortex-M3 (e.g., exclusive accesses, bit field processing) are not available on the Cortex-M0.

Some Cortex-M0 microcontrollers support a memory remapping feature. Applications that use the vector table relocation feature on the Cortex-M3 might able to use the memory remapping feature to handle vector table relocation.

Applications that require the user Thread mode or the MPU feature cannot be ported to the Cortex-M0 because these features are not supported in the Cortex-M0.

Porting Software between the Cortex-M0 and the Cortex-M4 Processor

The Cortex-M4 processor is based on the same architecture as that used for the Cortex-M3. It is similar to the Cortex-M3 in many aspects: it has the same Harvard bus architecture, approximately the same performance in terms of Dhrystone DMIPS/MHz, the same exception types, and so on.

Compared to the Cortex-M3, the Cortex-M4 has additional instructions such as single instruction, multiple data (SIMD) instructions, saturation arithmetic instructions, data packing and extraction instructions, and optional single precision floating point instructions if a floating point unit is implemented. The floating point support in the Cortex-M4 is optional; therefore, not all Cortex-M4 microcontrollers will support this feature. If the floating point unit is included, it includes an additional floating point register bank and additional registers, as well as extra bit fields in the xPSR and CONTROL special registers (Figure 21.6). The floating point unit can be turned on or off by software to reduce power consumption.

Apart from these additional instructions, the system features of the Cortex-M4 are similar to those of the Cortex-M3 processor. Therefore, the techniques for porting software between the Cortex-M0 and the Cortex-M3 processors can also be used on porting software between the Cortex-M0 and Cortex-M4 processors. However, because of the differences between the nature of the two processors, some applications developed for the Cortex-M4 processor (e.g., high-end audio processing or industrial applications that require floating point operations) are unsuitable for the Cortex-M0 processor.

Porting Software from 8-Bit/16-Bit Microcontrollers to the Cortex-M0

Common Modifications

Some application developers might need to port applications from 8-bit or 16-bit microcontrollers to the Cortex-M0. By moving from these architectures to the

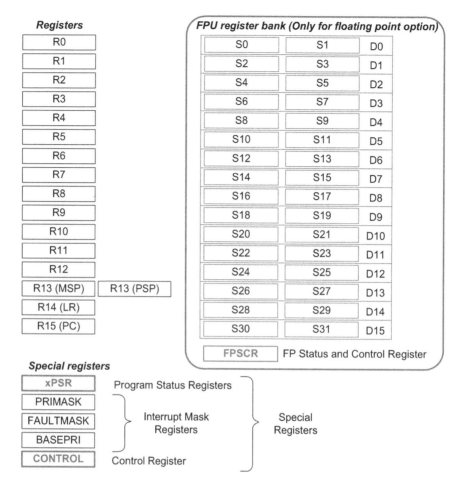

Figure 21.6:
Programmer's model of the Cortex-M4 with a floating point.

Cortex-M0, often you can get better code density, higher performance, and lower power consumption.

When porting applications from these microcontrollers to the Cortex-M0, the modifications of the software typically involve the following:

- *Startup code and vector table.* Different processor architectures have different startup code and interrupt vector tables. Usually the startup code and the vector table will have to be replaced.
- *Stack allocation adjustment.* With the Cortex-M processors, the stack size requirement can be very different from an 8-bit or 16-bit architecture. In addition, the methods to define stack location and stack size are also different from 8-bit and 16-bit development tools.

- *Architecture-specific/tool-chain-specific C language extensions.* Many of the C compilers for 8-bit and 16-bit microcontrollers support a number of C language extensions features. These include special data types like Special Function Registers (SFRs) and bit data in 8051, or various "#pragma" statements in various C compilers.
- *Interrupt control.* In 8-bit and 16-bit microcontroller programming, the interrupt configuration is usually done by directly writing to various interrupt control registers. When porting the applications to the ARM Cortex-M processor family, these codes should be converted to use the CMSIS interrupt control functions. For example, the enable and disable functions of interrupts can be converted to "__enable_irq()" and "__disable_irq ()". The configuration of individual interrupts can be handled by various NVIC functions in CMSIS.
- *Peripheral programming.* In 8-bit and 16-bit microcontroller programming, the peripherals control is usually handled by programming to registers directly. When using ARM microcontrollers, many microcontroller vendors provide device driver libraries to make use of the microcontroller easier. You can use these library functions to reduce software development time or write to the hardware registers directly if preferred. If you prefer to program the peripherals by accessing the registers directly, it is still beneficial to use the header files in the device driver library as these have all the peripheral registers defined and can save you time preparing and validating the code.
- *Assembly code and inline assembly.* Obviously all the assembly and inline assembly code needs to be rewritten. In many cases, you can rewrite the required function in C when the application is ported to the Cortex-M0.
- *Unaligned data.* Some 8-bit or 16-bit microcontrollers might support unaligned data. Because the Cortex-M0 does not support unaligned data, some data structures definitions or pointer manipulation codes might need to be changed. For data structures that require unaligned data handling, we can use the __packed attribute when defining the structure. However, the Cortex-M0 requires multiple instructions to access unaligned data. So it is best to convert the data structures so that all elements inside are aligned.
- *Be aware of data size differences.* The integers in most 8-bit and 16-bit processors are 16-bit, whereas in ARM architectures integers are 32-bit. This difference causes changes in behavior of overflow situations, it can also affect the memory size required for storing the data. For example, when a program file defines an array of integers from 8-bit or 16-bit architecture, we might want to change the code to use "short int" or "int16_t" (in "stdint.h," introduced in C99) when porting the code to ARM architecture so that the size remains unchanged.
- *Floating point.* Many 8-bit and 16-bit microcontrollers define "double" (double precision floating point) as 32-bit data. In ARM architecture, "double" is 64-bit. When porting applications containing floating point operations, you might need to change the

double precision floating point data to "float" (single precision floating point). Otherwise the processing speed would be reduced and the program size could increase because of the requirement to process the data in extra precision. For the same reason, some function calls for mathematical operation might need to be changed to ensure that the single precision version is used. For example, by default "cos ()" is the double precision version of the *cosine* function; for single precision operation, use "cosf()" instead.

- *Adding fault handlers.* In many 8-bit and 16-bit microcontrollers, there are no fault exceptions. Although embedded applications can operate without any fault handlers, the addition of fault handlers can help an embedded system to recover from error (e.g., data corruption caused by voltage drop or electromagnetic interference).

Memory Requirements

One of the points mentioned earlier is the stack size. After porting to the ARM architecture, the required stack size could increase or decrease, depending on the application. The stack size might increase for the following reasons:

- Each register push takes 4 bytes of memory in ARM, whereas in 16-bit or 8-bit models, each register push takes 2 bytes or 1 byte.
- In ARM programming, local variables are often stored in stack, whereas in some architectures local variables might be defined in a separate data memory area.

On the other hand, the stack size could decrease for the following reasons:

- With 8-bit or 16-bit architecture, multiple registers are required to hold large data, and often these architectures have fewer registers compared to ARM, so more stacking would be required.
- The more powerful addressing mode in ARM means address calculations can be carried out on the fly without taking up register space. The reduction of register space used for an operation can reduce the stacking requirement.

Overall, the total RAM size required could decrease significantly after porting because in some architectures, such as the 8051, local variables are defined statically in data memory space rather on the stack. So the memory space is used even when the function or subroutine is not running. On the other hand, in ARM processors, the local variables allocated on the stack only take up memory space when the function or subroutine is executing. Also, with more registers available in the ARM processor's register bank compared to some other architectures, some of the local variables might only need to be stored in the register bank instead of taking up memory space.

The program memory requirement in the ARM Cortex-M0 is normally much lower than it is for 8-bit microcontrollers, and it is often lower than that required for most 16-bit microcontrollers. So when you port your applications from these microcontrollers to the ARM Cortex-M0 microcontroller, you can use a device with smaller flash memory size. The reduction of the program memory size is often caused by the following:

- Better efficiency at handling 16-bit and 32-bit data (including integers and pointers)
- More powerful addressing modes
- Some memory access instructions can handle multiple data, including PUSH and POP

There can be exceptions. For applications that contains only a small amount of code, the code size in ARM Cortex-M0 microcontrollers could be larger compared to that for 8-bit or 16-bit microcontrollers for a couple of reasons:

- The ARM Cortex-M0 might have a much larger vector table because of more interrupts.
- The C startup code for ARM Cortex-M0 might be larger. If you are using ARM development tools like the Keil MDK or the RealView Development Suite, switching to the MicroLIB might help to reduce the code size.

Nonapplicable Optimizations for 8-Bit or 16-Bit Microcontrollers

Some optimization techniques used in 8-bit/16-bit microcontroller programming are not required on ARM processors. In some cases, these optimizations might result in extra overhead because of architectural differences. For example, many 8-bit microcontroller programmers use character data as loop counters for array accesses:

```
unsigned char i; /* use 8-bit data to avoid 16-bit processing */
char a[10] , b[10] ;
for (i=0;i<10;i++) a[i] = b[i];
```

When compiling the same program on ARM processors, the compiler will have to insert a UXTB instruction to replicate the overflow behavior of the array index ("i"). To avoid this extra overhead, we should declare "i" as integer "int", "int32_t", or "uint32_t" for best performance.

Another example is the unnecessary use of casting. For example, the following code uses casting to avoid the generation of a 16 × 16 multiply operation in an 8-bit processor:

```
unsigned int x, y, z;
z = ((char) x) * ((char) y); /* assumed both x and y must
                                be less than 256 */
```

Again, such a casting operation will result in extra instructions in ARM architecture. Since Cortex-M0 can handle a 32 × 32 multiply with a 32-bit result in a single instruction, the program code can be simplified:

```
unsigned int x, y, z;
z = x * y;
```

Example: Migrate from the 8051 to the ARM Cortex-M0

In general, because most applications can be programmed in C entirely on the Cortex-M0, the porting of applications from 8-bit/16-bit microcontrollers is usually straightforward and easy. Here we will see some simple examples of the modifications required.

Vector Table

In the 8051, the vector table contains a number of JMP instructions that branch to the start of the interrupt service routines. In some development environments, the compiler might create the vector table for you automatically. In ARM, the vector table contains the address of the main stack pointer initial values and starting addresses of the exception handlers. The vector table is part of the startup code, which is often provided by the development environment. For example, when creating a new project, the Keil MDK project wizard will offer to copy and add the default startup code, which contains the vector table (Table 21.6).

Table 21.6: Vector Table Porting

8051			Cortex-M0
org	00h		__Vectors DCD __initial_sp ; Top of Stack
	jmp	start	DCD Reset_Handler ; Reset Handler
org	03h ;	Ext Int0 vector	DCD NMI_Handler ; NMI Handler
	ljmp	handle_interrupt0	DCD HardFault_Handler ; Hard Fault
org	0Bh	;Timer 0 vector	DCD 0,0,0,0,0,0,0 ; Reserved
	ljmp	handle_timer0	DCD SVC_Handler ; SVCall Handler
org	13h ;	Ext Int1 vector	DCD 0,0 ; Reserved
	ljmp	handle_interrupt1	DCD PendSV_Handler ; PendSV Handler
org	1Bh	; Timer 1 vector	DCD SysTick_Handler ; SysTick
	ljmp	handle_timer1	Handler
org	23h	; Serial interrupt	; External Interrupts
	ljmp	handle_serial0	DCD WAKEUP_IRQHandler ; Wakeup
org	2bh	; Timer 2 vector	PIO0.0
	ljmp	handle_timer2	...

Data Type

In some cases, we need to modify the data type so as to maintain the same program behavior (Table 21.7).

Table 21.7: Data Type Change during Software Porting

8051	Cortex-M0

int my_data[20]; // array of 16-bit values
double pi;
	short int my_data[20]; // array of 16-bit values
	float pi;

Some function calls might also need to be changed if we want to ensure only single precision floating point is used (Table 21.8).

Table 21.8: Floating Point C Code Change during Software Porting

8051	Cortex-M0
Y = T*atan(T2*sin(Y)*cos(Y)/ (cos(X+Y)+cos(X-Y)-1.0));	Y = T*atanf(T2*sinf(Y)*cosf(Y)/ (cosf(X+Y)+cosf(X-Y)-1.0F));

Some special data types in 8051 are not available on the Cortex-M0: bit, sbit, sfr, sfr16, idata, xdata, and bdata.

Interrupt

Interrupt control code in 8051 are normally written as direct access to SFRs. They need to be changed to the CMSIS functions when ported to the ARM Cortex-M0. (Table 21.9).

Table 21.9: Interrupt Control Change during Software Porting

8051	Cortex-M0
EA = 0; /* Disable all interrupts */	__disable_irq(); /* Disable all interrupts */
EA = 1; /* Enable all interrupts */	__enable_irq(); /* Enable all interrupts */
EX0 = 1; /* Enable Interrupt 0 */	NVIC_EnableIRQ(Interrupt0_IRQn);
EX0 = 0; /* Disable Interrupt 0 */	NVIC_DisableIRQ(Interrupt0_IRQn);
PX0 = 1; /* Set interrupt 0 to high priority*/	NVIC_SetPriority(Interrupt0_IRQn, 0);

The interrupt service routine also requires minor modifications. Some of the special directives used by the interrupt service routine need to be removed when the application code is ported to the Cortex-M0 (Table 21.10).

Table 21.10: Interrupt Handler Change during Software Porting

8051	Cortex-M0
void timer1_isr(void) interrupt 1 using 2	__irq void timer1_isr(void)
{ /* Use register bank 2 */	{
…;	…;
return;	return;
}	}

Sleep Mode

Entering of sleep mode is different too (Table 21.11). In 8051, sleep mode can be entered by setting the IDL (idle) bit in PCON. In the Cortex-M0, you can use the WFI instruction, or use vendor-specific functions provided in the device driver library.

Table 21.11: Sleep Mode Control Change during Software Porting

8051	Cortex-M0
PCON = PCON | 1; /* Enter Idle mode */	__WFI(); /* Enter sleep mode */

## CHAPTER 22

Cortex-M0 Products

Overview

A number of ARM Cortex-M0 products are available, including microcontrollers, development boards, starter kits, and development suites. In this chapter we will have a quick glance at some of these products.

The descriptions here are based on information collected in the middle of 2010. Not every product is listed here, and additional products might be available that I am not aware off. By the time this book is printed, more products will have been released. For information about microcontroller devices based on the Cortex-M0 processor, the device database on www.onarm.com/devices provides a lot of useful information.

Microcontroller Products and Application-Specific Standard Products (ASSPs)

NXP Cortex-M0 Microcontrollers

NXP (www.nxp.com) provides a number of Cortex-M0 microcontrollers. The ARM Cortex-M0–based LPC1000 microcontroller family provides low-cost 32-bit MCU products targeted for traditional 8/16-bit MCU applications. They provide performance, low-power, and easy-to-use peripherals. Table 22.1 details the current range of products within the LPC1000, and more variants are planned.

Table 22.1: LPC1000 Product Family

Product	Features
LPC1111, LPC1112, LPC1113 and LPC1114	8KB to 32KB of flash memory, 2KB to 8KB of SRAM, GPIO, UART, SPI, SSP, I^2C, 16-bit timers and 32-bit timers, watchdog timer, serial wire debug, power management unit, 10-bit ADC, Brown-Out Detect, In System Programmable (ISP), and In Application Programmable (IAP); 2μA deep sleep current
LPC11C12, LPC11C14	All features of LPC1112/LPC1114, plus a CAN controller
LPC1102	Up to 32KB of flash memory, 8KB of SRAM. GPIO, UART, SPI, 16-bit timers and 32-bit timers, 10-bit ADC. 130 μA/MHz.
LPC1224, LPC1225, LPC1226, LPC1227	32KB to 128KB of flash memory, 4KB to 8KB of SRAM, micro-DMA controller, GPIO, UART(2), SSP/SPI, I2C Fast Mode+, RTC, 16-bit and 32-bit timers, windowed WDT, 10-bit ADC, analog comparators (2), same clocking and power features as LPC1112/LPC1114

The Definitive Guide to the ARM Cortex-M0. DOI: 10.1016/B978-0-12-385477-3.10022-9

Figure 22.1:
The LPC1102.

The LPC1102 microcontroller is the smallest 32-bit microcontroller device with a package size of only 5 mm^2 (Figure 22.1). It supports many peripherals you can find in the LPC1114 like UART, SPI, timers, I/O, and ADC. Device drivers and comprehensive example codes for all the LPC1000 products can be downloaded from the NXP web site.

NuMicro^{TM} Microcontroller Family

The NuMicro microcontroller family was developed by Nuvoton Technology Corp (www. nuvoton.com). The NuMicro devices included a number of product lines, as described in Table 22.2.

Table 22.2: NuMicro Product Family

Product Lines	Features
NUC140 Connectivity Line	CAN, LIN, USB, and NuMicro standard features
NUC130 Automotive Line	CAN, LIN, and NuMicro standard features
NUC120 USB Line	USB and NuMicro standard features
NUC100 Advance Line	NuMicro standard features

The common features of these product lines include the following:

- Up to 128KB flash, up to 32KB SRAM
- In System Programmable and In-Application Programmable
- Peripheral DMA mode

- GPIO
- 24-bit timers, watchdog timer, real-time clock
- Serial interfaces including UART, SPI, I^{2}C, I^{2}S
- ADC, analog comparator, temperature sensor
- Brown-out detector, LDO, low-voltage reset
- Up to 50MHz operation, with wide operating voltage range

Further high-density products will provide larger memory and additional functionality including Ethernet and motor control PWM.

Mocha-1 ARM Cortex-M0 Configurable Array

In Chapter 1 we emphasized that the Cortex-M0 processor is a good candidate for mixed-signal applications. The Mocha-1 ARM Cortex-M0 Configurable Array is a product of Triad Semiconductor (www.triadsemi.com). It is a flexible platform that supports a wide-range of mixed-signal applications (Figure 22.2).

Using Via-Configurable Array (VCA) technology in conjunction with the Mocha-1 platform, a system designer can specify custom analog and digital functions that complement the processing and control capabilities of the Cortex-M0 processor. VCA technology configures circuit building blocks ("tiles" of analog and digital resources) to form complex circuits, by placing vias at specific locations of an interconnecting metal fabric. Because only a single custom via layer and back-end metal processing is required, this approach produces custom ASICs that are typically less expensive and provide a quicker turnaround than fully custom ASICs (fewer IC processing steps are required to implement designs and make design changes).

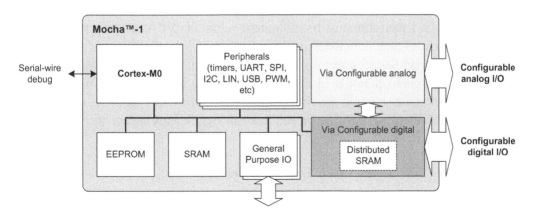

Figure 22.2:
The Mocha-1 product.

Implemented as an array of 99 digital tiles, the configurable digital logic portion of the Mocha-1 platform contains nearly 75,000 ASIC gates, 12.375 kB of distributed memory, and I/O capabilities. These resources are suitable for digital applications that range from simple state machines (PWMs, timers, etc.) to communications interfaces (e.g., SPI, I2C, USB, TCP/IP) and signal processing algorithms (e.g., FFT, audio and video processing).

Most of the analog VCA portion of the Mocha-1 platform contains three types of op-amp based general purpose tiles (including in total 8 low-noise, 12 high-speed single-ended, and 16 high-speed fully differential op-amps). Each tile also contains fundamental components, such as resistors, capacitors, switches, and transistors and distributed control logic. These tiles can be interconnected to create a variety of analog functions that include switched-capacitor circuits and filters, active filters, digital-to-analog converters (DACs), analog-to-digital converters (ADCs), instrumentation and programmable-gain amplifiers, multiplexers, sample-and-hold circuits, voltage and current references, power supply and temperature monitoring, and external sensor interfaces. Customizing each amplifier with respect to key features (power consumption, bandwidth, output drive, common-mode range, etc.) is accomplished via configuration as well. Other tiles are, such as the system PLL, a 10-bit ADC, two 10-bit DACs, six current-steering DAC tiles, and a band-gap circuit is also part of Mocha-1's analog VCA section and can be optimized for specific applications.

The processor subsystem contains a Cortex-M0 with 32 kB of EEPROM and 24 kB of SRAM. It also includes various peripherals (serial wire debug interface, watchdog timer, GPIO) and supports several power-saving modes. The SRAM memory provides a zero-wait state operation, whereas the EEPROM memory controller is optimized to minimize performance impact even when running at system clock speeds that require wait states.

Because the Mocha-1 platform must be configured before the VCA can be used, it is not available as an off-the-shelf component as are other microcontroller devices (evaluation versions, such as Triad Semiconductor's TSX-1001, are available that demonstrate the platform's abilities in a microcontroller-like format). But for engineers who work on mixed-signal applications where stock solutions are either too general-purpose or expensive, it is a desirable alternative to the full-custom ASIC approach that can significantly reduce development time and overall cost.

Melfas MCS™-7000 Series Touch Screen Controllers

The MCS-7000 Series Touch screen controllers are developed by Melfas Inc (www.melfas.com). The product family provides from 24 to 40 capacitive touch sensing channels (Table 22.3).

Table 22.3: MCS-7000 Product Family

Product	Features
MCS7024	I^2C, 24 capacitive touch channels, 32KB flash and 8KB SRAM
MCS7032	I^2C, 32 capacitive touch channels, 32KB flash and 8KB SRAM
MCS7040	I^2C, 40 capacitive touch channels, 32KB flash and 8KB SRAM

Compilers and Software Development Suites

Keil Microcontroller Development Kit (MDK)

The Keil MDK is a popular development suite. It consists of the following items:

- C/C++ compiler, assembler, linker and utilities supporting all Cortex-M, ARM7, and ARM9 microcontrollers
- Integrated debugger supporting in-circuit debuggers such as ULINK2, ULINK-Pro, and third-party products such as J-LINK and JtagJet. The debug environment also supports device-level simulation for users who want to test their program but do not have access to any hardware.
- Choices of run-time libraries: MicroLIB, a run-time library optimized for small memory footprint, and standard full-feature C libraries
- Keil RTX real-time kernel, which allows you to develop multitasking systems easily
- Flash programming algorithms

The μVision IDE allow easy access to various project options and features. Learning to use the Keil MDK is easy. You can download the evaluation version (limited to 32KB code size) of the Keil MDK from the Keil web site (www.keil.com/arm). Details of using the Keil MDK are covered in various chapters of this book starting from Chapter 14. For users who are migrating from 8051 or C166 designs and have used Keil μVision in the past, the switch is even easier because the same μVision environment can be used.

TASKING VX-Toolset for ARM

Apart from the development tools from ARM, there are a number of alternative choices. For instance, the TASKING VX-toolset for ARM supports the Cortex-M processors, including the Cortex-M0. It includes an Eclipse-based IDE and an integrated debugger, which can be used with the in-circuit debug interface adaptor from SEGGER (J-Link).

The main features included the following:

- ISO C++ Compiler, scalable to EC++
- C compiler, Linker, Assembler for Cortex-M processors
- C/C++ libraries, run-time libraries, floating point libraries

- Integrated static code analysis for CERT C secure coding standard
- MISRA C enhanced code checking

Details of the TASKING VX-toolset can be found on the TASKING web site (www.tasking. com).

IAR Embedded Workbench for ARM

The IAR Embedded Workbench is an integrated development environment for building and debugging ARM-based embedded applications. It includes the following main features:

- Optimizing C/C++ compiler
- ARM EABI and CMSIS compliant
- IAR C-SPY Debugger
- Extensive HW target system support
- Optional IAR J-Link and IAR J-Trace hardware debug probes

the details of the IAR Embedded Workbench for ARM and IAR KickStart Kit for LPC1114 can be found on www.iar.com.

CrossWorks for ARM

The CrossWorks for ARM is a C, C++, and assembly development suite from Rowley Associates (www.rowley.co.uk/arm/index.htm). It contains an IDE called CrossStudio with the GNU tool chain integrated. The source-level debugger in CrossStudio can work with a number of in-circuit debuggers including CrossConnect for ARM (from Rowley Associates) and third party in-circuit debugger hardware such as the SEGGER J-Link and Amontec JTAGkey.

The CrossWorks for ARM is available in various editions, including noncommercial low-cost packages (personal and educational licenses).

Red Suite

Red Suite from Code Red Technologies (www.code-red-tech.com) is a fully featured development suite for ARM-based microcontrollers, which includes all the tools necessary to develop high-quality software solutions in a timely and cost-effective fashion. The Red Suite IDE is based on the latest version of Eclipse with many ease-of-use and microcontroller-specific enhancements. It also features the industry standard GNU tool chain, allowing us to provide professional quality tools at low cost.

Features

The Red Suite integrated development environment (IDE) provides a comprehensive C/C++ programming environment, with syntax-coloring, source formatting, function folding, online and offline integrated help, extensive project management automation, and integrated source repository support (CVS integrated or Subversion via download).

It includes the following features:

- Wizards that create projects for all supported microcontrollers
- Automatic linker script generation including support for microcontroller memory maps
- Direct download to flash when debugging
- Inbuilt flash programmer
- Built-in datasheet browser
- Support for Cortex-M3, Cortex-M0, ARM7TDMI, and ARM926-EJ-based microcontrollers

With Cortex-M3-based microcontrollers, Red Suite can take advantage of its advanced features, including the following:

- Full support for serial wire viewing (SWV) through our Red Trace technology—unique in its class of tools
- No assembler required, even for startup code and interrupt handlers

Peripheral and Register Views

The peripheral viewer provides complete visibility of all registers and bit fields in all target peripherals in a simple tree-structured display. A powerful processor-register viewer is provided that gives access to all processors register and provides smart formatting for complex registers such as flags and status registers.

Red Trace

When used with Red Probe on Cortex-M3—based microcontrollers, the integrated Red Trace functionality gives the developer an unprecedented level of visibility into what is really happening on the target device. Unlike traditional trace solutions, Red Trace gathers trace data nonintrusively while the target application continues to run at full speed.

LabView C Code Generator

Besides C and assembly languages, there are other options to create program code for Cortex-M0 microcontrollers. One possible method is using the National Instruments LabVIEW graphical development environment, which works on PCs as well as ARM microcontrollers including Cortex-M0 microcontrollers.

The LabVIEW graphical programming language offers all of the features you expect in any programming language such as looping, conditional execution, and the handling of different data types. The main difference in working with LabVIEW is that you implement the design of the program in diagrams. For example, you can represent a simple loop to compute the sum of 1 to 10 by the for loop shown in Figure 22.3.

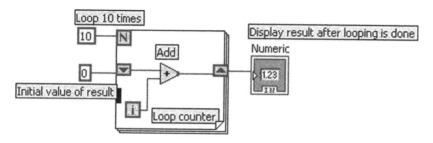

Figure 22.3:
A simple loop to add 1 to 10 in LabVIEW programming (image courtesy of National Instruments).

The LabVIEW programming environment provides a comprehensive library of functions including functions for digital signal processing (e.g., filter and spectral analysis), mathematic, array/matrix processing, and so on. These ready-to-use components allow application software to be developed without a in-depth knowledge of programming or algorithms. For complex applications, you can design the software into a hierarchy of modules called *virtual instruments* (VIs) and subVIs. For example, Figure 22.4 shows a LabVIEW subVI on the right, which finds the largest variables from four input variables, and this subVI is used by another VI.

What Is Needed for Using LabVIEW to Program ARM Microcontrollers?

To start using LabVIEW for the Cortex-M0 microcontroller, you need the LabVIEW C Code Generator. The LabVIEW C Code Generator takes LabVIEW graphical code and generates procedural C code from the diagram. Learn more about these products on the National Instruments web site (www.ni.com/embedded).

Before importing the LabVIEW-generated C code to the embedded platform, you can test your LabVIEW code by running the LabVIEW application on a PC. After the test is done and you are happy with the result, you can then use the LabVIEW C Code Generator to produce algorithm-level C code that you can integrate into another development environment and use to develop a full application. Figure 22.5 shows the typical development steps.

The generated C code can be used in various ARM microcontrollers and various tool chains.

Development Boards

LPCXpresso

The LPCXpresso (http://ics.nxp.com/lpcxpresso) is a low-cost development platform from NXP and Code Red Technologies. It consists of a low-cost development board and the

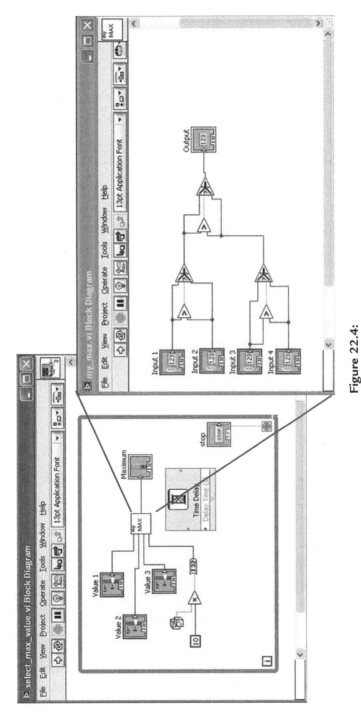

Figure 22.4:

Hierarchical software design in LabVIEW (image courtesy of National Instruments).

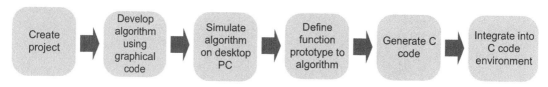

Figure 22.5:
Example design flow (diagram courtesy of National Instruments).

LPCXpresso IDE, a simplified Eclipse-based IDE. The LPCXpresso board is available for the NXP LPC1114 (Cortex-M0) product and LPC1343 (Cortex-M3). An LPC1700 (Cortex-M3) version will be available soon.

The LPCXpresso IDE contains C compiler, debugger, flash programming support, and examples. The C compiler in the LPCXpresso IDE is based on the GNU tool chain. The LPCXpresso IDE connects to the target board via the LPC-LINK, which is built in as part of the LPCXpresso board (Figure 22.6).

The LPCXpresso board is divided into two halves. The first half is a simple Cortex-M0 development board with an LPC1114. The other half is the LPC-LINK, an in-circuit debugger that allows the LPCXpresso IDE to connect to the LPC1114 via a USB connection. After the application is developed, you can use a USB connection to download your program to the Cortex-M0 microcontroller and test the application. The two halves of the LPCXpresso board can be separated by cutting the PCB, which would then allow the LPC-LINK to be used with other NXP Cortex-M microcontroller devices.

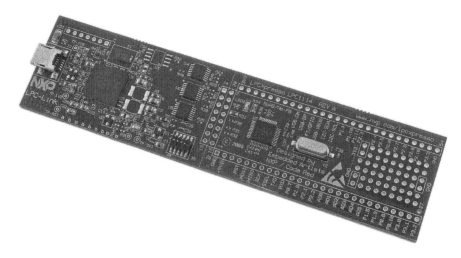

Figure 22.6:
An LPCXpresso board with an LPC1114.

IAR KickStart Kit for LPC1114

The IAR KickStart Kit for the LPC1114 includes an LPC114 evaluation board, an IAR Embedded Workbench (8KB KickStart edition), and an evaluation edition of IAR visual-STATE, a UML graphical state machine development tool (Figure 22.7).

Figure 22.7:
IAR KickStart Kit for LPC1114.

LPC1114 Cortex-M0 Stamp Module

The LPC1114 Cortex-M0 Stamp module is developed by Steinert Technologies (www.steitec. net). It provides easy access to I/O pins via 2.54mm pitch headers, making it attractive to hobbyists and students working on educational projects (Figure 22.8).

Keil Cortex-M0 Boards

Keil provides a number of development boards including products for ARM7, ARM9, Cortex-M3, and Cortex-M0 microcontrollers. For example, the MCBNUC1xx evaluation board

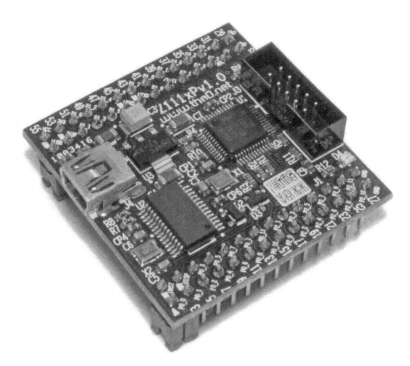

Figure 22.8:
LPC1114 stamp module from Steinert Technologies.

contains the NUC140VE3AN device (128KB flash, 16KB SRAM) in an easy-to-use form factor (Figure 22.9).

In addition, the MCB1000 board for NXP LPC1114 became available from Keil in the fourth quarter of 2010 (Figure 22.10).

Figure 22.9:
Keil MCBNUC1xx evaluation board.

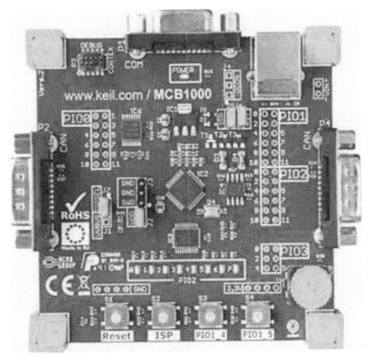

Figure 22.10:
Keil MCB1000 evaluation board.

Instruction Set Summary

The instructions supported on the Cortex-M0 processor include those shown in the following table:

Syntax (Unified Assembly Language)	Description
ADCS <Rd>, <Rm>	ADD with carry and update APSR
ADDS <Rd>, <Rn>, <Rm>	ADD registers and update APSR
ADDS <Rd>, <Rn>, #immed3	ADD register and a 3-bit immediate value
ADDS <Rd>, #immed8	ADD register and an 8-bit immediate value
ADD <Rd>, <Rm>	ADD two registers without update APSR
ADD <Rd>, SP, <Rd>	ADD the stack pointer to a register
ADD SP, <Rm>	ADD a register to the stack pointer
ADD <Rd>, SP, #immed8	ADD a stack pointer with an immediate value; Rd = SP + ZeroExtend(#immed8 <<2)
ADD SP, SP, #immed7	ADD an immediate value to the stack pointer; SP = SP + ZeroExtend(#immed7 <<2)
ADR <Rd>, <label>	Put an address to a register. Alternative syntax:ADD <Rd>, PC, #immed8
ANDS <Rd>, <Rd>, <Rm>	Logical AND between two registers
ASRS <Rd>, <Rd>, <Rm>	Arithmetic Shift Right
ASRS <Rd>, <Rd>, #immed5	Arithmetic Shift Right
BICS <Rd>, <Rd>, <Rm>	Logical Bitwise Clear
B <label>	Branch to an address (unconditional)
B <cond> <label>	Conditional branch
BL <label>	Branch and Link (return address store in LR)
BX <Rm>	Branch to address in register with exchange (LSB of target register should be set to 1 to indicate Thumb state)
BLX <Rm>	Branch to address in register and link (return address store in LR) with exchange (LSB of target register should be set to 1 to indicate Thumb state)
BKPT #immed8	Software breakpoint; immediate value of 0xAB is reserved for semi hosting
CMP <Rn>, <Rm>	Compare two registers and update APSR
CMP <Rn>, #immed8	Compare a register and an 8-bit immediate value and update APSR
CMN <Rn>, <Rm>	Compare negative (effectively an ADD operation)
CPSIE I	Clear PRIMASK (enable interrupt); in a CMSIS-compliant device driver, you can use the "__enable_irq()" CMSIS function for "CPSIE I"
CPSID I	Set PRIMASK (disable interrupt); in a CMSIS-compliant device driver, you can use the "__disable_irq()" CMSIS function for "CPSIE I"

(Continued)

Syntax (Unified Assembly Language)	Description
DMB	Data Memory Barrier; ensures that all memory accesses are completed before new memory access is committed; in a CMSIS-compliant device driver, you can use the "__DMB()" CMSIS function for DMB
DSB	Data Synchronization Barrier; ensures that all memory accesses are completed before next instruction is executed; in a CMSIS-compliant device driver you can use the "__DSB()" CMSIS function for DSB
EORS <Rd>, <Rd>, <Rm>	Logical Exclusive OR between two registers
ISB	Instruction Synchronization Barrier; flushes the pipeline and ensures that all previous instructions are completed before executing new instructions; in a CMSIS-compliant device driver, you can use the "__ISB()" CMSIS function for ISB
LDM <Rn>, {<Ra>, <Rb>,....}	Load multiple registers from memory; <Rn> is in the destination register list and gets updated by load
LDMIA <Rn>, {<Ra>, <Rb>,....}	Load multiple registers from memory; <Rn> is not in the destination register list and gets updated by address increment; alternative syntax: LDMFD <Rn>, {<Ra>, <Rb>,....}
LDR <Rt>, [<Rn>, <Rm>]	Load word from memory. <Rt> = memory[<Rn>+<Rm>]
LDR <Rt>, [<Rn>, #immed5]	Load word from memory; <Rt> = memory[<Rn> + #immed5<<2]
LDR <Rt>, [PC, #immed8]	Load word (literal data) from memory; <Rt> = memory[PC+ #immed8<<2]
LDR <Rt>, [SP, #immed8]	Load word from memory; <Rt> = memory[SP+ #immed8<<2]
LDRH <Rt>, [<Rn>, <Rm>]	Load half word from memory; <Rt> = memory[<Rn>+<Rm>]
LDRH <Rt>, [<Rn>, #immed5]	Load half word from memory; <Rt> = memory[<Rn> + #immed5<<1]
LDRB <Rt>, [<Rn>, <Rm>]	Load byte from memory; <Rt> = memory[<Rn>+<Rm>]
LDRB <Rt>, [<Rn>, #immed5]	Load byte from memory; <Rt> = memory[<Rn> + #immed5]
LDRSH <Rt>, [<Rn>, <Rm>]	Load signed half word from memory; <Rt> = signed_extend (memory[<Rn>+<Rm>])
LDRSB <Rt>, [<Rn>, <Rm>]	Load signed byte from memory; <Rt> = signed_extend (memory[<Rn>+<Rm>])
LSLS <Rd>, <Rd>, <Rm>	Logical shift left
LSLS <Rd>, <Rm>, #immed5	Logical shift left
LSRS <Rd>, <Rd>, <Rm>	Logical shift right
LSRS <Rd>, <Rm>, #immed5	Logical shift right
MOV <Rd>, <Rm>	Move register into register
MOVS <Rd>, <Rm>	Move register into register and update APSR
MOVS <Rd>, #immed8	Move immediate data (sign extended) into register
MRS <Rd>, <SpecialReg>	Move Special Register into register; in a CMSIS-compliant device driver library, a number of functions are available for special register accesses (see Appendix C)

(Continued)

Syntax (Unified Assembly Language)		Description
MSR	\<SpecialReg\>, \<Rd\>	Move register into Special Register; in a CMSIS-compliant device driver library, a number of functions are available for special register accesses (see Appendix C)
MVNS	\<Rd\>, \<Rm\>	Logical Bitwise NOT. Rd = NOT(Rm)
MULS	\<Rd\>, \<Rm\>, \<Rd\>	Multiply
NOP		No Operation; in a CMSIS-compliant device driver, you can use the "__NOP()" CMSIS function for NOP
ORRS	\<Rd\>, \<Rd\>, \<Rm\>	Logical OR
POP	{\<Ra\>, \<Rb\>,....}	Read single or multiple registers from stack memory and update
POP	{\<Ra\>, \<Rb\>, ..., PC}	the stack pointer
PUSH	{\<Ra\>, \<Rb\>,....}	Store single or multiple register to stack memory and update the
PUSH	{\<Ra\>, \<Rb\>, ..., LR}	stack pointer
REV	\<Rd\>, \<Rm\>	Byte Order Reverse
REV16	\<Rd\>, \<Rm\>	Byte Order Reverse within half word
REVSH	\<Rd\>, \<Rm\>	Byte order reverse within lower half word, then signed extend result
RORS	\<Rd\>, \<Rd\>, \<Rm\>	Rotate Right
RSBS	\<Rd\>, \<Rn\>, #0	Reverse Subtract (negative).
SBCS	\<Rd\>, \<Rd\>, \<Rm\>	Subtract with carry (borrow)
SEV		Send event to all processors in multiprocessing environment (including itself); in a CMSIS-compliant device driver, you can use the "__SEV()" CMSIS function for SEV
STMIA	\<Rn\>!, {\<Ra\>, \<Rb\>,....}	Store multiple registers to memory. \<Rn\> gets updated by address increment.
STR	\<Rt\>, [\<Rn\>, \<Rm\>]	Write word to memory; memory[\<Rn\>+\<Rm\>] = \<Rt\>
STR	\<Rt\>, [\<Rn\>, #immed5]	Write word to memory; memory[\<Rn\> + #immed5\<\<2] = \<Rt\>
STR	\<Rt\>, [SP, #immed8]	Write word to memory; memory[SP+ #immed8\<\<2] = \<Rt\>
STRH	\<Rt\>, [\<Rn\>, \<Rm\>]	Write half word to memory; memory [\<Rn\>+\<Rm\>] = \<Rt\>
STRH	\<Rt\>, [\<Rn\>, #immed5]	Write half word to memory; memory[\<Rn\> + #immed5\<\<1] = \<Rt\>
STRB	\<Rt\>, [\<Rn\>, \<Rm\>]	Write byte to memory; memory [\<Rn\>+\<Rm\>] = \<Rt\>
STRB	\<Rt\>, [\<Rn\>, #immed5]	Write byte to memory; memory[\<Rn\> + #immed5] = \<Rt\>
SUBS	\<Rd\>, \<Rn\>, \<Rm\>	Subtract two registers
SUBS	\<Rd\>, \<Rn\>, #immed3	Subtract a register with a 3-bit immediate data value
SUBS	\<Rd\>, #immed8	Subtract a register with an 8-bit immediate data value
SUB	SP, SP, #immed7	Subtract SP by an immediate data value; SP = SP − ZeroExtend (#immed7 \<\<2)
SVC	#\<immed8\>	Supervisor call; alternative syntax: SVC \<immed8\>
SXTB	\<Rd\>, \<Rm\>	Signed Extend lowest byte in a word data item
SXTH	\<Rd\>, \<Rm\>	Signed Extend lower half word in a word data item
TST	\<Rn\>, \<Rm\>	Test (bitwise AND)
UXTB	\<Rd\>, \<Rm\>	Extend lowest byte in a word data item

(Continued)

Syntax (Unified Assembly Language)	Description
UXTH <Rd>, <Rm>	Extend lower half word in a word data item
WFE	Wait for Event; if no record of previous event, enter sleep mode; if there is previous event, clear event latch register and continue; in a CMSIS-compliant device driver, you can use the "__WFE()" CMSIS function for WFE, but you might get better power optimization using vendor-specific sleep functions
WFI	Wait For Interrupt; enter sleep mode; in a CMSIS-compliant device driver, you can use the "__WFI()" CMSIS function for WFI, but you might get better power optimization using vendor-specific sleep functions
YIELD	Hint for thread switching and indicate task is stalled; execute as NOP on the Cortex-M0 processor

Cortex-M0 Exception Type Quick Reference

Exception Types

The exception types and corresponding control registers are listed in Table B.1.

Stack Contents after Exception Stacking

Table B.2 describes the layout of a stack frame in the stack memory after an exception stacking sequence is carried out. This information is useful for extracting stacked data within the exception handler.

Table B.1: Exception Types and Associated Control Registers

Exception Type	Name	Priority (Word address)	Enable
1	Reset	−3	Always
2	NMI	−2	Always
3	HardFault	−1	Always
11	SVC	Programmable (0xE000ED1C, byte 3)	Always
14	PendSV	Programmable (0xE000ED20, byte 2)	Always
15	SYSTICK	Programmable (0xE000ED20, byte 3)	SYSTICK Control and Status Register (SysTick->CTRL)
16	Interrupt #0	Programmable (0xE000E400, byte 0)	NVIC SETENA0 (0xE000E100, bit 0)
17	Interrupt #1	Programmable (0xE000E400, byte 1)	NVIC SETENA0 (0xE000E100, bit 1)
18	Interrupt #2	Programmable (0xE000E400, byte 2)	NVIC SETENA0 (0xE000E100, bit 2)
19	Interrupt #3	Programmable (0xE000E400, byte 3)	NVIC SETENA0 (0xE000E100, bit 3)
20	Interrupt #4	Programmable (0xE000E404, byte 0)	NVIC SETENA0 (0xE000E100, bit 4)
21	Interrupt #5	Programmable (0xE000E404, byte 1)	NVIC SETENA0 (0xE000E100, bit 5)
22-31	Interrupt #6—#31	Programmable (0xE000E404 − 0xE000E41C)	NVIC SETENA0 (0xE000E100, bit 6 − bit 31)

Table B.2: Stack Contents after Exception Stacking

Address	Data
(N+36)	(Previous stacked data)
(N+32)	(Previous stacked data/padding)
(N+28)	Stacked xPSR
(N+24)	Stacked PC (return address)
(N+20)	Stacked LR
(N+16)	Stacked R12
(N+12)	Stacked R3
(N+8)	Stacked R2
(N+4)	Stacked R1
New SP (N) →	Stacked R0

Depending on the SP value before the exception has taken place, the previous SP can be either the new SP value plus 32 or the new SP value plus 36. If the previous SP was aligned to a double word aligned address boundary, then the previous SP is new SP + 32. Otherwise, a padding word would be allocated before stacking and therefore the previous SP is new SP + 36.

CMSIS Quick Reference

The Cortex Microcontroller Software Interface Standard (CMSIS) contains a number of standardized functions:

- Core peripheral access functions
- Core register access functions
- Special instruction access functions

This appendix covers the basic information about these functions and other information related to using the CMSIS.

Data Type

The CMSIS uses standard data types defined in "stdint.h" (Table C.1).

Table C.1: Standard Data Types Used in CMSIS

Type	Data
uint32_t	Unsigned 32-bit integer
uint16_t	Unsigned 16-bit integer
uint8_t	Unsigned 8-bit integer

Exception Enumeration

Instead of using integer values for exception types, the CMSIS uses the IRQn enumeration to identify exceptions. The CMSIS defines the following enumeration and handler names for system exceptions:

Exception Type	Exception	CMSIS Handler Name	CMSIS IRQn Enumeration (Value)
1	Reset	Reset_Handler	—
2	NMI	NMI_Handler	NonMaskableInt_IRQn (-14)
3	HardFault	HardFault_Handler	HardFault_IRQn (-13)
11	SVC	SVC_Handler	SVCall_IRQn (-5)
14	PendSV	PendSV_Handler	PendSV_IRQn (-2)
15	SYSTICK	SysTick_Handler	SysTick_IRQn (-1)

The exception type 16 and above are device specific. In the case of NXP LPC11xx, these peripheral exceptions are defined as follows:

Exception Type	Exception	CMSIS Handler Name	CMSIS IRQn Enumeration (Value)
16	I/O wakeup 0	WAKEUP_IRQHandler	WAKEUP0_IRQn (0)
17	I/O wakeup 1	WAKEUP_IRQHandler	WAKEUP1_IRQn (1)
18	I/O wakeup 2	WAKEUP_IRQHandler	WAKEUP2_IRQn (2)
19	I/O wakeup 3	WAKEUP_IRQHandler	WAKEUP3_IRQn (3)
20	I/O wakeup 4	WAKEUP_IRQHandler	WAKEUP4_IRQn (4)
21	I/O wakeup 5	WAKEUP_IRQHandler	WAKEUP5_IRQn (5)
22	I/O wakeup 6	WAKEUP_IRQHandler	WAKEUP6_IRQn (6)
23	I/O wakeup 7	WAKEUP_IRQHandler	WAKEUP7_IRQn (7)
24	I/O wakeup 8	WAKEUP_IRQHandler	WAKEUP8_IRQn (8)
25	I/O wakeup 9	WAKEUP_IRQHandler	WAKEUP9_IRQn (9)
26	I/O wakeup 10	WAKEUP_IRQHandler	WAKEUP10_IRQn (10)
27	I/O wakeup 11	WAKEUP_IRQHandler	WAKEUP11_IRQn (11)
29	I/O wakeup 12	WAKEUP_IRQHandler	WAKEUP12_IRQn (12)
29	Reserved	—	—
30	SSP1	SSP1_IRQHandler	SSP1_IRQn (14)
31	I2C	I2C_IRQHandler	I2C_IRQn (15)
32	16-bit Timer0	TIMER16_0_IRQHandler	TIMER_16_0_IRQn (16)
33	16-bit Timer1	TIMER16_1_IRQHandler	TIMER_16_1_IRQn (17)
34	32-bit Timer0	TIMER32_0_IRQHandler	TIMER_32_0_IRQn (18)
35	32-bit Timer1	TIMER32_1_IRQHandler	TIMER_32_1_IRQn (19)
36	SSP0	SSP0_IRQHandler	SSP0_IRQn (20)
37	UART	UART_IRQHandler	UART_IRQn (21)
38	Reserved	—	—
39	Reserved	—	—
40	A/C converter	ADC_IRQHandler	ADC_IRQn (24)
41	Watchdog	WDT_IRQHandler	WDT_IRQn (25)
42	Brown Out Detect	BOD_IRQHandler	BOD_IRQn (26)
43	Reserved	—	—
44	External IRQ 3	PIOINT3_IRQHandler	EINT3_IRQn (28)
45	External IRQ 2	PIOINT2_IRQHandler	EINT2_IRQn (29)
46	External IRQ 1	PIOINT1_IRQHandler	EINT1_IRQn (30)
47	External IRQ 0	PIOINT0_IRQHandler	EINT0_IRQn (31)

NVIC Access Functions

The following functions are available for interrupt control:

Function Name	void NVIC_EnableIRQ(IRQn_Type IRQn)
Description	Enable Interrupt in NVIC Interrupt Controller
Parameter	IRQn_Type IRQn specifies the interrupt number (IRQn enum); this function does not support system exceptions
Return	None

Function Name	void NVIC_DisableIRQ(IRQn_Type IRQn)
Description	Disable Interrupt in NVIC Interrupt Controller
Parameter	IRQn_Type IRQn is the positive number of the external interrupt; this function does not support system exceptions
Return	None

Function Name	uint32_t NVIC_GetPendingIRQ(IRQn_Type IRQn)
Description	Read the interrupt pending bit for a device-specific interrupt source
Parameter	IRQn_Type IRQn is the number of the device specific interrupt; this function does not support system exceptions
Return	1 if pending interrupt else 0

Function Name	void NVIC_SetPendingIRQ(IRQn_Type IRQn)
Description	Set the pending bit for an external interrupt
Parameter	IRQn_Type IRQn is the number of the interrupt; this function does not support system exceptions
Return	None

Function Name	void NVIC_ClearPendingIRQ(IRQn_Type IRQn)
Description	Clear the pending bit for an external interrupt
Parameter	IRQn_Type IRQn is the number of the interrupt; this function does not support system exceptions
Return	None

Function Name	void NVIC_SetPriority(IRQn_Type IRQn, uint32_t priority)
Description	Set the priority for an interrupt or system exceptions with a programmable priority level
Parameter	IRQn_Type IRQn is the number of the interrupt unint32_t priority is the priority for the interrupt; this function automatically shifts the input priority value left to put priority value in implemented bits
Return	None

Function Name	uint32_t NVIC_GetPriority(IRQn_Type IRQn)
Description	Read the priority for an interrupt or system exceptions with programmable priority level
Parameter	IRQn_Type IRQn is the number of the interrupt
Return	uint32_t priority is the priority for the interrupt; this function automatically shifts the input priority value right to remove unimplemented bits in the priority value register

System and SysTick Access Functions

The following functions are available for system control and SysTick setup:

Function Name	void NVIC_SystemReset(void)
Description	Initiate a system reset request
Parameter	None
Return	None

Function Name	uint32_t SysTick_Config(uint32_t ticks)
Description	Initialize and start the SysTick counter and its interrupt; this function programs the SysTick to generate SysTick exception for every "ticks" number of core clock cycles.
Parameter	ticks is the number of clock ticks between two interrupts
Return	Always return 0

Function Name	void SystemInit (void)
Description	Initialize the system; device specific—this function is implemented in system_<device>.c (e.g., system_LPC11xx.c)
Parameter	None
Return	None

Function Name	void SystemCoreClockUpdate (void)
Description	Update the SystemCoreClock variable; this function is available from CMSIS version 1.3 and is device specific—this function is implemented in system_<device>.c (e.g., system_LPC11xx.c); it should be used every time after the clock settings have been changed
Parameter	None
Return	None

Core Registers Access Functions

The following functions are available for accessing core registers:

Function Name	Descriptions
uint32_t __get_MSP(void)	Get MSP value
void __set_MSP(uint32_t topOfMainStack)	Change MSP value
uint32_t __get_PSP(void)	Get PSP value
void __set_PSP(uint32_t topOfProcStack)	Change PSP value
uint32_t __get_CONTROL(void)	Get CONTROL value
void __set_CONTROL(uint32_t control)	Change CONTROL value

Special Instructions Access Functions

The following special instructions access functions are available in CMSIS:

Functions for System Features

Function Name	Instruction	Descriptions
void __WFI(void)	WFI	Wait for interrupt (sleep)
void __WFE(void)	WFE	Wait for event (sleep)
void __SEV(void)	SEV	Send event
void __enable_irq(void)	CPSIE i	Enable interrupt (clear PRIMASK)
void __disable_irq(void)	CPSID i	Disable interrupt (set PRIMASK)
void __NOP(void)	NOP	No operation
void __ISB(void)	ISB	Instruction synchronization barrier
void __DSB(void)	DSB	Data synchronization barrier
void __DMB(void)	DMB	Data memory barrier

Functions for Data Processing

Function Name	Instruction	Descriptions
uint32_t __REV(uint32_t value)	REV	Reverse byte order inside a word
uint32_t __REV16(uint32_t value)	REV16	Reverse byte order inside each of the two half word Note: early versions of CMSIS define input value as uint16_t
uint32_t __REVSH(uint32_t value)	REVSH	Reverse byte order in the lower half word and then signed extend the result to 32-bit Note: early versions of CMSIS define input value as uint16_t

NVIC, SCB, and SysTick Registers Quick Reference

NVIC Register Summary

Address	Name	CMSIS Symbol	Full Name
0xE000E100	ISER	NVIC->ISER	Interrupt Set Enable Register
0xE000E180	ICER	NVIC->ICER	Interrupt Clear Enable Register
0xE000E200	ISPR	NVIC->ISPR	Interrupt Set Pending Register
0xE000E280	ICPR	NVIC->ICPR	Interrupt Clear Pending Register
0xE000E400	IPR0-7	NVIC->IPR[0] to NVIC->IPR[7]	Interrupt Priority Register

Interrupt Set Enable Register (NVIC -> ISER)

To enable an interrupt with a CMSIS-compliant device driver library, please use the `NVIC_EnableIRQ` function:

Address	Name	Type	Reset Value	Descriptions
0xE000E100	SETENA	R/W	0x00000000	Set enable for Interrupts 0 to 31; write 1 to set bit to 1, write 0 has no effect Bit[0] for Interrupt #0 (exception #16) Bit[1] for Interrupt #1 (exception #17) … Bit[31] for Interrupt #31 (exception #47) Read value indicates the current enable status

Interrupt Clear Enable Register (NVIC -> ICER)

To disable an interrupt with a CMSIS-compliant device driver library, please use the `NVIC_DisableIRQ` function:

Address	Name	Type	Reset Value	Descriptions
0xE000E180	CLRENA	R/W	0x00000000	Clear enable for Interrupts 0 to 31;write 1 to clear bit to 0, write 0 has no effect Bit[0] for Interrupt #0 (exception #16) …

(Continued)

Address	Name	Type	Reset Value	Descriptions
				Bit[31] for Interrupt #31 (exception #47) Read value indicates the current enable status

Interrupt Set Pending Register (NVIC -> ISPR)

For setting pending status with a CMSIS-compliant device driver library, please use the `NVIC_SetPendingIRQ` function:

Address	Name	Type	Reset Value	Descriptions
0xE000E200	SETPEND	R/W	0x00000000	Set pending for Interrupts 0 to 31; write 1 to set bit to 1, write 0 has no effect. Bit[0] for Interrupt #0 (exception #16) Bit[1] for Interrupt #1 (exception #17) … Bit[31] for Interrupt #31 (exception #47) Read value indicates the current pending status

Interrupt Clear Pending Register (NVIC -> ICPR)

For clearing pending status with CMSIS-compliant device driver library, please use the `NVIC_ClearPendingIRQ` function:

Address	Name	Type	Reset Value	Descriptions
0xE000E280	CLRPEND	R/W	0x00000000	Clear pending for interrupt 0 to 31; write 1 to clear bit to 0, write 0 has no effect Bit[0] for Interrupt #0 (exception #16) … Bit[31] for Interrupt #31 (exception #47) Read value indicates the current pending status

Interrupt Priority Registers (NVIC -> IPR[0] to NVIC -> IPR[7])

For programming of Interrupt Priority with CMSIS-compliant device driver library, please use the `NVIC_SetPriority` function:

Address	Name	Type	Reset Value	Descriptions
0xE000E400	PRIORITY0	R/W	0x00000000	Priority level for interrupt 0 to 3 [31:30] Interrupt priority 3 [23:22] Interrupt priority 2 [15:14] Interrupt priority 1 [7:6] Interrupt priority 0

(Continued)

Address	Name	Type	Reset Value	Descriptions
0xE000E404	PRIORITY1	R/W	0x00000000	Priority level for interrupt 4 to 7
0xE000E408	PRIORITY2	R/W	0x00000000	Priority level for interrupt 8 to 11
0xE000E40C	PRIORITY3	R/W	0x00000000	Priority level for interrupt 12 to 15
0xE000E410	PRIORITY4	R/W	0x00000000	Priority level for interrupt 16 to 19
0xE000E414	PRIORITY5	R/W	0x00000000	Priority level for interrupt 20 to 23
0xE000E418	PRIORITY6	R/W	0x00000000	Priority level for interrupt 24 to 27
0xE000E41C	PRIORITY7	R/W	0x00000000	Priority level for interrupt 28 to 31

SCB Register Summary

Address	Name	CMSIS Symbol	Full Name
0xE000ED00	CPUID	SCB->CPUID	CPU ID (Identity) Base register
0xE000ED04	ICSR	SCB->ICSR	Interrupt Control State Register
0xE000ED0C	AIRCR	SCB->AIRCR	Application Interrupt and Reset Control Register
0xE000ED10	SCR	SCB->SCR	System Control Register
0xE000ED14	CCR	SCB->CCR	Configuration Control Register
0xE000ED1C	SHPR2	SCB->SHP[0]	System Handler Priority Register 2
0xE000ED20	SHPR3	SCB->SHP[1]	System Handler Priority Register 3
0xE000ED24	SHCSR	SCB->SHCSR	System Handler Control and State Register (accessible from debugger only)

CPU ID Base Register (SCB -> CPUID)

This register's value can be used to determine CPU type and revision:

Bits	Field	Type	Reset Value	Descriptions
31:0	CPU ID	RO	0x410CC200 (r0p0)	CPU ID value; used by debugger as well as application code to determine processor type and revision [31:24] Implementer [23:20] Variant (0x0) [19:16] Constant (0xC) [15:4] Part number (0xC20) [3:0] Revision (0x0)

Interrupt Control State Register (SCB -> ICSR)

Bits	Field	Type	Reset Value	Descriptions
31	NMIPENDSET	R/W	0	Write 1 to pend NMI, write 0 has no effect. On reads return pending state of NMI.
30:29	Reserved	—	—	Reserved.

(Continued)

Bits	Field	Type	Reset Value	Descriptions
28	PENDSVSET	R/W	0	Write 1 to set PendSV, write 0 has no effect. On reads return the pending state of PendSV.
27	PENDSVCLR	R/W	0	Write 1 to clear PendSV, write 0 has no effect. On reads return the pending state of PendSV.
26	PENDSTSET	R/W	0	Write 1 to pend SysTick, write 0 has no effect. On reads return the pending state of SysTick.
25	PENDSTCLR	R/W	0	Write 1 to clear SysTick pending, write 0 has no effect. On reads return the pending state of SysTick.
24	Reserved	—	—	Reserved.
23	ISRPREEMPT	RO	—	During debugging, this bit indicates that an exception will be served in the next running cycle, unless it is suppressed by debugger by C_MASKINTS in Debug Control and Status Register.
22	ISRPENDING	RO	—	During debugging, this bit indicates that an exception is pended.
21:18	Reserved	—	—	Reserved.
17:12	VECTPENDING	RO	—	Indicates the exception number of the highest priority pending exception. If it is read as 0, it means no exception is currently pended.
11:6	Reserved	—	—	Reserved.
5:0	VECTACTIVE	RO	—	Current active exception number, same as IPSR. If the processor is not serving an exception (Thread mode), this field read as 0.

Application Interrupt and Control State Register (SCB -> AIRCR)

Bits	Field	Type	Reset value	Descriptions
31:16	VECTKEY (during write operation)	WO	—	Register access key. When writing to this register, the VECTKEY field need to be set to 0x05FA, otherwise the write operation would be ignored.
31:16	VECTKEYSTAT (during read operation)	RO	0xFA05	Read as 0xFA05.
15	ENDIANESS	RO	0 or 1	1 indicates the system is big endian. 0 indicates the system is little endian.
14:3	Reserved	—	—	Reserved.
2	SYSRESETREQ	WO	—	Write 1 to this bit cause the external signal SYSRESETREQ to be asserted.
1	VECTCLRACTIVE	WO	—	Write 1 to this bit causes: Exception active status to be cleared Processor return to Thread mode IPSR to be cleared This bit can be only be used by debugger.
0	Reserved	—	—	Reserved.

System Control Register (SCB -> SCR)

Bits	Field	Type	Reset Value	Descriptions
31:5	Reserved	—	—	Reserved.
4	SEVONPEND	R/W	0	When set to 1, an event is generated for each new pending of an interrupt. This can be used to wake up the processor if Wait-for-Event sleep is used.
3	Reserved	—	—	Reserved.
2	SLEEPDEEP	R/W	0	When set to 1, deep sleep mode is selected when sleep mode is entered. When this bit is zero, normal sleep mode is selected when sleep mode is entered.
1	SLEEPONEXIT	R/W	0	When set to 1, enter sleep mode (Wait-for-Interrupt) automatically when exiting an exception handler and returning to thread level. When set to 0 this feature is disabled.
0	Reserved	—	—	Reserved.

Configuration Control Register (SCB -> CCR)

This register is read only and has fixed value. It is implemented to maintain compatibility between ARMv6-M and ARMv7-M architectures:

Bits	Field	Type	Reset Value	Descriptions
31:10	Reserved	—	—	Reserved.
9	STKALIGN	RO	1	Double word exception stacking alignment behavior is always used.
8:4	Reserved	—	—	Reserved.
3	UNALIGN_TRP	RO	1	Instruction trying to carry out an unaligned access always causes a fault exception.
2:0	Reserved	—	—	Reserved.

System Handler Priority Register 2 (SCB -> SHR[0])

For programming the Interrupt Priority with the CMSIS-compliant device driver library, please use the NVIC_SetPriority function rather than directly accessing the CMSIS register symbol. This ensures software compatibility between various Cortex-M processors:

Address	Name	Type	Reset Value	Descriptions
0xE000ED1C	SHPR2	R/W	0x00000000	System Handler Priority Register 2 [31:30] SVC priority

System Handler Priority Register 3 (SCB -> SHR[1])

For programming the Interrupt Priority with the CMSIS-compliant device driver library, please use the `NVIC_SetPriority` function rather than directly access the CMSIS register symbol. This ensures software compatibility between various Cortex-M processors:

Address	Name	Type	Reset Value	Descriptions
0xE000ED20	SHPR3	R/W	0x00000000	System Handler Priority Register 3 [31:30] SysTick priority [23:22] PendSV priority

System Handler Control and State Register

This register is only accessible from a debugger. Application software cannot access this register:

Bits	Field	Type	Reset Value	Descriptions
31:16	Reserved	—	—	Reserved.
15	SVCALLPENDED	RO	0	1 indicates SVC execution is pended. Accessible from debugger only.
14:0	Reserved	—	—	Reserved.

SysTick Register Summary

Address	Name	CMSIS Symbol	Full Name
0xE000E010	SYST_CSR	SysTick->CTRL	SysTick Control and Status Register
0xE000E014	SYST_RVR	SysTick->LOAD	SysTick Reload Value Register
0xE000E018	SYST_CVR	SysTick->VAL	SysTick Current Value Register
0xE000E01C	SYST_CALIB	SysTick->CALIB	SysTick Calibration Register

SysTick Control and Status Register (SysTick -> CTRL)

Bits	Field	Type	Reset Value	Descriptions
31:17	Reserved	—	—	Reserved.
16	COUNTFLAG	RO	0	Set to 1 when the SysTick timer reach zero. Clear to 0 by reading of this register.
15:3	Reserved	—	—	Reserved.

(Continued)

Bits	Field	Type	Reset Value	Descriptions
2	CLKSOURCE	R/W	0	Value of 1 indicates that the core clock is used for the SysTick timer. Otherwise a reference clock frequency (depending on MCU design) is used.
1	TICKINT	R/W	0	SysTick interrupt enable. When this bit is set, the SysTick exception is generated when the SysTick timer count down to 0.
0	ENABLE	R/W	0	When set to 1 the SysTick timer is enabled. Otherwise the counting is disabled.

SysTick Reload Value Register (SysTick -> LOAD)

Bits	Field	Type	Reset Value	Descriptions
31:24	Reserved	—	—	Reserved.
23:0	RELOAD	R/W	Undefined	Specify the reload value of the SysTick Timer.

SysTick Current Value Register (SysTick -> VAL)

Bits	Field	Type	Reset Value	Descriptions
31:24	Reserved	—	—	Reserved.
23:0	CURRENT	R/W	Undefined	On read returns the current value of the SysTick timer. Write to this register with any value to clear the register and the COUNTFLAG to 0. (This does not cause SysTick exception.)

SysTick Calibration Value Register (SysTick -> CALIB)

Bits	Field	Type	Reset Value	Descriptions
31	NOREF	RO	—	If it is read as 1, it indicates SysTick always use core clock for counting as no external reference clock is available. If it is 0, then an external reference clock is available and can be used. The value is MCU design dependent.
30	SKEW	RO	—	If set to 1, the TENMS bit field is not accurate. The value is MCU design dependent.
29:24	Reserved	—	—	Reserved.
23:0	TENMS	RO	—	Ten millisecond calibration value. The value is MCU design dependent.

Debug Registers Quick Reference

Overview

The Cortex-M0 debug system contains a number of programmable registers. These registers can be accessed by an in-circuit debuggers only and cannot be accessed by the application software. This quick reference is intended for tools developers, or if you are using a debugger that supports debug scripts (e.g., RealView Debugger), you can use debug scripts to access to these registers to carry out testing operations automatically.

The debug system in the Cortex-M0 is partitioned into the following segments:

- Debug support in the processor core
- Breakpoint unit
- Data watchpoint unit
- ROM table.

System-on-chip developers can add debug support components if required. If additional debug components are added, another ROM table unit can also be added to the system so that a debugger can identify available debug components included in the system.

The debug support is configurable; some Cortex-M0 based products might not have any debug support.

Core Debug Registers

The processor core contains a number of registers for debug purpose.

Address	Name	Descriptions
0xE000ED24	SHCSR	System Handler Control and State Register—indicate system exception status
0xE000ED30	DFSR	Debug Fault Status Register—allow debugger to determine the cause of halting
0xE000EDF0	DHCSR	Debug Halting Control and Status Register—control processor debug activities like halting, single stepping, restart
0xE000EDF4	DCRSR	Debug Core Register Selector Register—control read and write of core registers during halt

(Continued)

461

Address	Name	Descriptions
0xE000EDF8	DCRDR	Debug Core Register Data Register—data transfer register for reading or writing core registers during halt
0xE000EDFC	DEMCR	Debug Exception Monitor Control Register—for enabling of data watchpoint unit and vector catch feature; vector catch allows the debugger to halt the processor if the processor is reset or if a hard fault exception is triggered
0xE000EFD0 to 0xE000EFFC	PIDs, CIDs	ID registers

System Handler Control and State Register (0xE000ED24)

Bits	Field	Type	Reset Value	Descriptions
31:16	Reserved	—	—	Reserved
15	SVCALLPENDED	RO	0	1 indicates SVC execution is pended; accessible from debugger only
14:0	Reserved	—	—	Reserved

Debug Fault Status Register (0xE000ED30)

Bits	Field	Type	Reset Value	Descriptions
31:5	Reserved	—	—	Reserved
4	EXTERNAL	RWc	0	EDBGRQ was asserted
3	VCATCH	RWc	0	Vector catch occurred
2	DWTTRAP	RWc	0	Data watchpoint occurred
1	BKPT	RWc	0	Breakpoint occurred
0	HALTED	RWc	0	Halted by debugger or single stepping

Debug Halting Control and Status Register (0xE000EDF0)

Bits	Field	Type	Reset Value	Descriptions
31:16	DBGKEY (during write)	WO	—	Debug Key. During write, the value of 0xA05F must be used on the top 16-bit. Otherwise the write is ignored.
25	S_RESET_ST (during read)	RO	—	Reset status flag (sticky). Core has been reset or being reset; this bit is clear on read.
24	S_RETIRE_ST (during read)	RO	—	Instruction is completed since last read; this bit is clear on reset.

(Continued)

Bits	Field	Type	Reset Value	Descriptions
19	S_LOCKUP	RO	—	When this bit is 1, the core is in lockup state.
18	S_SLEEP	RO	—	When this bit is 1, the core is sleeping.
17	S_HALT (during read)	RO	—	When this bit is 1, the core is halted.
16	S_REGRDY_ST	RO	—	When this bit is 1, the core completed a register read or register write operation.
15:4	Reserved	—	—	Reserved.
3	C_MASKINTS	R/W	0	Mask exceptions while stepping (does not affect NMI and hard fault); valid only if C_DEBUGEN is set.
2	C_STEP	R/W	0	Single step control. Set this to 1 to carry out single step operation; valid only if C_DEBUGEN is set.
1	C_HALT	R/W	0	Halt control. This bit is only valid when C_DEBUGEN is set.
0	C_DEBUGEN	R/W	0	Debug enable. Set this bit to 1 to enable debug.

Debug Core Register Selector Register (0xE000EDF4)

Bits	Field	Type	Reset Value	Descriptions
31:17	Reserved	—	—	Reserved
16	REGWnR	WO	—	Set to 1 to write value to register Set to 0 to read value from register
15:5	Reserved	—	—	Reserved
4:0	REGSEL	WO	0	Register select

Debug Core Register Data Register (0xE000EDF8)

Bits	Field	Type	Reset Value	Descriptions
31:0	DBGTMP	RW	0	Data value for the core register transfer

Debug Exception and Monitor Control Register (0xE000EDFC)

Bits	Field	Type	Reset Value	Descriptions
31:25	Reserved	—	—	Reserved
24	DWTENA	RW	0	Data watchpoint unit enable
23:11	Reserved	—	—	Reserved

(Continued)

Bits	Field	Type	Reset Value	Descriptions
10	VC_HARDERR	RW	0	Debug trap at hard fault exception
9:1	Reserved	—	—	Reserved
0	VC_CORERESET	RW	0	Halt processor after system reset and before the first instruction executed

Breakpoint Unit

The breakpoint unit contains up to four comparators for instruction breakpoints. Each comparator can produce a breakpoint for up to two instructions (if the two instructions are located in the same word address). Additional breakpoints can be implemented by inserting breakpoint instructions in the program image if the program memory can be modified.

The breakpoint unit design is configurable. Some microcontrollers might contain no breakpoint unit or a breakpoint unit with fewer than four comparators.

Address	Name	Descriptions
0xE0002000	BP_CTRL	Breakpoint Control Register—for enabling the breakpoint unit and provide information about the breakpoint unit
0xE0002008	BP_COMP0	Breakpoint Comparator Register 0
0xE000200C	BP_COMP1	Breakpoint Comparator Register 1
0xE0002010	BP_COMP2	Breakpoint Comparator Register 2
0xE0002014	BP_COMP3	Breakpoint Comparator Register 3
0xE0002FD0 to 0xE0002FFC	PIDs, CIDs	ID registers

Breakpoint Control Register (0xE0002000)

Bits	Field	Type	Reset Value	Descriptions
31:17	Reserved	—	—	Reserved
7:4	NUM_CODE	RO	0 to 4	Number of comparators
3:2	Reserved	—	—	Reserved
1	KEY	WO	—	Write Key—when there is a write operation to this register, this bit should be set to 1, otherwise the write operation is ignored
0	ENABLE	RW	0	Enable control

Breakpoint Comparator Registers (0xE0002008−0xE0002014)

Bits	Field	Type	Reset Value	Descriptions
31:30	BP_MATCH	RW	—	Breakpoint setting: 00: No breakpoint 01: Breakpoint at lower half word address 10: Breakpoint at upper half word address 11: Breakpoint at both lower and upper half word
29	Reserved	—	—	Reserved
28:2	COMP	RW	—	Compare instruction address
1	Reserved	—	—	Reserved
0	ENABLE	RW	0	Enable control for this comparator

Data Watchpoint Unit

The data watchpoint unit has two main functions:

- Setting data watchpoints
- Providing a PC sampling register for basic profiling

Before accessing the DWT, the TRCENA bit in Debug Exception and Monitor Control Register (DEMCR, address 0xE000EDFC) must be set to 1 to enable the DWT. Unlike the Data Watchpoint and Trace unit in the Cortex-M3/M4, the DWT in the Cortex-M0 does not support trace. But the programming models of its registers are mostly compatible to the DWT in ARMv7-M.

The DWT design is configurable. Some microcontrollers might contain no DWT or a DWT with just 1 comparator.

Address	Name	Descriptions
0xE0001000	DWT_CTRL	DWT Control Register—provide information about the data watchpoint unit
0xE000101C	DWT_PCSR	Program Counter Sample Register—provide current program address
0xE0001020	DWT_COMP0	Comparator Register 0
0xE0001024	DWT_MASK0	Mask Register 0
0xE0001028	DWT_FUNCTION0	Function Register 0
0xE0001030	DWT_COMP1	Comparator Register 1
0xE0001034	DWT_MASK1	Mask Register 1
0xE0001038	DWT_FUNCTION1	Function Register 1
0xE0001FD0 to 0xE0001FFC	PIDs, CIDs	ID registers

DWT Control Register (0xE0001000)

Bits	Field	Type	Reset Value	Descriptions
31:28	NUMCOMP	RO	0 to 2	Number of comparator implemented
27:0	Reserved	—	—	Reserved

Program Counter Sample Register (0xE000101C)

Bits	Field	Type	Reset Value	Descriptions
31:0	EIASAMPLE	RO	—	Execution instruction address sample; read as 0xFFFFFFFF if core is halted or if DWTENA is 0

DWT COMP0 Register and DWT COMP1 Registers (0xE0001020, 0xE0001030)

Bits	Field	Type	Reset Value	Descriptions
31:0	COMP	RW	—	Address value to compare to; the value must be aligned to the compare address range defined by the compare mask register

DWT MASK0 Register and DWT MASK1 Registers (0xE0001024, 0xE0001034)

Bits	Field	Type	Reset Value	Descriptions
31:4	Reserved	—	—	Reserved
3:0	MASK	RW	—	Mask pattern: 0000: compare mask = 0xFFFFFFFF 0001: compare mask = 0xFFFFFFFE ... 1110: compare mask = 0xFFFFC000 1111: compare mask = 0xFFFF8000

DWT FUNC0 Register and DWT FUNC1 Registers (0xE0001028, 0xE0001038)

Bits	Field	Type	Reset Value	Descriptions
31:4	Reserved	—	—	Reserved
3:0	FUNC	RW	0	Function: 0000: Disable 0100: Watchpoint on PC match 0101: Watchpoint on read address 0110: Watchpoint on write address 0111: Watchpoint on read or write address Other values: Reserved

ROM Table Registers

The ROM table is used to allow a debugger to identify available components in the system. The lowest two bits of each entry are used to indicate if the debug component is present and if there is another valid entry following in the next address in the ROM table. The rest of the bits in the ROM table contain the address offset of the debug unit from the ROM table base address:

Address	Value	Name	Descriptions
0xE00FF000	0xFFF0F003	SCS	Points to System Control Space base address 0xE000E000
0xE00FF004	0xFFF02003	DWT	Points to DW base address 0xE0001000
0xE00FF008	0xFFF03003	BPU	Points to BPU base address 0xE0002000
0xE00FF00C	0x00000000	end	End of table marker
0xE00FFFCC	0x00000001	MEMTYPE	Indicates that system memory is accessible on this memory map
0xE00FFFD0 to 0xE00FFFFC	0x000000–	IDs	Peripheral ID and component ID values (values dependent on the design versions)

Using the ROM table, the debugger can identify the debug components available as shown in Figure E.1.

The ROM table lookup can be divided into multiple stages if a system-on-chip design contains additional debug components and an extra ROM table. In such cases, the ROM table lookup can be cascaded so that the debugger can identify all the debug components available (Figure E.2).

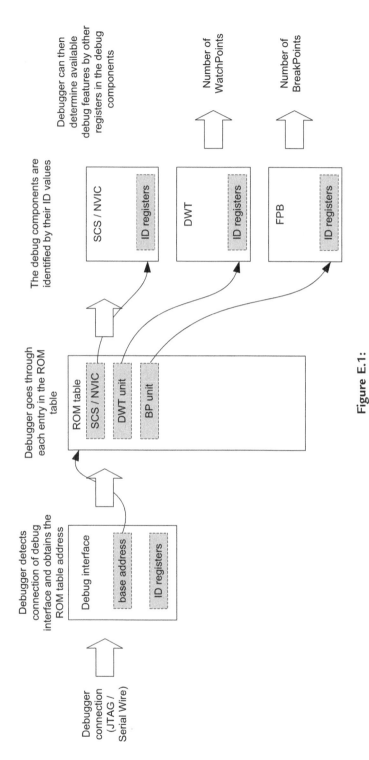

Figure E.1:

The debugger can use the ROM table to detect available debug components automatically.

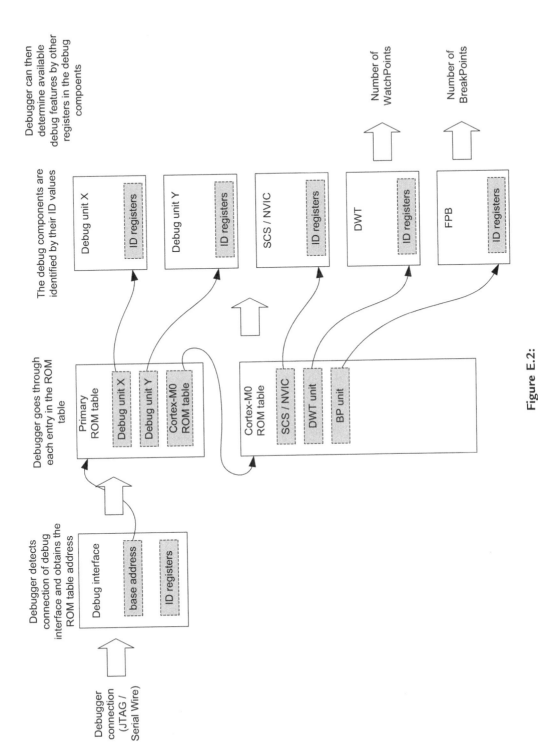

Figure E.2:
Multistage ROM table lookup when additional debug components are present.

Debug Connector Arrangement

A number of standard debug connector configurations are defined to allow in-circuit debuggers to connect to target boards easily. Most of the Cortex-M0 development boards use these standard pin-out arrangements. If you are designing your own Cortex-M0 microcontroller board, you should use one of these connector arrangements to make connection to the in-circuit debugger easier.

The 10-Pin Cortex Debug Connector

For PCB design with small size, the 0.05" pitch Cortex debug connector is ideal (Figures F.1 and F.2). The board space required is approximately 10 mm × 3 mm (the PCB header size is smaller, only 5 mm × 6 mm) and is based on the Samtec micro header.

Figure F.1:
The 10-pin Cortex debug connector.

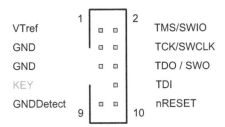

Figure F.2:
The pin out in the 10-pin Cortex debug connector.

The 10-pin Cortex debug connector supports both JTAG and serial wire protocols. The VTref is normally connected to VCC (e.g., 3.3 volt) and the nRESET signal can usually be ignored (the debugger normally resets the microcontroller using the System Reset Request feature in the AIRCR of System Control Block). The GNDDetect signal allows the in-circuit debugger to detect that it is connected to a target board. This connector arrangement is also called the CoreSight debug connector in some ARM documentation.

The 20-Pin Cortex Debug + ETM Connector

In some cases you might also find a 20-pin 0.05" pitch pin debug connector (Figures F.3 and F4). It is used in some Cortex-M3/M4 board where instruction trace is required. The header (Samtec FTSH-120) includes addition signals for trace information transfer. Although the Cortex-M0 does not support trace, some in-circuit debuggers might use this connector arrangement.

When using a Cortex-M0 microcontroller with this debug connection arrangement, you can ignore the trace signals. Both JTAG and serial wire debug protocol can be used with this debug connection arrangement.

The Legacy 20-Pin IDC Connector Arrangement

Many existing in-circuit debuggers and development boards still use the larger 20 pin IDC connector arrangement (Figures F.5 and F.6). Using a 0.1" pitch, it is easy for hobbyists to use (easy for soldering) and provides stronger mechanical support.

Figure F.3:
The 20 pin Cortex debug+ETM connector.

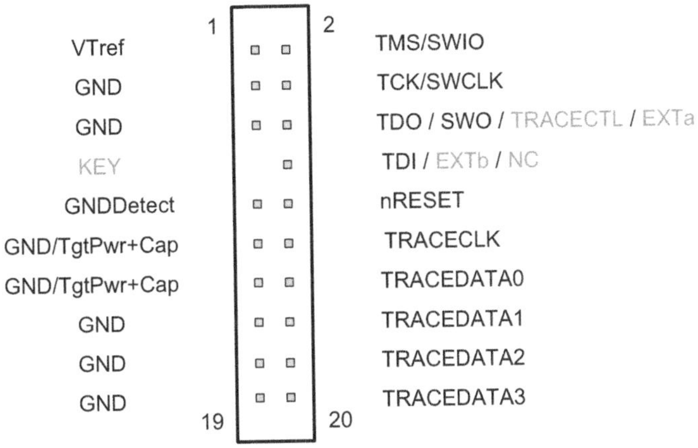

Figure F.4:
Pin out assignment for the 20-pin Cortex debug + ETM connector.

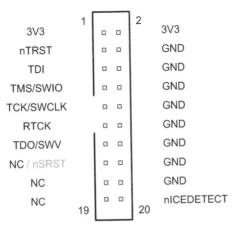

Figure F.5:
20-pin IDC connector.

Figure F.6:
Pin out assignment of 20 pin IDC debug connector.

Linker Script for CodeSourcery G++ Lite

The following linker script is modified from a generic linker script included in Sourcery G++ Lite installation (original file is generic-m.ld). This linker script requires a CS3 startup mechanism and therefore is tool chain specific. If you are using another GNU tool chain, you will have to modify the linker script to fit the requirement of the tool chain you are using.

The memory map arrangement in this script is targeted to the NXP LPC1114 microcontroller. This script is used in examples in chapter 20.

LPC1114.ld

```
/* Linker script for generic-m
 *
 * Version:Sourcery G++ Lite 2009q1-161
 * BugURL:https://support.codesourcery.com/GNUToolchain/
 *
 * Copyright 2007, 2008 CodeSourcery, Inc.
 *
 * The authors hereby grant permission to use, copy, modify, distribute,
 * and license this software and its documentation for any purpose, provided
 * that existing copyright notices are retained in all copies and that this
 * notice is included verbatim in any distributions.  No written agreement,
 * license, or royalty fee is required for any of the authorized uses.
 * Modifications to this software may be copyrighted by their authors
 * and need not follow the licensing terms described here, provided that
 * the new terms are clearly indicated on the first page of each file where
 * they apply.
 * */
OUTPUT_FORMAT ("elf32-littlearm", "elf32-bigarm", "elf32-littlearm")
ENTRY(_start)
SEARCH_DIR(.)
GROUP(-lgcc -lc -lcs3 -lcs3unhosted -lcs3micro)

MEMORY
{
  /* ROM is a readable (r), executable region (x)    */
  rom (rx)   : ORIGIN = 0, LENGTH = 32k

  /* RAM is a readable (r), writable (w) and         */
  /* executable region (x)                           */
  ram (rwx)  : ORIGIN = 0x10000000, LENGTH = 8k
}

/* These force the linker to search for particular symbols from
 * the start of the link process and thus ensure the user's
```

(Continued)

LPC1114.ld—Cont'd

```
 * overrides are picked up
 */
EXTERN(__cs3_reset_generic_m)
INCLUDE micro-names.inc
EXTERN(__cs3_interrupt_vector_micro)
EXTERN(__cs3_start_c main __cs3_stack __cs3_heap_end)

PROVIDE(__cs3_heap_start = _end);
PROVIDE(__cs3_heap_end = __cs3_region_start_ram + __cs3_region_size_ram);
PROVIDE(__cs3_region_num = (__cs3_regions_end - __cs3_regions) / 20);
PROVIDE(__cs3_stack = __cs3_region_start_ram + __cs3_region_size_ram);

SECTIONS
{

  .text :
  {
   CREATE_OBJECT_SYMBOLS
   __cs3_region_start_rom = .;
   *(.cs3.region-head.rom)
   ASSERT (. == __cs3_region_start_rom, ".cs3.region-head.rom not permitted");

   /* Vector table */
   __cs3_interrupt_vector = __cs3_interrupt_vector_micro;
   *(.cs3.interrupt_vector)  /* vector table */
   /* Make sure we pulled in an interrupt vector.  */
   ASSERT (. != __cs3_interrupt_vector_micro, "No interrupt vector");

   /* Map CS3 vector symbols to handler names in C */
   _start = __cs3_start_c;
   __cs3_reset = __cs3_start_c;

   *(.text .text.* .gnu.linkonce.t.*)
   *(.plt)
   *(.gnu.warning)
   *(.glue_7t) *(.glue_7) *(.vfp11_veneer)

   *(.ARM.extab* .gnu.linkonce.armextab.*)
   *(.gcc_except_table)
  } >rom
  .eh_frame_hdr : ALIGN (4)
  {
   KEEP (*(.eh_frame_hdr))
  } >rom
  .eh_frame : ALIGN (4)
  {
   KEEP (*(.eh_frame))
  } >rom
  /* .ARM.exidx is sorted, so has to go in its own output section.  */
```

```
  __exidx_start = .;

.ARM.exidx :
{
  *(.ARM.exidx* .gnu.linkonce.armexidx.*)
} >rom
__exidx_end = .;
.rodata : ALIGN (4)
{
  *(.rodata .rodata.* .gnu.linkonce.r.*)

. = ALIGN(4);
KEEP(*(.init))

. = ALIGN(4);
__preinit_array_start = .;
KEEP (*(.preinit_array))
__preinit_array_end = .;

. = ALIGN(4);
__init_array_start = .;
KEEP (*(SORT(.init_array.*)))
KEEP (*(.init_array))
__init_array_end = .;

. = ALIGN(4);
KEEP(*(.fini))

. = ALIGN(4);
__fini_array_start = .;
KEEP (*(.fini_array))
KEEP (*(SORT(.fini_array.*)))
__fini_array_end = .;

. = ALIGN(0x4);
KEEP (*crtbegin.o(.ctors))
KEEP (*(EXCLUDE_FILE (*crtend.o) .ctors))
KEEP (*(SORT(.ctors.*)))
KEEP (*crtend.o(.ctors))

. = ALIGN(0x4);
KEEP (*crtbegin.o(.dtors))
KEEP (*(EXCLUDE_FILE (*crtend.o) .dtors))
KEEP (*(SORT(.dtors.*)))
KEEP (*crtend.o(.dtors))

/* Add debug information
. = ALIGN(4);
__my_debug_regions = .;
LONG (__cs3_heap_start)
```

(*Continued*)

LPC1114.ld—Cont'd

```
   LONG (__cs3_heap_end)
   LONG (__cs3_stack) */

   . = ALIGN(4);
   __cs3_regions = .;
   LONG (0)
   LONG (__cs3_region_init_ram)
   LONG (__cs3_region_start_ram)
   LONG (__cs3_region_init_size_ram)
   LONG (__cs3_region_zero_size_ram)
   __cs3_regions_end = .;

   . = ALIGN (8);
   *(.rom)
   *(.rom.b)
   _etext = .;
} >rom

.data : ALIGN (8)
{
   __cs3_region_start_ram = .;
   _data = .;
   *(.cs3.region-head.ram)
   KEEP(*(.jcr))
   *(.got.plt) *(.got)
   *(.shdata)
   *(.data .data.* .gnu.linkonce.d.*)
   . = ALIGN (8);
   *(.ram)
   _edata = .;
} >ram AT>rom
.bss :
{
   _bss = .;
   *(.shbss)
   *(.bss .bss.* .gnu.linkonce.b.*)
   *(COMMON)
   . = ALIGN (8);
   *(.ram.b)
   _ebss = .;
   _end = .;
   __end = .;
} >ram AT>rom
__cs3_region_init_ram = LOADADDR (.data);
__cs3_region_init_size_ram = _edata - ADDR (.data);
__cs3_region_zero_size_ram = _end - _edata;
__cs3_region_size_ram = LENGTH(ram);


.stab 0 (NOLOAD) : { *(.stab) }
.stabstr 0 (NOLOAD) : { *(.stabstr) }
/* DWARF debug sections.
 * Symbols in the DWARF debugging sections are relative to the beginning
```

```
   * of the section so we begin them at 0.   */
  /* DWARF 1 */
  .debug           0 : { *(.debug) }
  .line            0 : { *(.line) }
  /* GNU DWARF 1 extensions */
  .debug_srcinfo  0 : { *(.debug_srcinfo) }
  .debug_sfnames  0 : { *(.debug_sfnames) }
  /* DWARF 1.1 and DWARF 2 */
  .debug_aranges  0 : { *(.debug_aranges) }
  .debug_pubnames 0 : { *(.debug_pubnames) }
  /* DWARF 2 */
  .debug_info     0 : { *(.debug_info .gnu.linkonce.wi.*) }
  .debug_abbrev   0 : { *(.debug_abbrev) }
  .debug_line     0 : { *(.debug_line) }
  .debug_frame    0 : { *(.debug_frame) }
  .debug_str      0 : { *(.debug_str) }
  .debug_loc      0 : { *(.debug_loc) }
  .debug_macinfo  0 : { *(.debug_macinfo) }
  /* DWARF 2.1 */
  .debug_ranges   0 : { *(.debug_ranges) }
  /* SGI/MIPS DWARF 2 extensions */
  .debug_weaknames 0 : { *(.debug_weaknames) }
  .debug_funcnames 0 : { *(.debug_funcnames) }
  .debug_typenames 0 : { *(.debug_typenames) }
  .debug_varnames  0 : { *(.debug_varnames) }

  .note.gnu.arm.ident 0 : { KEEP (*(.note.gnu.arm.ident)) }
  .ARM.attributes 0 : { KEEP (*(.ARM.attributes)) }
  /DISCARD/ : { *(.note.GNU-stack) }
}
```

Example Code Files

This appendix contains a number of source code files which are used in various projects in this book:

- system_LPC11xx.c (NXP LPC11xx system initialization code for CMSIS)
- system_LPC11xx.h (NXP LPC11xx system initialization code header for CMSIS)
- LPC11xx.hs (assembly header file for the assembly projects in Chapter 16)
- uart_test.s (assembly code for the UART test example in Chapter 16)
- RTX-Config.c (RTX Kernel configuration file used in examples in Chapter 18)

Appendix H.1 system_LPC11xx.c

When including the system_LPC11xx.c, you might need to edit a number of parameters to fit your design requirements. Typically these include CLOCK_SETUP, SYSCLK_SETUP, SYSOSC_SETUP, SYSPLLCTRL_Val, __XTAL, and so on.

```
system_LPC11xx.c
/*****************************************************************************//**
 * @file     system_LPC11xx.c
 * @brief    CMSIS Cortex-M0 Device Peripheral Access Layer Source File
 *           for the NXP LPC11xx Device Series
 * @version  V1.00
 * @date     17. November 2009
 *
 * @note
 * Copyright (C) 2009 ARM Limited. All rights reserved.
 *
 * @par
 * ARM Limited (ARM) is supplying this software for use with Cortex-M
 * processor based microcontrollers.  This file can be freely distributed
 * within development tools that are supporting such ARM based processors.
 *
 * @par
 * THIS SOFTWARE IS PROVIDED "AS IS".  NO WARRANTIES, WHETHER EXPRESS, IMPLIED
 * OR STATUTORY, INCLUDING, BUT NOT LIMITED TO, IMPLIED WARRANTIES OF
 * MERCHANTABILITY AND FITNESS FOR A PARTICULAR PURPOSE APPLY TO THIS SOFTWARE.
 * ARM SHALL NOT, IN ANY CIRCUMSTANCES, BE LIABLE FOR SPECIAL, INCIDENTAL, OR
 * CONSEQUENTIAL DAMAGES, FOR ANY REASON WHATSOEVER.
 *
 *****************************************************************************/
```

(Continued)

system_LPC11xx.c—Cont'd

```c
#include <stdint.h>
#include "LPC11xx.h"


/*
//-------- <<< Use Configuration Wizard in Context Menu >>> -------------------
*/


/*-------------------- Clock Configuration -----------------------------------
//
// <e> Clock Configuration
//    <e1> System Clock Setup
//      <e2> System Oscillator Enable
//        <o3.1> Select System Oscillator Frequency Range
//                    <0=> 1 - 20 MHz
//                    <1=> 15 - 25 MHz
//      </e2>
//      <e4> Watchdog Oscillator Enable
//        <o5.0..4> Select Divider for Fclkana
//                    <0=>    2 <1=>    4 <2=>    6 <3=>    8
//                    <4=>  10 <5=>  12 <6=>  14 <7=>  16
//                    <8=>  18 <9=>  20 <10=> 22 <11=> 24
//                    <12=> 26 <13=> 28 <14=> 30 <15=> 32
//                    <16=> 34 <17=> 36 <18=> 38 <19=> 40
//                    <20=> 42 <21=> 44 <22=> 46 <23=> 48
//                    <24=> 50 <25=> 52 <26=> 54 <27=> 56
//                    <28=> 58 <29=> 60 <30=> 62 <31=> 64
//        <o5.5..8> Select Watchdog Oscillator Analog Frequency (Fclkana)
//                    <0=> Disabled
//                    <1=>  0.5 MHz
//                    <2=>  0.8 MHz
//                    <3=>  1.1 MHz
//                    <4=>  1.4 MHz
//                    <5=>  1.6 MHz
//                    <6=>  1.8 MHz
//                    <7=>  2.0 MHz
//                    <8=>  2.2 MHz
//                    <9=>  2.4 MHz
//                    <10=> 2.6 MHz
//                    <11=> 2.7 MHz
//                    <12=> 2.9 MHz
//                    <13=> 3.1 MHz
//                    <14=> 3.2 MHz
//                    <15=> 3.4 MHz
//      </e4>
//      <o6> Select Input Clock for sys_pllclkin (Register: SYSPLLCLKSEL)
//                    <0=> IRC Oscillator
//                    <1=> System Oscillator
//                    <2=> WDT Oscillator
//                    <3=> Invalid
```

```
//      <e7> Use System PLL
//                    <i> F_pll = M * F_in
//                    <i> F_in must be in the range of 10 MHz to 25 MHz
//      <o8.0..4>   M: PLL Multiplier Selection
//                    <1-32><#-1>
//      <o8.5..6>   P: PLL Divider Selection
//                    <0=> 2
//                    <1=> 4
//                    <2=> 8
//                    <3=> 16
//      <o8.7>      DIRECT: Direct CCO Clock Output Enable
//      <o8.8>      BYPASS: PLL Bypass Enable
//      </e7>
//      <o9> Select Input Clock for Main clock (Register: MAINCLKSEL)
//                    <0=> IRC Oscillator
//                    <1=> Input Clock to System PLL
//                    <2=> WDT Oscillator
//                    <3=> System PLL Clock Out
//    </e1>
//    <o10.0..7> System AHB Divider <0-255>
//                    <i> 0 = is disabled
//    <o11.0>   SYS Clock Enable
//    <o11.1>   ROM Clock Enable
//    <o11.2>   RAM Clock Enable
//    <o11.3>   FLASHREG Flash Register Interface Clock Enable
//    <o11.4>   FLASHARRAY Flash Array Access Clock Enable
//    <o11.5>   I2C Clock Enable
//    <o11.6>   GPIO Clock Enable
//    <o11.7>   CT16B0 Clock Enable
//    <o11.8>   CT16B1 Clock Enable
//    <o11.9>   CT32B0 Clock Enable
//    <o11.10>  CT32B1 Clock Enable
//    <o11.11>  SSP0 Clock Enable
//    <o11.12>  UART Clock Enable
//    <o11.13>  ADC Clock Enable
//    <o11.15>  WDT Clock Enable
//    <o11.16>  IOCON Clock Enable
//    <o11.18>  SSP1 Clock Enable
//
//    <o12.0..7> SSP0 Clock Divider <0-255>
//                    <i> 0 = is disabled
//    <o13.0..7> UART Clock Divider <0-255>
//                    <i> 0 = is disabled
//    <o14.0..7> SSP1 Clock Divider <0-255>
//                    <i> 0 = is disabled
// </e>
*/
```

(*Continued*)

system_LPC11xx.c—Cont'd

```
#define CLOCK_SETUP            1
#define SYSCLK_SETUP           1
#define SYSOSC_SETUP           1
#define SYSOSCCTRL_Val         0x00000000
#define WDTOSC_SETUP           0
#define WDTOSCCTRL_Val         0x000000A0
#define SYSPLLCLKSEL_Val       0x00000001
#define SYSPLL_SETUP           1
#define SYSPLLCTRL_Val         0x00000023
#define MAINCLKSEL_Val         0x00000003
#define SYSAHBCLKDIV_Val       0x00000001
#define AHBCLKCTRL_Val         0x0001005F
#define SSP0CLKDIV_Val         0x00000001
#define UARTCLKDIV_Val         0x00000001
#define SSP1CLKDIV_Val         0x00000001


/*-------------------- Memory Mapping Configuration ------------------------
//
// <e> Memory Mapping
//   <o1.0..1> System Memory Remap (Register: SYSMEMREMAP)
//                     <0=> Bootloader mapped to address 0
//                     <1=> RAM mapped to address 0
//                     <2=> Flash mapped to address 0
//                     <3=> Flash mapped to address 0
// </e>
*/
#define MEMMAP_SETUP           0
#define SYSMEMREMAP_Val        0x00000001


/*
//-------- <<< end of configuration section >>> -------------------------------
*/


/*------------------------------------------------------------------------------
  Check the register settings
 *-----------------------------------------------------------------------------*/
#define CHECK_RANGE(val, min, max)                   ((val < min) || (val > max))
#define CHECK_RSVD(val, mask)                        (val & mask)

/* Clock Configuration -------------------------------------------------------*/
#if (CHECK_RSVD((SYSOSCCTRL_Val),  ~0x00000003))
   #error "SYSOSCCTRL: Invalid values of reserved bits!"
#endif


#if (CHECK_RSVD((WDTOSCCTRL_Val),  ~0x000001FF))
   #error "WDTOSCCTRL: Invalid values of reserved bits!"
#endif


#if (CHECK_RANGE((SYSPLLCLKSEL_Val), 0, 2))
   #error "SYSPLLCLKSEL: Value out of range!"
```

```
#endif
#if (CHECK_RSVD((SYSPLLCTRL_Val),  ~0x000001FF))
   #error "SYSPLLCTRL: Invalid values of reserved bits!"
#endif


#if (CHECK_RSVD((MAINCLKSEL_Val),  ~0x00000003))
   #error "MAINCLKSEL: Invalid values of reserved bits!"
#endif


#if (CHECK_RANGE((SYSAHBCLKDIV_Val), 0, 255))
   #error "SYSAHBCLKDIV: Value out of range!"
#endif


#if (CHECK_RSVD((AHBCLKCTRL_Val),  ~0x0001FFFF))
   #error "AHBCLKCTRL: Invalid values of reserved bits!"
#endif


#if (CHECK_RANGE((SSP0CLKDIV_Val), 0, 255))
   #error "SSP0CLKDIV: Value out of range!"
#endif


#if (CHECK_RANGE((UARTCLKDIV_Val), 0, 255))
   #error "UARTCLKDIV: Value out of range!"
#endif


#if (CHECK_RANGE((SSP1CLKDIV_Val), 0, 255))
   #error "SSP1CLKDIV: Value out of range!"
#endif


#if (CHECK_RSVD((SYSMEMREMAP_Val), ~0x00000003))
   #error "SYSMEMREMAP: Invalid values of reserved bits!"
#endif



/*-------------------------------------------------------------------------
  DEFINES
 *-----------------------------------------------------------------------*/


/*-------------------------------------------------------------------------
  Define clocks
 *-----------------------------------------------------------------------*/
#define __XTAL            (12000000UL)    /* Oscillator frequency          */
#define __SYS_OSC_CLK     (    __XTAL)    /* Main oscillator frequency     */
#define __IRC_OSC_CLK     (12000000UL)    /* Internal RC oscillator frequency */


#define __FREQSEL   ((WDTOSCCTRL_Val >> 5) & 0x0F)
#define __DIVSEL    (((WDTOSCCTRL_Val & 0x1F) << 1) + 2)

#if (CLOCK_SETUP)                         /* Clock Setup                 */
  #if (SYSCLK_SETUP)                      /* System Clock Setup          */
    #if (WDTOSC_SETUP)                    /* Watchdog Oscillator Setup*/
```

(*Continued*)

system_LPC11xx.c—Cont'd

```c
      #if  (__FREQSEL ==   0)
        #define __WDT_OSC_CLK           ( 400000 / __DIVSEL)
      #elif (__FREQSEL ==   1)
        #define __WDT_OSC_CLK           ( 500000 / __DIVSEL)
      #elif (__FREQSEL ==   2)
        #define __WDT_OSC_CLK           ( 800000 / __DIVSEL)
      #elif (__FREQSEL ==   3)
        #define __WDT_OSC_CLK           (1100000 / __DIVSEL)
      #elif (__FREQSEL ==   4)
        #define __WDT_OSC_CLK           (1400000 / __DIVSEL)
      #elif (__FREQSEL ==   5)
        #define __WDT_OSC_CLK           (1600000 / __DIVSEL)
      #elif (__FREQSEL ==   6)
        #define __WDT_OSC_CLK           (1800000 / __DIVSEL)
      #elif (__FREQSEL ==   7)
        #define __WDT_OSC_CLK           (2000000 / __DIVSEL)
      #elif (__FREQSEL ==   8)
        #define __WDT_OSC_CLK           (2200000 / __DIVSEL)
      #elif (__FREQSEL ==   9)
        #define __WDT_OSC_CLK           (2400000 / __DIVSEL)
      #elif (__FREQSEL ==  10)
        #define __WDT_OSC_CLK           (2600000 / __DIVSEL)
      #elif (__FREQSEL ==  11)
        #define __WDT_OSC_CLK           (2700000 / __DIVSEL)
      #elif (__FREQSEL ==  12)
        #define __WDT_OSC_CLK           (2900000 / __DIVSEL)
      #elif (__FREQSEL ==  13)
        #define __WDT_OSC_CLK           (3100000 / __DIVSEL)
      #elif (__FREQSEL ==  14)
        #define __WDT_OSC_CLK           (3200000 / __DIVSEL)
      #else
        #define __WDT_OSC_CLK           (3400000 / __DIVSEL)
      #endif
#else
        #define __WDT_OSC_CLK           (1600000 / 2)
#endif  // WDTOSC_SETUP

/* sys_pllclkin calculation */
#if   ((SYSPLLCLKSEL_Val & 0x03) == 0)
  #define __SYS_PLLCLKIN                (__IRC_OSC_CLK)
#elif ((SYSPLLCLKSEL_Val & 0x03) == 1)
  #define __SYS_PLLCLKIN                (__SYS_OSC_CLK)
#elif ((SYSPLLCLKSEL_Val & 0x03) == 2)
  #define __SYS_PLLCLKIN                (__WDT_OSC_CLK)
#else
  #define __SYS_PLLCLKIN                (0)
#endif

#if (SYSPLL_SETUP)                      /* System PLL Setup        */
  #define __SYS_PLLCLKOUT      (__SYS_PLLCLKIN * ((SYSPLLCTRL_Val & 0x01F)
```

```
 + 1))
  #else
    #define __SYS_PLLCLKOUT          (__SYS_PLLCLKIN * (1))
  #endif  // SYSPLL_SETUP

  /* main clock calculation */
  #if   ((MAINCLKSEL_Val & 0x03) == 0)
    #define __MAIN_CLOCK            (__IRC_OSC_CLK)
  #elif ((MAINCLKSEL_Val & 0x03) == 1)
    #define __MAIN_CLOCK            (__SYS_PLLCLKIN)
  #elif ((MAINCLKSEL_Val & 0x03) == 2)
    #define __MAIN_CLOCK            (__WDT_OSC_CLK)
  #elif ((MAINCLKSEL_Val & 0x03) == 3)
    #define __MAIN_CLOCK            (__SYS_PLLCLKOUT)
  #else
    #define __MAIN_CLOCK            (0)
  #endif

  #define __SYSTEM_CLOCK             (__MAIN_CLOCK / SYSAHBCLKDIV_Val)

  #else  // SYSCLK_SETUP
    #if (SYSAHBCLKDIV_Val == 0)
      #define __SYSTEM_CLOCK        (0)
    #else
      #define __SYSTEM_CLOCK        (__XTAL / SYSAHBCLKDIV_Val)
    #endif
  #endif  // SYSCLK_SETUP

#else
  #define __SYSTEM_CLOCK            (__XTAL)
#endif  // CLOCK_SETUP

/*----------------------------------------------------------------------
  Clock Variable definitions
 *---------------------------------------------------------------------*/
uint32_t SystemCoreClock = __SYSTEM_CLOCK;/*!< System Clock Frequency (Core
Clock)*/


/*----------------------------------------------------------------------
  Clock functions
 *---------------------------------------------------------------------*/
void SystemCoreClockUpdate (void)                /* Get Core Clock Frequency    */
{
  uint32_t wdt_osc = 0;

  /* Determine clock frequency according to clock register values            */
  switch ((LPC_SYSCON->WDTOSCCTRL >> 5) & 0x0F) {
    case 0:  wdt_osc =  400000; break;
    case 1:  wdt_osc =  500000; break;
    case 2:  wdt_osc =  800000; break;
    case 3:  wdt_osc = 1100000; break;
```

(*Continued*)

system_LPC11xx.c—Cont'd

```
  case 4:  wdt_osc = 1400000; break;
  case 5:  wdt_osc = 1600000; break;
  case 6:  wdt_osc = 1800000; break;
  case 7:  wdt_osc = 2000000; break;
  case 8:  wdt_osc = 2200000; break;
  case 9:  wdt_osc = 2400000; break;
  case 10: wdt_osc = 2600000; break;
  case 11: wdt_osc = 2700000; break;
  case 12: wdt_osc = 2900000; break;
  case 13: wdt_osc = 3100000; break;
  case 14: wdt_osc = 3200000; break;
  case 15: wdt_osc = 3400000; break;
}
wdt_osc /= ((LPC_SYSCON->WDTOSCCTRL & 0x1F) << 1) + 2;

switch (LPC_SYSCON->MAINCLKSEL & 0x03) {
  case 0:                             /* Internal RC oscillator          */
    SystemCoreClock = __IRC_OSC_CLK;
    break;
  case 1:                             /* Input Clock to System PLL       */
    switch (LPC_SYSCON->SYSPLLCLKSEL & 0x03) {
        case 0:                       /* Internal RC oscillator          */
          SystemCoreClock = __IRC_OSC_CLK;
          break;
        case 1:                       /* System oscillator               */
          SystemCoreClock = __SYS_OSC_CLK;
          break;
        case 2:                       /* WDT Oscillator                  */
          SystemCoreClock = wdt_osc;
          break;
        case 3:                       /* Reserved                        */
          SystemCoreClock = 0;
          break;
    }
   break;
  case 2:                             /* WDT Oscillator                  */
    SystemCoreClock = wdt_osc;
    break;
  case 3:                             /* System PLL Clock Out            */
    switch (LPC_SYSCON->SYSPLLCLKSEL & 0x03) {
        case 0:                       /* Internal RC oscillator          */
          if (LPC_SYSCON->SYSPLLCTRL & 0x180) {
            SystemCoreClock = __IRC_OSC_CLK;
          } else {
            SystemCoreClock = __IRC_OSC_CLK * ((LPC_SYSCON->SYSPLLCTRL & 0x01F) +
1);
          }
          break;
        case 1:                       /* System oscillator               */
          if (LPC_SYSCON->SYSPLLCTRL & 0x180) {
            SystemCoreClock = __SYS_OSC_CLK;
```

```
                SystemCoreClock = __SYS_OSC_CLK;
            } else {
                SystemCoreClock = __SYS_OSC_CLK * ((LPC_SYSCON->SYSPLLCTRL & 0x01F) +
1);
            }
          break;
        case 2:                            /* WDT Oscillator                */
          if (LPC_SYSCON->SYSPLLCTRL & 0x180) {
            SystemCoreClock = wdt_osc;
          } else {
            SystemCoreClock = wdt_osc * ((LPC_SYSCON->SYSPLLCTRL & 0x01F) + 1);
          }
          break;
        case 3:                            /* Reserved                      */
          SystemCoreClock = 0;
          break;
      }
      break;
  }

  SystemCoreClock /= LPC_SYSCON->SYSAHBCLKDIV;

}

/**
 * Initialize the system
 *
 * @param  none
 * @return none
 *
 * @brief  Setup the microcontroller system.
 *         Initialize the System.
 */
void SystemInit (void)
{
#if (CLOCK_SETUP)                                  /* Clock Setup            */
#if (SYSCLK_SETUP)                                 /* System Clock Setup     */
#if (SYSOSC_SETUP)                                 /* System Oscillator Setup */
  uint32_t i;

  LPC_SYSCON->PDRUNCFG       &= ~(1 << 5);         /* Power-up System Osc    */
  LPC_SYSCON->SYSOSCCTRL     = SYSOSCCTRL_Val;
  for (i = 0; i < 200; i++) __NOP();
  LPC_SYSCON->SYSPLLCLKSEL   = SYSPLLCLKSEL_Val;   /* Select PLL Input       */
  LPC_SYSCON->SYSPLLCLKUEN   = 0x01;               /* Update Clock Source    */
  LPC_SYSCON->SYSPLLCLKUEN   = 0x00;               /* Toggle Update Register */
  LPC_SYSCON->SYSPLLCLKUEN   = 0x01;
  while (!(LPC_SYSCON->SYSPLLCLKUEN & 0x01));      /* Wait Until Updated     */
#if (SYSPLL_SETUP)                                 /* System PLL Setup       */
  LPC_SYSCON->SYSPLLCTRL     = SYSPLLCTRL_Val;
  LPC_SYSCON->PDRUNCFG       &= ~(1 << 7);         /* Power-up SYSPLL        */
```

(Continued)

```
system_LPC11xx.c—Cont'd
  while (!(LPC_SYSCON->SYSPLLSTAT & 0x01));          /* Wait Until PLL Locked    */
#endif
#endif
#if (WDTOSC_SETUP)                                   /* Watchdog Oscillator Setup*/
  LPC_SYSCON->WDTOSCCTRL     = WDTOSCCTRL_Val;
  LPC_SYSCON->PDRUNCFG      &= ~(1 << 6);            /* Power-up WDT Clock       */
#endif
  LPC_SYSCON->MAINCLKSEL     = MAINCLKSEL_Val;       /* Select PLL Clock Output  */
  LPC_SYSCON->MAINCLKUEN     = 0x01;                 /* Update MCLK Clock Source */
  LPC_SYSCON->MAINCLKUEN     = 0x00;                 /* Toggle Update Register   */
  LPC_SYSCON->MAINCLKUEN     = 0x01;
  while (!(LPC_SYSCON->MAINCLKUEN & 0x01));          /* Wait Until Updated       */
#endif

  LPC_SYSCON->SYSAHBCLKDIV  = SYSAHBCLKDIV_Val;
  LPC_SYSCON->SYSAHBCLKCTRL = AHBCLKCTRL_Val;
  LPC_SYSCON->SSP0CLKDIV    = SSP0CLKDIV_Val;
  LPC_SYSCON->UARTCLKDIV    = UARTCLKDIV_Val;
  LPC_SYSCON->SSP1CLKDIV    = SSP1CLKDIV_Val;
#endif


#if (MEMMAP_SETUP || MEMMAP_INIT)     /* Memory Mapping Setup              */
  LPC_SYSCON->SYSMEMREMAP = SYSMEMREMAP_Val;
#endif
}
```

Appendix H.2 system_LPC11xx.h

This file is required for declaration of the system initialization function. It is normally included in the Keil MDK-ARM installation.

```
system_LPC11xx.h
/**********************************************************************//**
 * @file     system_LPC11xx.h
 * @brief    CMSIS Cortex-M0 Device Peripheral Access Layer Header File
 *           for the NXP LPC11xx Device Series
 * @version  V1.00
 * @date     17. November 2009
 *
 * @note
 * Copyright (C) 2009 ARM Limited. All rights reserved.
 *
 * @par
 * ARM Limited (ARM) is supplying this software for use with Cortex-M
 * processor based microcontrollers.  This file can be freely distributed
```

```
 * within development tools that are supporting such ARM based processors.
 *
 * @par
 * THIS SOFTWARE IS PROVIDED "AS IS".  NO WARRANTIES, WHETHER EXPRESS, IMPLIED
 * OR STATUTORY, INCLUDING, BUT NOT LIMITED TO, IMPLIED WARRANTIES OF
 * MERCHANTABILITY AND FITNESS FOR A PARTICULAR PURPOSE APPLY TO THIS SOFTWARE.
 * ARM SHALL NOT, IN ANY CIRCUMSTANCES, BE LIABLE FOR SPECIAL, INCIDENTAL, OR
 * CONSEQUENTIAL DAMAGES, FOR ANY REASON WHATSOEVER.
 *
 ******************************************************************************/


#ifndef __SYSTEM_LPC11xx_H
#define __SYSTEM_LPC11xx_H

#ifdef __cplusplus
extern "C" {
#endif

#include <stdint.h>

extern uint32_t SystemCoreClock;     /*!< System Clock Frequency (Core Clock)  */


/**
 * Initialize the system
 *
 * @param  none
 * @return none
 *
 * @brief  Setup the microcontroller system.
 *         Initialize the System and update the SystemCoreClock variable.
 */
extern void SystemInit (void);

/**
 * Update SystemCoreClock variable
 *
 * @param  none
 * @return none
 *
 * @brief  Updates the SystemCoreClock with current core Clock
 *         retrieved from cpu registers.
 */
extern void SystemCoreClockUpdate (void);
#ifdef __cplusplus
}
#endif

#endif /* __SYSTEM_LPC11x_H */
```

Appendix H.3 LPC11xx.hs

This file is for the assembly programming example in Chapter 16. It provides register name definitions for the processor peripherals as well as LPC11xx peripherals.

(The syntax of this assembly header file is for the KEIL MDK-ARM or the ARM RealView Development Suite.)

```
LPC11xx.hs
; Base Address Constant and register offset definition
; Cortex-M0
SCS_BASE                EQU     0xE000E000
SysTick_BASE            EQU     0xE000E010
NVIC_BASE               EQU     0xE000E100
SCB_BASE                EQU     0xE000ED00


; Core peripheral address offset
; 1) SysTick (use SysTick_BASE)
SysTick_CTRL            EQU     0x000
SysTick_LOAD            EQU     0x004
SysTick_VAL             EQU     0x008
SysTick_CALIB           EQU     0x00C


; 2) NVIC (use NVIC_BASE)
NVIC_ISER               EQU     0x000   ; 0xE000E100
NVIC_ICER               EQU     0x080   ; 0xE000E180
NVIC_ISPR               EQU     0x100   ; 0xE000E200
NVIC_ICPR               EQU     0x180   ; 0xE000E280
NVIC_IPR0               EQU     0x300   ; 0xE000E400


; 3) SCB (use SCB_BASE)
SCB_CPUID               EQU     0x000
SCB_ICSR                EQU     0x004
SCB_AIRCR               EQU     0x00C
SCB_SCR                 EQU     0x010
SCB_CCR                 EQU     0x014
SCB_SHP0                EQU     0x01C
SCB_SHP1                EQU     0x020
SCB_SHCSR               EQU     0x024
SCB_DFSR                EQU     0x030


; LPC111x specific peripheral base address
; APB0 peripherals
LPC_I2C_BASE            EQU     0x40000000
LPC_WDT_BASE            EQU     0x40004000
LPC_UART_BASE           EQU     0x40008000
LPC_CT16B0_BASE         EQU     0x4000C000
LPC_CT16B1_BASE         EQU     0x40010000
LPC_CT32B0_BASE         EQU     0x40014000
LPC_CT32B1_BASE         EQU     0x40018000
LPC_ADC_BASE            EQU     0x4001C000
```

```
LPC_PMU_BASE          EQU    0x40038000
LPC_SSP0_BASE         EQU    0x40040000
LPC_GPIO1_BASE        EQU    0x50010000
LPC_GPIO2_BASE        EQU    0x50020000
LPC_GPIO3_BASE        EQU    0x50030000


; AHB peripherals
LPC_GPIO_BASE         EQU    0x50000000
LPC_GPIO0_BASE        EQU    0x50000000
LPC_IOCON_BASE        EQU    0x40044000
LPC_SYSCON_BASE       EQU    0x40048000
LPC_SSP1_BASE         EQU    0x40058000


; Register offset
; 1) SYSCON
SYSMEMREMAP           EQU    0x000
PRESETCTRL            EQU    0x004
SYSPLLCTRL            EQU    0x008
SYSPLLSTAT            EQU    0x00C
SYSOSCCTRL            EQU    0x020
WDTOSCCTRL            EQU    0x024
IRCCTRL               EQU    0x028
SYSRESSTAT            EQU    0x030
SYSPLLCLKSEL          EQU    0x040
SYSPLLCLKUEN          EQU    0x044
MAINCLKSEL            EQU    0x070
MAINCLKUEN            EQU    0x074
SYSAHBCLKDIV          EQU    0x078
SYSAHBCLKCTRL         EQU    0x080
SSP0CLKDIV            EQU    0x094
UARTCLKDIV            EQU    0x098
SSP1CLKDIV            EQU    0x09C
SYSTICKCLKDIV         EQU    0x0B0
WDTCLKSEL             EQU    0x0D0
WDTCLKUEN             EQU    0x0D4
WDTCLKDIV             EQU    0x0D8
CLKOUTCLKSEL          EQU    0x0E0
CLKOUTUEN             EQU    0x0E4
CLKOUTDIV             EQU    0x0E8
PIOPORCAP0            EQU    0x100
PIOPORCAP1            EQU    0x104
BODCTRL               EQU    0x150
SYSTCKCAL             EQU    0x158
STARTAPRP0            EQU    0x200
STARTERP0             EQU    0x204
STARTRSRP0CLR         EQU    0x208
STARTSRP0             EQU    0x20C

PDSLEEPCFG            EQU    0x230
PDAWAKECFG            EQU    0x234
PDRUNCFG              EQU    0x238

DEVICE_ID             EQU    0x3F4
```

(Continued)

LPC11xx.hs—Cont'd

```
; 2) IOCON register offset
PIO2_6                     EQU      0x000
PIO2_0                     EQU      0x008
PIO0_1                     EQU      0x010
PIO1_8                     EQU      0x014
PIO0_2                     EQU      0x01C

PIO2_7                     EQU      0x020
PIO2_8                     EQU      0x024
PIO2_1                     EQU      0x028
PIO0_3                     EQU      0x02C
PIO0_4                     EQU      0x030
PIO0_5                     EQU      0x034
PIO1_9                     EQU      0x038
PIO3_4                     EQU      0x03C

PIO2_4                     EQU      0x040
PIO2_5                     EQU      0x044
PIO3_5                     EQU      0x048
PIO0_6                     EQU      0x04C
PIO0_7                     EQU      0x050
PIO2_9                     EQU      0x054
PIO2_10                    EQU      0x058
PIO2_2                     EQU      0x05C

PIO0_8                     EQU      0x060
PIO0_9                     EQU      0x064
JTAG_TCK_PIO0_10           EQU      0x068
PIO1_10                    EQU      0x06C
PIO2_11                    EQU      0x070
JTAG_TDI_PIO0_11           EQU      0x074
JTAG_TMS_PIO1_0            EQU      0x078
JTAG_TDO_PIO1_1            EQU      0x07C
JTAG_nTRST_PIO1_2          EQU      0x080
PIO3_0                     EQU      0x084
PIO3_1                     EQU      0x08C
PIO2_3                     EQU      0x08C
ARM_SWDIO_PIO1_3           EQU      0x090
PIO1_4                     EQU      0x094
PIO1_11                    EQU      0x098
PIO3_2                     EQU      0x09C

PIO1_5                     EQU      0x0A0
PIO1_6                     EQU      0x0A4
PIO1_7                     EQU      0x0A8
PIO3_3                     EQU      0x0AC
SCK_LOC                    EQU      0x0B0
DSR_LOC                    EQU      0x0B4
DCD_LOC                    EQU      0x0B8
RI_LOC                     EQU      0x0BC
```

```
; 3) PMU
PMU_PCON                EQU     0x000
PMU_GPREG0              EQU     0x004
PMU_GPREG1              EQU     0x008
PMU_GPREG2              EQU     0x00C
PMU_GPREG3              EQU     0x010
PMU_GPREG4              EQU     0x014


; 4) GPIO
; Most control registers are start from offset 0x8000
LPC_GPIO0_REGBASE       EQU     LPC_GPIO0_BASE + 0x8000
LPC_GPIO1_REGBASE       EQU     LPC_GPIO1_BASE + 0x8000
LPC_GPIO2_REGBASE       EQU     LPC_GPIO2_BASE + 0x8000
LPC_GPIO3_REGBASE       EQU     LPC_GPIO3_BASE + 0x8000


; Since data register offset is large, use fixed constant
; instead.
LPC_GPIO0_DATA          EQU     0x50000000 + 0x3FFC
LPC_GPIO1_DATA          EQU     0x50010000 + 0x3FFC
LPC_GPIO2_DATA          EQU     0x50020000 + 0x3FFC
LPC_GPIO3_DATA          EQU     0x50030000 + 0x3FFC


; Masked access use LPC_GPIO0/1/2/3_BASE
GPIO_MASKED_ACCESS_BASE    EQU     0x000


; Following are control register offset
; (Use LPC_GPIO0/1/2/3_REGBASE)
GPIO_DIR                EQU     0x000 ; 0x8000
GPIO_IS                 EQU     0x004 ; 0x8004
GPIO_IBE                EQU     0x008 ; 0x8008
GPIO_IEV                EQU     0x00C ; 0x800C
GPIO_IE                 EQU     0x010 ; 0x8010
GPIO_RIS                EQU     0x014 ; 0x8014
GPIO_MIS                EQU     0x018 ; 0x8018
GPIO_IC                 EQU     0x01C ; 0x801C

; 5) Timer register offset
TIMER_IR                EQU     0x000
TIMER_TCR               EQU     0x004
TIMER_TC                EQU     0x008
TIMER_PR                EQU     0x00C
TIMER_PC                EQU     0x010
TIMER_MCR               EQU     0x014
TIMER_MR0               EQU     0x018
TIMER_MR1               EQU     0x01C
TIMER_MR2               EQU     0x020
TIMER_MR3               EQU     0x024
TIMER_CCR               EQU     0x028
TIMER_CR0               EQU     0x02C
TIMER_EMR               EQU     0x03C
TIMER_CTCR              EQU     0x070
TIMER_PWMC              EQU     0x074
```

(*Continued*)

LPC11xx.hs—Cont'd

```
; 6) UART register offset
UART_RBR              EQU       0x000
UART_THR              EQU       0x000
UART_DLL              EQU       0x000
UART_DLM              EQU       0x004
UART_IER              EQU       0x004
UART_IIR              EQU       0x008
UART_FCR              EQU       0x008
UART_LCR              EQU       0x00C
UART_MCR              EQU       0x010
UART_LSR              EQU       0x014
UART_MSR              EQU       0x018
UART_SCR              EQU       0x01C
UART_ACR              EQU       0x020
UART_FDR              EQU       0x028
UART_TER              EQU       0x030
UART_RS485CTRL        EQU       0x04C
UART_ADRMATCH         EQU       0x050
UART_RS485DLY         EQU       0x054
UART_FIFOLVL          EQU       0x058


; 7) SSP register offset
SSP_CR0               EQU       0x000
SSP_CR1               EQU       0x004
SSP_DR                EQU       0x008
SSP_SR                EQU       0x00C
SSP_CPSR              EQU       0x010
SSP_IMSC              EQU       0x014
SSP_RIS               EQU       0x018
SSP_MIS               EQU       0x01C
SSP_ICR               EQU       0x020


; 8) I2C register offset
I2C_CONSET            EQU       0x000
I2C_STAT              EQU       0x004
I2C_DAT               EQU       0x008
I2C_ADR0              EQU       0x00C
I2C_SCLH              EQU       0x010
I2C_SCLL              EQU       0x014
I2C_CONCLR            EQU       0x018
I2C_MMCTRL            EQU       0x01C
I2C_ADR1              EQU       0x020
I2C_ADR2              EQU       0x024
I2C_ADR3              EQU       0x028
I2C_DATA_BUFFER       EQU       0x02C
I2C_MASK0             EQU       0x030
I2C_MASK1             EQU       0x034
I2C_MASK2             EQU       0x038
I2C_MASK3             EQU       0x03C
```

```
; 9) Watchdog
WDT_MOD                 EQU     0x000
WDT_TC                  EQU     0x004
WDT_FEED                EQU     0x008
WDT_TV                  EQU     0x00C

; 10) ADC register offset
ADC_CR                  EQU     0x000
ADC_GDR                 EQU     0x004
ADC_INTEN               EQU     0x00C
ADC_DR0                 EQU     0x010
ADC_DR1                 EQU     0x014
ADC_DR2                 EQU     0x018
ADC_DR3                 EQU     0x01C
ADC_DR4                 EQU     0x020
ADC_DR5                 EQU     0x024
ADC_DR6                 EQU     0x028
ADC_DR7                 EQU     0x02C
ADC_STAT                EQU     0x030
        END
```

Appendix H.4 uart_test.s

The code that follows is a full assembly listing for the UART project in Chapter 16.

```
uart_test.s

        PRESERVE8 ; Indicate the code here preserve
                  ; 8 byte stack alignment
        THUMB     ; Indicate THUMB code is used
        AREA    |.text|, CODE, READONLY   ; Start of CODE area

        INCLUDE LPC11xx.hs
        EXPORT UartTest
UartTest   FUNCTION
        BL Set48MHzClock       ; Switch Clock to 48MHz
        BL UartConfig          ; Initialize UART

        ; UartPuts ("Hello\n");
        LDR     R0,=Hello_message ; start address of hello message
        BL      UartPuts       ; print message
UartTest_loop
        BL      UartGetRxDataAvail ; check if data is received
        CMP     R0, #0
        BEQ     UartTest_loop ; Return value =0 (no data), try again
        BL      UartGetRxData ; Get receive data in R0
        BL      UartPutc      ; output receive data
```

(Continued)

uart_test.s—Cont'd

```
        B          UartTest_loop
Hello_message
        DCB        "Hello\n", 0
        ALIGN
        ENDFUNC
; -------------------------------------------------------
UartGetRxDataAvail FUNCTION
        ; Return 1 if data is received
        LDR        R0,=LPC_UART_BASE
        LDR        R0, [R0, #UART_LSR]
        MOVS       R1, #1
        ANDS       R0, R0, R1
        BX         LR
        ENDFUNC
; -------------------------------------------------------
UartGetRxData FUNCTION
        ; Return received data in R0
        LDR        R0,=LPC_UART_BASE
        LDR        R0, [R0, #UART_RBR]
        BX         LR
        ENDFUNC
; -------------------------------------------------------
UartPutc FUNCTION
        ; Print a character to UART
        ; Input R0 - value of character to be printed
        LDR        R1,=LPC_UART_BASE
        MOVS       R3, #0x20  ; LSR.transmit holding register empty
        CMP        R0,#10 ; new line character
        BNE        UartPutc_wait_2

        ; while ((LPC_UART->LSR & (1<<5))==0);
UartPutc_wait_1
        LDR        R2, [R1,#UART_LSR]
        TST        R2, R3 ; Check transmit empty
        BEQ        UartPutc_wait_1
        ; LPC_UART->THR = 13 (for correct display on Hyperterminal)
        MOVS       R2, #13 ; carriage return
        STR        R2, [R1, #UART_THR]

UartPutc_wait_2
        LDR        R2, [R1,#UART_LSR]
        TST        R2, R3 ; Check transmit empty
        BEQ        UartPutc_wait_2
        ; LPC_UART->THR = data
        STR        R0, [R1, #UART_THR]

        CMP        R0,#13 ; new line character
        BNE        UartPutc_end
        ; while ((LPC_UART->LSR & (1<<5))==0);
UartPutc_wait_3
```

```
            LDR     R2, [R1,#UART_LSR]
            TST     R2, R3 ; Check transmit empty
            BEQ     UartPutc_wait_3
            ; LPC_UART->THR = 10 (for correct display on Hyperterminal)
            MOVS    R2, #10 ; new line
            STR     R2, [R1, #UART_THR]
UartPutc_end
            BX      LR
            ENDFUNC
; --------------------------------------------------
UartPuts FUNCTION
            ; Print a text string to UART
            ; Input R0 - starting address of string
            PUSH    {R4, LR}
            MOV     R4, R0
UartPuts_loop
            LDRB    R0, [R4] ; Read one character
            CMP     R0, #0
            BEQ     UartPuts_end
            BL      UartPutc
            ADDS    R4, R4, #1
            B       UartPuts_loop
UartPuts_end
            POP     {R4, PC}
            ENDFUNC
; --------------------------------------------------
UartConfig FUNCTION
            ; UART interface are : PIO1_7 (TXD) and PIC1_6 (RXD)
            ; Other UART signals (DTR, DSR, CTS, RTS, RI) are not used

            ; Enable clock to IO configuration block
            ; (bit[16] of AHBCLOCK Control register)
            ; LPC_SYSCON->SYSAHBCLKCTRL = LPC_SYSCON->SYSAHBCLKCTRL | (1<<16);
            LDR     R0,=(LPC_SYSCON_BASE+SYSAHBCLKCTRL)
            LDR     R2,=0x10000   ; (1<<16)
            LDR     R1,[R0]
            ORRS    R1, R1, R2
            STR     R1,[R0]

            ; PIO1_7 IO output config
            ; bit[5]   - Hysteresis (0=disable, 1 =enable)
            ; bit[4:3] - MODE(0=inactive, 1 =pulldown, 2=pullup, 3=repeater)
            ; bit[2:0] - Function (0 = IO, 1=TXD, 2=CT32B0_MAT1)
            ; LPC_IOCON->PIO1_7 = (0x1) + (0<<3) + (0<<5);
            LDR     R0,=LPC_IOCON_BASE
            MOVS    R1,#PIO1_7
            MOVS    R2,#0x01
            STR     R2,[R0, R1]

            ; PIO1_6 IO output config
            ; bit[5]   - Hysteresis (0=disable, 1 =enable)
            ; bit[4:3] - MODE(0=inactive, 1 =pulldown, 2=pullup, 3=repeater)
```

(Continued)

uart_test.s—Cont'd

```
        ; bit[2:0] - Function (0 = IO, 1=RXD, 2=CT32B0_MAT0)
        ; LPC_IOCON->PIO1_6 = (0x1) + (2<<3) + (1<<5);
        MOVS    R1,#PIO1_6
        MOVS    R2,#0x31
        STR     R2,[R0, R1]


        ; Enable clock to UART (bit[12] of AHBCLOCK Control register
        ;  LPC_SYSCON->SYSAHBCLKCTRL = LPC_SYSCON->SYSAHBCLKCTRL | (1<<12);
        LDR     R0,=(LPC_SYSCON_BASE+SYSAHBCLKCTRL)
        LDR     R2,=0x1000  ; (1<<12)
        LDR     R1,[R0]
        ORRS    R1, R1, R2
        STR     R1,[R0]


        ; UART_PCLK divide ratio = 1
        ;  LPC_SYSCON->UARTCLKDIV = 1;
        LDR     R0,=(LPC_SYSCON_BASE+UARTCLKDIV)

        MOVS    R1,#1
        STR     R1,[R0]

        ; UART_PCLK = 48MHz, Baudrate = 38400, divide ratio = 1250
        ; Line Control Register
        ; LPC_UART->LCR = (1<<7) |   // Enable access to Divisor Latches
        ;    (0<<6) |   // Disable Break Control
        ;    (0<<4) |   // Bit[5:4] parity select (odd, even, sticky-1, sticky-0)
        ;    (0<<3) |   // parity disabled
        ;    (0<<2) |   // 1 stop bit
        ;    (3<<0);    // 8-bit data
        LDR     R0,=LPC_UART_BASE
        MOVS    R1,#0x83
        STR     R1,[R0,#UART_LCR]


        ; LPC_UART->DLL = 78;  // Divisor Latch Least Significant Byte
        ;                      // 48MHz/38400/16 = 78.125
        MOVS    R1,#78
        STR     R1,[R0,#UART_DLL]


        ; LPC_UART->DLM = 0;   // Divisor Latch Most Significant Byte  : 0
        MOVS    R1,#0
        STR     R1,[R0,#UART_DLM]


        ; LPC_UART->LCR = (0<<7) |   // Disable access to Divisor Latches
        ;    (0<<6) |   // Disable Break Control
        ;    (0<<4) |   // Bit[5:4] parity select (odd, even, sticky-1, sticky-0)
        ;    (0<<3) |   // parity disabled
        ;    (0<<2) |   // 1 stop bit
        ;    (3<<0);    // 8-bit data
        MOVS    R1,#0x3
        STR     R1,[R0,#UART_LCR]


        ; LPC_UART->FCR = 1; // Enable FIFO
```

```
        MOVS    R1,#1
        STR     R1,[R0,#UART_FCR]

        BX      LR
        ENDFUNC
; --------------------------------------------------
Set48MHzClock FUNCTION
        LDR     R0,=LPC_SYSCON_BASE

        ; Power up the PLL and System oscillator
        ; (clear the powerdown bits for PLL and System oscillator)
        ; LPC_SYSCON->PDRUNCFG = LPC_SYSCON->PDRUNCFG & 0xFFFFFF5F;
        LDR     R3, =PDRUNCFG
        LDR     R1,[R0, R3]
        MOVS    R2, #0xA0
        BICS    R1, R1, R2
        STR     R1,[R0, R3]

        ; Select PLL source as crystal oscillator
        ;   0 - IRC oscillator
        ;   1 - System oscillator
        ;   2 - WDT oscillator
        ; LPC_SYSCON->SYSPLLCLKSEL = 1;
        MOVS    R1, #0x1
        STR     R1,[R0, #SYSPLLCLKSEL]


        ; Update SYSPLL setting (0->1 sequence)
        ; LPC_SYSCON->SYSPLLCLKUEN = 0;
        MOVS    R1, #0x0
        STR     R1,[R0, #SYSPLLCLKUEN]


        ; LPC_SYSCON->SYSPLLCLKUEN = 1;
        MOVS    R1, #0x1
        STR     R1,[R0, #SYSPLLCLKUEN]


        ; Set PLL to 48MHz generate from 12MHz
        ;   M = 48/12 = 4 (MSEL = 3)
        ;   FCCO (must be between 156 to 320MHz, and is 2x, 4x, 8x or 16x of Clock)
        ;   Clock freq out selected as 192MHz
        ;   P = 192MHz/48MHz/2 = 2 (PSEL = 1)
        ;    bit[8]   - BYPASS
        ;    bit[7]   - DIRECT
        ;    bit[6:5] - PSEL  (1,2,4,8)
        ;    bit[4:0] - MSEL  (1-32)
        ; LPC_SYSCON->SYSPLLCTRL = (3 + (1<<5)); // M = 4, P = 2
        MOVS    R1, #0x23
        STR     R1,[R0, #SYSPLLCTRL]

        ; wait until PLL is locked
        ; while(LPC_SYSCON->SYSPLLSTAT == 0);
Set48MHzClock_waitloop1
        LDR     R1,[R0,#SYSPLLSTAT]
```

(Continued)

```
uart_test.s—Cont'd
        CMP     R1, #0
        BEQ     Set48MHzClock_waitloop1

        ; Switch main clock to PLL clock
        ;    0 - IRC
        ;    1 - Input clock to system PLL
        ;    2 - WDT clock
        ;    3 - System PLL output
        ; LPC_SYSCON->MAINCLKSEL = 3;
        MOVS    R1, #0x3
        STR     R1,[R0, #MAINCLKSEL]

        ; Update Main Clock Select setting (0->1 sequence)
        ; LPC_SYSCON->MAINCLKUEN = 0;
        MOVS    R1, #0x0
        STR     R1,[R0, #MAINCLKUEN]

        ; LPC_SYSCON->MAINCLKUEN = 1;
        MOVS    R1, #0x1
        STR     R1,[R0, #MAINCLKUEN]
        BX      LR
        ENDFUNC
; -------------------------------------------------------
        END
; -------------------------------------------------------
```

Appendix H.5 RTX_config.c

The following file is a configuration file for the Keil RTX Kernel for examples in Chapter 18. This file might need modifying depending on your project requirements. Please refer to Chapter 18 for details.

```
RTX_config.c
/*----------------------------------------------------------------------
 *      RL-ARM - RTX
 *----------------------------------------------------------------------
 *      Name:    RTX_CONFIG.C
 *      Purpose: Configuration of RTX Kernel for Cortex-M
 *      Rev.:    V4.10
 *----------------------------------------------------------------------
 *      This code is part of the RealView Run-Time Library.
 *      Copyright (c) 2004-2010 KEIL - An ARM Company. All rights reserved.
 *----------------------------------------------------------------------*/
```

```
#include <RTL.h>

/*----------------------------------------------------------------------
 *      RTX User configuration part BEGIN
 *--------------------------------------------------------------------*/

//-------- <<< Use Configuration Wizard in Context Menu >>> ------------------
//
// <h>Task Configuration
// =======================
//
//    <o>Number of concurrent running tasks <0-250>
//    <i> Define max. number of tasks that will run at the same time.
//    <i> Default: 6
#ifndef OS_TASKCNT
 #define OS_TASKCNT      6
#endif

//    <o>Number of tasks with user-provided stack <0-250>
//    <i> Define the number of tasks that will use a bigger stack.
//    <i> The memory space for the stack is provided by the user.
//    <i> Default: 0
#ifndef OS_PRIVCNT
 #define OS_PRIVCNT      0
#endif

//    <o>Task stack size [bytes] <20-4096:8><#/4>
//    <i> Set the stack size for tasks which is assigned by the system.
//    <i> Default: 200
#ifndef OS_STKSIZE
 #define OS_STKSIZE      100
#endif

// <q>Check for the stack overflow
// ================================
// <i> Include the stack checking code for a stack overflow.
// <i> Note that additional code reduces the Kernel performance.
#ifndef OS_STKCHECK
 #define OS_STKCHECK     1
#endif

// <q>Run in privileged mode
// ==========================
// <i> Run all Tasks in privileged mode.
// <i> Default: Unprivileged
#ifndef OS_RUNPRIV
 #define OS_RUNPRIV      0

#endif
```

(Continued)

RTX_config.c—Cont'd

```
// </h>
// <h>SysTick Timer Configuration
// ===============================
//    <o>Timer clock value [Hz] <1-1000000000>
//    <i> Set the timer clock value for selected timer.
//    <i> Default: 6000000  (6MHz)
#ifndef OS_CLOCK
 #define OS_CLOCK        48000000
#endif


//    <o>Timer tick value [us] <1-1000000>
//    <i> Set the timer tick value for selected timer.
//    <i> Default: 10000  (10ms)
#ifndef OS_TICK
 #define OS_TICK         10000
#endif


// </h>

// <h>System Configuration
// =======================
// <e>Round-Robin Task switching
// =============================
// <i> Enable Round-Robin Task switching.
#ifndef OS_ROBIN
 #define OS_ROBIN        1
#endif


//    <o>Round-Robin Timeout [ticks] <1-1000>
//    <i> Define how long a task will execute before a task switch.
//    <i> Default: 5
#ifndef OS_ROBINTOUT
 #define OS_ROBINTOUT    5
#endif

// </e>

//    <o>Number of user timers <0-250>
//    <i> Define max. number of user timers that will run at the same time.
//    <i> Default: 0  (User timers disabled)
#ifndef OS_TIMERCNT
 #define OS_TIMERCNT     0
#endif

//    <o>ISR FIFO Queue size<4=>   4 entries  <8=>   8 entries
//                          <12=> 12 entries  <16=> 16 entries
//                          <24=> 24 entries  <32=> 32 entries
//                          <48=> 48 entries  <64=> 64 entries
//                          <96=> 96 entries
```

```
//    <i> ISR functions store requests to this buffer,
//    <i> when they are called from the iterrupt handler.
//    <i> Default: 16 entries
#ifndef OS_FIFOSZ
 #define OS_FIFOSZ       16
#endif


// </h>
//------------- <<< end of configuration section >>> -----------------------

// Standard library system mutexes
// ================================
//   Define max. number system mutexes that are used to protect
//   the arm standard runtime library. For microlib they are not used.
#ifndef OS_MUTEXCNT
 #define OS_MUTEXCNT     8
#endif

/*----------------------------------------------------------------------------
 *      RTX User configuration part END
 *---------------------------------------------------------------------------*/


#define OS_TRV          ((U32)(((double)OS_CLOCK*(double)OS_TICK)/1E6)-1)


/*----------------------------------------------------------------------------
 *      Global Functions
 *---------------------------------------------------------------------------*/

/*-------------------------- os_idle_demon ----------------------------------*/

__task void os_idle_demon (void) {
  /* The idle demon is a system task, running when no other task is ready */
  /* to run. The 'os_xxx' function calls are not allowed from this task.  */

  for (;;) {
  /* HERE: include optional user code to be executed when no task runs.*/
  }
}



/*-------------------------- os_tmr_call ------------------------------------*/

void os_tmr_call (U16 info) {
  /* This function is called when the user timer has expired. Parameter   */
  /* 'info' holds the value, defined when the timer was created.          */

  /* HERE: include optional user code to be executed on timeout. */
}
/*-------------------------- os_error ---------------------------------------*/
```

(Continued)

RTX_config.c—Cont'd

```
void os_error (U32 err_code) {
  /* This function is called when a runtime error is detected. Parameter */
  /* 'err_code' holds the runtime error code (defined in RTL.H).          */

  /* HERE: include optional code to be executed on runtime error. */
  for (;;);
}



/*-------------------------------------------------------------------------
 *        RTX Configuration Functions
 *------------------------------------------------------------------------*/

#include <RTX_lib.c>

/*-------------------------------------------------------------------------
 * end of file
 *------------------------------------------------------------------------*/
```

Troubleshooting

Chapter 12 of this book covered various techniques for locating problems in program code. In this section, we will summarize the most common mistakes and problems that software developers might find when preparing software for the Cortex-M0.

I.1. Program Does Not Run/Start

There can be many possible reasons.

I.1.1. Vector Table Missing or in Wrong Place

Depending in the tool chain, you might need to create a vector table. If you do have a vector table in the project, make sure it is suitable for the Cortex-M0 (e.g., vector table code for the ARM7TDMI cannot be used). It is also possible for the vector table to be removed during the link stage or to be placed into the wrong address location.

You should generate a disassembled listing of the compiled image or a linker report to see if the vector table is present and if it is correctly placed at the start of the memory.

I.1.2. Incorrect C Startup Code Being Used

In addition to reviewing compiler options, make sure you are specifying the correct linker options as well. Otherwise a linker might pull in incorrect C startup code. For example, it might end up using startup code for another ARM processor, which contains instructions not supported by the Cortex-M0, or it could use startup code for a debug environment with semihosting, which might contain a breakpoint instruction (BKPT) or supervisor call (SVC). This can cause an unexpected hard fault or software exception.

I.1.3. Incorrect Value in Reset Vector

Make sure the reset vector is really pointing to the intended reset handler. Also, you should check that the exception vectors in the vector table have the LSB set to 1 to indicate Thumb code.

I.1.4. Program Image not Programmed in Flash Correctly

Most flash programming tools automatically verify the flash memory after programming. If not, after the program image is programmed into the flash, you might need to double-check to ensure that the flash memory has been updated correctly. In some cases, you might need to erase the flash first, and then program the program image.

I.1.5. Incorrect Tool Chain Configurations

Some other tool chain configurations can also cause problems with the startup sequence—for example, memory map settings, CPU options, endianness settings, and the like.

I.1.6. Incorrect Stack Pointer Initialization Value

This involves two parts. First, the initial stack pointer value (the first word on the vector table) needs to point to a valid memory address. Second, the C startup code might have a separate stack setup step. Try getting the processor to halt at the startup sequence, and single step through it to make sure the stack pointer is not changed to point to an invalid address value.

I.1.7. Incorrect Endian Setting

Most ARM microcontrollers are using little endian, but there is a chance that someday you may use an ARM Cortex-M0 microcontroller in big endian. If this is the case, make sure the C compiler options, assembler options, and linker options are set up correctly to support big endian mode.

I.2. Program Started, but Entered a Hard Fault

I.2.1. Invalid Memory Access

One of the most common problems is accidentally accessing an invalid memory region. Usually you can trace the faulting memory access instruction following the instructions in Chapter 12. Using the method described there, you can locate the program code that caused the fault.

I.2.2. Unaligned Data Access

If you directly manipulate a pointer, or if you have assembly code, you can generate code that attempts to carry out an unaligned access. If the faulting instruction is a memory access instruction, determine if the address value used for the transfer is aligned or not.

I.2.3. Bus Slave Return Error

Some peripherals might return an error response if it has not been initialized or if the clock to the peripheral is disabled. In some less common cases, a peripheral might only be able to accept 32-bit transfers and return error responses for byte or half-word transfers.

I.2.4. Stack Corruption in Exception Handler

If the program crashes after an interrupt handler execution, it might be a stack frame corruption problem. Because local variables can be stored on the stack memory, if a data array is defined inside an exception handler and the array index being used exceeds the array size, the stack frame of the exception could become corrupted. As a result, the program could crash after exiting the exception.

I.2.5. Program Crash at Some C Functions

Please check if you have reserved sufficient stack space and heap space. By default, the heap space defined in the default startup code for NXP LPC111x in Keil MDK-ARM is zero bytes. You will need to modify this if you are using C functions like malloc, printf, and so on.

Another possible reason for this problem is an incorrect C library function being pulled in by the linker. The linker can normally output verbosely to show the user what library functions were pulled in, which is something a user should check under such circumstances.

I.2.6. Accidentally Trying to Switch to ARM State

After a hard fault is entered, if the T bit in the stacked xPSR is 0, the fault was triggered by switching to ARM state. This can be caused by, for example, an invalid function pointer value, the LSB of a vector in vector table not being set to 1, corruption of the stack frame during exception, or even an incorrect linker setting that ends up causing an incorrect C library being used.

I.2.7. SVC Executed at Incorrect Priority Level

If the SVC instruction is executed inside an SVC handler, or any other exception handlers that have same or higher priority than the SVC exception, it will trigger a fault. If an SVC is used in an NMI handler or the hard fault handler, it will result in a lockup.

I.3. Sleep Problems

I.3.1. Execute of WFE Does not Enter Sleep

Execution of a WFE instruction does not always result in entering of sleep mode. If a past event has occurred, the internal event latch inside the Cortex-M0 processor will be set. In this situation, execution of a WFE instruction will clear the event latch and continue to the next instruction. Therefore, a WFE instruction is usually used in a conditional idle loop so that it can be executed again if sleep did not occur in the first WFE execution.

I.3.2. Sleep-on-Exit Triggers Sleep Too Early

If you enable the Sleep-on-Exit feature too early during the initialization stage of a program, the processor will enter sleep mode as soon as the first exception handler is completed.

I.3.3. SEVONPEND does not Work for Interrupt that is Already in a Pending State

The Send Event on Pending (SEVONPEND) feature generates an event when an idle interrupt changes into the pending state if the feature is enabled. The event can be used to wake up the Cortex-M0 if it has been entering sleep mode by WFE instruction. However, if the pending status of the interrupt was already set before entering sleep, a new interrupt request that arrives during sleep will not trigger an event. In this case, the Cortex-M0 processor will not be awakened.

I.3.4. Processor Cannot Wake up Because Sleep Mode Might Disable Some Clocks

Depending on the microcontroller you are using and the chosen sleep mode, the peripherals or the processor clock might be stopped and you might not be able to wake up the processor unless some special wakeup signal is used. Please refer to documentation from your micro-controller vendors for details.

I.3.5. Race Condition

Sometimes we need to pass software flags from interrupt handlers to thread level codes. However, the following code has a race condition:

```
volatile int irq_flag=0;

while (1){
  if (irq_flag==0) {
    __WFI(); // enter sleep
  }
  else {
    process_a(); // Execute if IRQ_Handler had executed
  }
}
void IRQ_Handler(void){
  irq_flag=1;
  return;
}
```

If the IRQ takes place after the "irq_flag" checking and before the WFI, the process will enter sleep mode and will not execute "process_a()." To solve this problem, the WFE instruction should be used. The execution of IRQ_Handler causes the internal event latch to set. As a result, the next execution of WFE will only cause the event latch to be cleared and will not enter sleep.

If a microcontroller with Cortex-M3 r2p0 or earlier versions is used for the same operation, an __SEV() instruction needs to be included inside the "IRQ_Handler." This is because of errata in the processor design that prevent the event latch from being set correctly in an interrupt event. Therefore, the code should be changed to the following:

```
volatile int irq_flag=0;

while (1){
  if (irq_flag==0){
    __WFE(); // enter sleep if event latch is 0
    }
  else{
    process_a(); // Execute if IRQ_Handler had executed
    }
  }
void IRQ_Handler(void){
  irq_flag=1;
  __SEV(); // required for Cortex-M3 r2p0 or earlier versions
  return;
  }
```

I.4. Interrupt Problem

I.4.1. Extra Interrupt Handler Executed

In some microcontrollers, the peripherals are connected to a peripheral bus running at a different speed from the processor system bus, and the data transfer through the bus bridge might have a delay (depending on the design of the bus bridge). If the interrupt request of the peripheral is cleared at the end of an interrupt service routine and the exception is exited immediately, the interrupt signal connected to the processor might still be high when the exception exit takes place. This results in another execution of the same exception handler. To solve the problem, you can clear the interrupt request earlier in the interrupt service routine, or add an extra access to the peripheral after clearing the interrupt request. In most cases, these arrangements can solve this problem.

I.4.2. Additional SysTick Handler Execution

If you set up the SysTick timer for a single shot arrangement with a short delay, a second SysTick interrupt event could be generated during the SysTick handler execution. In such cases, in addition to disabling the SysTick interrupt generation, you should also clear the SysTick interrupt pending status before exiting the SysTick handler. Otherwise the SysTick handler will be entered again.

I.4.3. Disabling of Interrupt Within the Interrupt Handler

If you are porting application code from an ARM7TDMI microcontroller, you might need to update some interrupt handlers if they disable the interrupts during interrupt handling to ensure that the interrupts are reenabled before the exception exit. In the ARM7TDMI, interrupts can be reenabled at the same time as the exception return because the I-bit in CPSR is restored during the process. In the Cortex-M0, reenabling of the interrupt (clearing of PRIMASK) has to be done separately.

I.4.4. Incorrect Interrupt Return Instructions

If you are porting software from the ARM7TDMI, make sure that all interrupt handlers are updated to remove wrapper code for nested interrupt support, and make sure the correct instruction is used for exception return. In the Cortex-M0, exception return must be carried out using BX or POP instructions.

I.4.5. Exception Priority Setup Values

Although the Exception/Interrupt Priority Level Registers contain 8 bits for the priority level of each exception or interrupt, only the top 2 bits are implemented. As a result, the priority leve values can only be 0x00, 0x40, 0x80, and 0xC0. If you are using NVIC functions from CMSIS-compliant device driver libraries, the priority setup function "NVIC_SetPriority()" automatically shift the values 0 to 3 to the implemented bits.

I.5. Other Issues

I.5.1. Incorrect SVC Parameter Passing Method

Unlike traditional ARM processors, the parameters pass on to the SVC exception and the return value from SVC handler must be transferred using exception stack frame. Otherwise the parameter could become corrupted. Please refer to Chapter 18 for details.

I.5.2. Debug Connection Affected by I/O Setting or Low-power Modes

If you change the I/O settings of pins that are used for a debug connection, you might be unable to debug your application or update the flash because the debug connection is affected by the I/O usage configuration changes. Similarly, low-power features might also disable debugger connections. In some microcontroller products, there is a special boot mode to allow you to disable the execution of your program during bootup. Chapter 17 covered the recovery method you can use on NXP LPC111x.

I.5.3. Debug Protocol Selection

Some Cortex-M0 microcontrollers use serial wire debug protocol and some other use JTAG debug protocol. If incorrect debug protocol is selected in the configuration of a debug environment, the debugger will not be able to connect to the microcontrollers.

I.5.4. Using Event Output as Pulse I/O

Some Cortex-M0 microcontrollers allow an I/O pin to be configured as an event output. When the SEV instruction is executed, a single cycle pulse is generated from the processor and this can be useful for external latch control.

When a sequence of multiple pulses is required, additional instructions need to be placed between the SEV instructions. Otherwise the pulses could be merged. For example, the following sequence might result in one pulse (of two clock cycles) or two pulses (of one cycle) depending on the wait state of the memory system:

```
    __SEV(); // First pulse
    __SEV(); // Second pulse, could be merged with first pulse
By changing the code to
    __SEV(); // First pulse
    __NOP(); // Produce timing gap between the two pulses.
    __SEV(); // Second pulse
```

If the C compiler you use can optimize away NOPs, an __ISB() could be used instead.

Index

CPSIA information can be obtained at www.ICGtesting.com
Printed in the USA
LVOW020509141112

307213LV00007B/34/P